THE MANAGEMENT OF ENGINEERING

THE MANAGEMENT OF ENGINEERING

Human, Quality, Organizational, Legal, and Ethical Aspects of Professional Practice

F. Lawrence Bennett, P. E.
University of Alaska Fairbanks, School of Engineering

John Wiley & Sons, Inc.
New York   Chichester   Brisbane   Toronto   Singapore

ACQUISITIONS EDITOR   Charity Robey
MARKETING MANAGER   Debra Riegert
SENIOR PRODUCTION EDITORS   Nancy Prinz, Tracey Kuehn
DESIGNER   Pedro Noa
MANUFACTURING MANAGER   Mark Cirillo
ILLUSTRATION COORDINATOR   Eugene Aiello
cover photo © Fourmy/REA/SABA; Photo has been altered.

Library of Congress Cataloging in Publication Data:

Bennett, F. Lawrence.
    The management of engineering: human, quality, organizational,
legal, and ethical aspects of professional practice / F. Lawrence
Bennett.
        p.   cm.
    Includes bibliographical references and index.
    ISBN 0–471–59329–X (alk. paper)
    1. Engineering—Management.   I. Title.
TA190.B36   1996
620′.0068—dc20
                                                        95-34032
                                                        CIP

10   9   8   7   6

To Margaret Ann
With All My Love

Preface

The Management of Engineering was written to fulfill the need for a single comprehensive textbook for advanced undergraduate or beginning graduate courses in engineering management. Although many engineering programs provide, or even require, more than one course in engineering management for undergraduates, many others are limited to a single course. This book is intended primarily for students in the single course. To date, no individual text has covered the topics included in this volume.

This book is first and foremost a textbook, but it will also serve as a handy reference or study aid for the practicing engineering professional. The topics are treated in a way that is both practical and easy to apply to professional practice.

I have tried to take the word *comprehensive* to heart. As the book's title implies, a broad range of topics of interest to those who will manage (or currently manage) engineering personnel, projects, and organizations have been integrated into this volume.

Chapter 1 offers a brief introduction to the engineering profession and discusses the typical transition from engineer to engineering manager. Chapter 2 presents various organizational structures. The next two chapters are devoted to the increasingly important topic of quality management. Chapters 5 and 6 deal, respectively, with personnel management and communications, both essential to successful engineering management. Chapter 7 distills the key fundamentals of project management, including organization, supervision, scheduling, cost planning, and control. The next chapters present an essential introduction of legal matters for the engineering manager, including contract law, agency, intellectual property, forms of business organizations, and the issue of professional liability. Numerous actual case studies support and enhance student understanding of these potentially delicate issues.

Following the legal material, Chapter 13 is devoted to a comprehensive discussion of ethical issues that are likely to arise during an engineer's career. Chapter 14 concludes the text with an appeal for professionalism and a plea that the essence of an engineer's career, in the final analysis, is really about his or her service to humanity.

The intent of this book is to summarize some essential material to assist the engineering manager. However, there still may be too much to cover in a one-semester or one-quarter course. The text is modular enough to permit sections to be omitted or reordered as needed. Otherwise, a two-course sequence might benefit from using this book. The extensive reference lists assist the reader interested in pursuing in greater depth individual topics brought up in each chapter.

Augmenting the text are numerous figures and tables, developed from personal experience and from many other acknowledged sources. Each chapter contains abundant problems and discussion questions. Case studies also add a touch of realism. A teacher's manual is available from the publisher for the use of instructors.

F. LAWRENCE BENNETT, P.E.

Fairbanks, Alaska
August 1995

Acknowledgments

No book like this is the work of a single individual. So many persons have made contributions that it is difficult to remember them all.

I am indebted to a generation of ESM 450 students at the University of Alaska Fairbanks (UAF) for their active interest in the topics included here, for their willingness to try the several combinations of texts we have used over the past 25 years, and for their struggles with, and suggestions for improving, draft versions of this text during the past several semesters.

I am especially grateful to the reviewers of my work: Colin Benjamin, University of Missouri—Rolla; Mike Deisenroth, Virginia Polytechnic Institute and State University; Floyd Miller, University of Illinois at Chicago; Ted Eschenbach, head of Engineering Management at the University of Alaska Anchorage and editor of *Engineering Management Journal;* Hal A. Rumsey, Washington State University—Spokane; Jerry D. Westbrook, University of Alabama at Huntsville; Elizabeth Paté-Cornell, Stanford University; E. Douglas Harris, Southern Methodist University; Michael L. Donnell, The George Washington University; Vince Haneman, dean emeritus of the UAF School of Engineering; E. R. Baker, my faculty colleague in Engineering Management at UAF; Andy Scott, retired manager of engineering for Tennessee Eastman; my neighbor and good buddy Wendell Shiffler, vice president of TRAF Construction; and the Reverend Chuck Young, another good buddy and cabin-building fanatic.

Those who provided materials on a variety of topics include Bert Bell, president of Ghemm Company Contractors; William O. Frank, president of Heartland Professional Construction Services; Jim Poirot, chairman emeritus of CH2M-Hill; Mel Hensey, president of Hensey Associates; D. L. Richter, vice president of Caltex Pacific Indonesia; G. F. Sandison, assistant division manager, Texaco Onshore Producing Division; Art Schwartz, counsel for the National Society of Professional Engineers; Tony Baez, technical manager for the American Society of Civil Engineers; Ed Cridge of the UAF Media Department; Baxter Burton, assistant director of UAF's Polar Ice Coring Office; and Dave Spell, another faculty colleague at UAF and a member of Alaska's registration board.

Sally Lowery provided helpful legal research, Marilyn West contributed typing assistance, and graduate students Lei Chen and Zhu Ye handled library research and other support efforts with aplomb. Personnel at UAF's Rasmuson Library also provided superb assistance.

I am grateful to the University of Alaska Fairbanks for granting a semester-long sabbatical leave, during which the early phases of this writing project were completed. My thanks also go to the generous folks at Sweden's Luleå University of Technology (especially Professors Sven Knutsson and Kennet Axelsson), where colleagueship and facilities were made available in abundance during that sabbatical leave.

The staff at Wiley have been most helpful; my editor, Charity Robey, provided that critical and proper balance between guidance and independence that makes her one of her industry's best. Sean Culhane assisted in paring the manuscript to proper size.

Kim Fisher deserves a paragraph of her own. Through many long and sometimes difficult years, she has provided clerical and administrative assistance to our little department and to me personally in a fashion that can only be considered professional in every sense of that word. I deeply appreciate the many ways in which Kim has contributed to this project.

Lastly, I express profound love and appreciation to my beloved wife Margaret for her many years of support and devotion. It was she who sat at home while I made one last run to the office at midnight to look up yet another reference. It was she who was always there to help overcome the hard times and celebrate the good ones. Her patience, faith, and love are a major reason for the completion of this volume, which I dedicate to her with lasting thanks and love.

I owe a great debt of gratitude to all those mentioned above, and to many others. I accept full responsibility for any errors that may be found in this book and for decisions to include some materials and omit others. Comments and suggestions are always welcome. Please write to

Dr. F. Lawrence Bennett, P.E., Head,
Engineering and Science Management
University of Alaska Fairbanks
P. O. Box 755900
Fairbanks, AK 99775-5900
FFFLB@aurora.alaska.edu

F. L. B.

Contents

1 The Engineer as Manager   **1**

   1.1 Engineering as a Profession and as a Career, 2

   1.2 Engineering Versus Management, 4

   1.3 The Transition to Engineering Manager, 7

   1.4 Maintaining Technical Competence, 13

   1.5 The Engineering Manager's Functions and Activities, 14

   1.6 A Definition and Some Distinctions, 16

   1.7 Discussion Questions, 19

   1.8 References, 20

2 The Engineering Organization   **22**

   2.1 The Effective Organization, 22

      2.1.1 Teamwork in Practice, 22

      2.1.2 A Descriptive Model, 24

      2.1.3 Principles of Organizational Structure, 25

      2.1.4 Organizational Design: An Approach, 27

   2.2 Organizational Structures, 28

      2.2.1 The Organizational Chart Can't Say It All!, 28

      2.2.2 Functional, 28

      2.2.3 Product, 29

      2.2.4 Project, 31

      2.2.5 Matrix, 34

      2.2.6 Self-Directed Teams, 36

      2.2.7 Clusters, 38

      2.2.8 Geographical, 39

      2.2.9 The Throwaway Organization, 40

      2.2.10 The Informal Organization, 40

   2.3 Relationships Within the Organization, 41

   2.4 Relationships Outside the Organization, 45

   2.5 The Future of Technical Organizations, 46

   2.6 Discussion Questions, 46

   2.7 References, 47

3 Total Quality Management: Principles and Approaches   **49**

   3.1 Some History, 50

3.2   Some Definitions, 54
3.3   Some Primary Principles, 56
       3.3.1   Continuous Improvement, 56
       3.3.2   Focus on the Customer, 58
       3.3.3   The Importance of Teams, 59
       3.3.4   Management Involvement and Support, 60
3.4   The Gurus and Their Approaches, 62
       3.4.1   Philip B. Crosby, 62
       3.4.2   W. Edwards Deming, 63
       3.4.3   Armand V. Feigenbaum, 64
       3.4.4   Kaoru Ishikawa, 65
       3.4.5   Joseph M. Juran, 65
       3.4.6   Genichi Taguchi, 66
3.5   Applications to Service Industries, 66
3.6   Cost of Quality, 69
3.7   Standards of Excellence, 71
       3.7.1   ISO 9000, 71
       3.7.2   Baldrige Award, 72
       3.7.3   Deming Prize (Japan), 73
       3.7.4   State Awards, 75
       3.7.5   President's Award, 76
3.8   Implementing Total Quality Management in the Engineering
       Organization, 76
       3.8.1   Culture Change, 76
       3.8.2   Approaches to Implementation, 77
3.9   Concluding Remarks, 77
3.10   Discussion Questions, 78
3.11   References, 79

4   Total Quality Management: Techniques and Applications   82
4.1   A Sampling of Numeric Tools, 82
       4.1.1   Control Charts, 82
       4.1.2   Histograms, 86
       4.1.3   Pareto Charts, 87
       4.1.4   Run Charts (Timeline Charts), 91
       4.1.5   Scatter Diagrams, 91
4.2   A Sampling of Nonnumeric Tools, 92
       4.2.1   Brainstorming, 92
       4.2.2   Cause-and-Effect Diagrams (Ishikawa or Fishbone
                Diagrams), 93
       4.2.3   Flowcharting, 93
       4.2.4   Force Fields, 94

        4.2.5   Nominal Group Technique, 96

   4.3   Benchmarking, 97

   4.4   A Comment on Total Quality Management Tools, 98

   4.5   Total Quality Management Applications in Engineering, 100

        4.5.1   Gee & Jenson: Design Consultants, 100

        4.5.2   Heartland Professional Construction Services, 103

        4.5.3   Texaco: Process Industry, 107

   4.6   The Quality Road Ahead, 113

   4.7   Problems and Discussion Questions, 115

   4.8   References, 118

5   The Human Element in Engineering Management   120

   5.1   Motivation, 120

   5.2   Leadership, 128

   5.3   Delegation, 134

   5.4   Personnel Development, 136

   5.5   Incentive, Recognition, and Reward Systems, 137

   5.6   The Personnel Administration Function, 139

   5.7   Discussion Questions and Cases, 142

   5.8   References, 145

6   Communication in the Engineering Organization   147

   6.1   The Communication Process, 147

   6.2   Communication in Organizations, 148

   6.3   Oral Communication, 152

   6.4   Written Communication, 156

   6.5   Choosing the Appropriate Method, 159

   6.6   Meetings, 161

   6.7   Negotiation, 162

   6.8   Some Final Thoughts on Improving the Communication Process, 163

   6.9   Discussion Questions and Case, 164

   6.10 References, 167

7   Management of Engineering Projects   169

   7.1   The Nature of Projects and Project Management, 169

        7.1.1   Some Definitions, 169

        7.1.2   The "Essence" of Successful Project Management, 171

        7.1.3   The Project Life Cycle, 172

   7.2   Project Organizations, 172

        7.2.1   An Important Decision: Project Organization or Not, 175

7.2.2   The Basic Three: Functional, Projectized, and Matrix, 176

    7.2.2.1   The Functional Organization, 176

    7.2.2.2   The Projectized Organization, 177

    7.2.2.3   The Matrix Organization, 178

7.2.3   Role Conflict in the Project Organization, 182

7.2.4   Don't Forget the Informal Structure!, 186

7.2.5   The Place of the Project Manager in the Organization, 186

7.3   The Project Manager, 188

7.3.1   A Different Kind of Position, 188

7.3.2   Roles and Responsibilities, 188

7.3.3   What Kind of Person Can Fulfill That Role?, 189

7.4   Project Planning, Scheduling, and Control, 190

7.4.1   Work Breakdown Structure, 191

7.4.2   Bar Charting Methods, 193

7.4.3   Network Scheduling Methods, 195

    7.4.3.1   The Network Approach, 195

    7.4.3.2   Activity-on-Node Method, 197

    7.4.3.3   Activity-on-Arrow Method, 203

    7.4.3.4   Precedence Method, 206

7.4.4   Resource and Cost Analysis and Control Based on a Network Schedule, 207

7.4.5   Computer Applications for Planning and Controlling the Schedule, 208

7.5   Project Budgeting and Cost Control, 210

7.5.1   Budgeting, 210

    7.5.1.1   Cost Estimating, 210

    7.5.1.2   The Project Budget, 213

7.5.2   Cost Control, 215

7.6   Some Final Thoughts on Project Control, 219

7.7   Discussion Questions and Problems, 219

7.8   References, 222

8   Engineers and the Law   **224**

8.1   Four Basic Types of Law, 225

8.1.1   Constitutional Law, 226

8.1.2   Administrative Law, 226

8.1.3   Statute Law, 226

8.1.4   Common Law, 227

8.2   Civil Law: Two Meanings, 228

8.3   Law Versus Equity, 228

8.4   The United States Court System, 229

8.4.1   Federal Courts, 229

       8.4.2 State Courts, 231

       8.4.3 Jurisdiction, 231

  8.5 Trial Procedure, 233

       8.5.1 Pretrial, 233

       8.5.2 Trial, 234

       8.5.3 Evidence, 235

  8.6 Citation of Cases, 237

  8.7 Agency, 238

  8.8 Intellectual Property, 241

       8.8.1 Patents, 241

       8.8.2 Trademarks, 242

       8.8.3 Copyrights, 243

       8.8.4 Trade Secrets, 243

  8.9 Regulatory Compliance, 244

  8.10 Alternate Forms of Dispute Resolution, 247

       8.10.1 Negotiation, 247

       8.10.2 Mediation, 247

       8.10.3 Mini-Trial, 248

       8.10.4 Arbitration, 249

  8.11 Topics Not Covered, 251

  8.12 Discussion Questions, 251

  8.13 References, 253

9 Contract Law Principles   254

  9.1 Definitions, 254

  9.2 Formation Principles, 259

       9.2.1 Competent Parties, 260

       9.2.2 Proper Subject Matter, 261

       9.2.3 Meeting of the Minds, 261

       9.2.4 Consideration, 264

       9.2.5 Form, 265

  9.3 Interpretation of Written Contracts, 266

  9.4 Third-Party Rights, 269

       9.4.1 Assignment of Contract Rights, 269

       9.4.2 Third-Party Beneficiaries, 270

  9.5 Contract Discharge, 272

       9.5.1 Satisfactory Performance/Completion, 272

       9.5.2 Contractual Provision, 273

       9.5.3 Breach of Contract, 273

       9.5.4 Mutual Agreement of the Parties, 274

       9.5.5 Impossibility of Performance, 274

       9.5.6 Impracticability of Performance, 275

9.5.7 Operation of Law, 276
9.6 Remedies for Breach of Contract, 276
   9.6.1 Damages, 277
   9.6.2 Restitution, 278
   9.6.3 Equity Remedies, 278
9.7 Discussion Questions, 279
9.8 References, 283

10 Engineering and Construction Contracts   **285**

10.1 Engineering and Architectural Services Contracts, 285
   10.1.1 Contract Elements, 286
   10.1.2 Scope of Services, 286
   10.1.3 Fee Arrangements, 288
   10.1.4 Relationships with Other Parties, 289
   10.1.5 Standard Documents, 290
10.2 Construction Contract Elements, 290
   10.2.1 Competent Parties, 290
   10.2.2 Proper Subject Matter, 291
   10.2.3 Meeting of the Minds, 291
   10.2.4 Consideration, 292
   10.2.5 Form, 293
10.3 Types of Contract, 293
   10.3.1 Fixed Price, 293
   10.3.2 Unit Price, 294
   10.3.3 Cost Plus, 297
   10.3.4 Design/Build Contracts, 298
   10.3.5 Turnkey, 299
   10.3.6 Construction Manager, 300
   10.3.7 Fast Track, 301
   10.3.8 End-Result Contracts, 301
10.4 The Contracting Process, 302
   10.4.1 Advertising, 302
   10.4.2 Proposals, 303
   10.4.3 Bid Opening, 305
   10.4.4 Evaluation and Award, 306
   10.4.5 The Work Proceeds, 307
   10.4.6 Completion, 310
10.5 Construction Documents, 311
   10.5.1 Agreement, 311
   10.5.2 Drawings, 311
   10.5.3 General Conditions, 312
   10.5.4 Special Condition, 312

10.5.5 Technical Specifications, 312

10.5.6 Other Documents, 313

10.6 Contract Provisions of Special Interest, 313

10.6.1 Bonds, 314

10.6.2 Incorporation by Reference, 315

10.6.3 Hierarchy of Documents, 315

10.6.4 Indemnity, 316

10.6.5 Time of Completion and Liquidated Damages, 316

10.6.6 Delay and No Damages for Delay, 317

10.6.7 Termination, 318

10.6.8 Differing Site Conditions, 319

10.7 Discussion Questions, 320

10.8 References, 323

11 Legal Structures of Business Organizations   **325**

11.1 Sole Proprietorship, 325

11.2 Partnership, 326

11.2.1 General Partnership, 326

11.2.2 Limited Partnership and Subpartnership, 332

11.2.3 Partnering, 332

11.3 Corporation, 334

11.3.1 Limited Liability of Owners, 335

11.3.2 Formation of Corporation, 335

11.3.3 Participation and Control, 337

11.3.4 Financing the Corporation, 338

11.3.5 Taxation: C Corporation and S Corporations, 339

11.3.6 Other Corporation Characteristics, 340

11.3.7 Professional Corporations, 340

11.3.8 Termination, 343

11.4 Joint Venture, 343

11.5 Other Organizations, 344

11.5.1 Not-for-Profit Corporation, 344

11.5.2 Franchise, 345

11.5.3 Limited Liability Company, 345

11.5.4 Unincorporated Association, 345

11.5.5 Share-Office Arrangement, 345

11.6 Discussion Questions, 346

11.7 References, 347

12 Professional Liability   **349**

12.1 Introduction to Tort Law, 349

12.1.1 Intentional Torts, 350

          12.1.2 Negligence, 351
          12.1.3 Strict Liability, 356
     12.2 Liability Trends of Importance to the Professional, 357
          12.2.1 Historical Overview, 357
          12.2.2 Demise of the Privity Defense, 361
          12.2.3 Standards of Care, 362
          12.2.4 Product Liability, 364
          12.2.5 Services Liability, 366
          12.2.6 Statutes of Limitation and Statutes of Repose, 368
          12.2.7 Review of Shop Drawings and Other Submittals, 369
          12.2.8 Job Site Safety, 369
          12.2.9 Computer-Aided Design, 371
     12.3 Strategies for Protection of the Professional, 373
     12.4 Tort Reform, 377
     12.5 Discussion Questions, 378
     12.6 References, 381

13 Professional Ethics   **383**
     13.1 Morality and Ethics, 385
     13.2 Codes of Ethics, 387
     13.3 Obligation in Three Sometimes Conflicting Directions, 397
     13.4 Some Ethical Issues in Engineering, 399
          13.4.1 Conflict of Interest, 399
          13.4.2 Confidentiality and Employee Loyalty, 401
          13.4.3 Contributions and Kickbacks, 404
          13.4.4 Whistleblowing, 405
          13.4.5 Professional Conduct, 410
          13.4.6 Global Issues, 411
          13.4.7 A Comment and a Concluding Case, 413
     13.5 Ethics and the Engineering Manager, 414
     13.6 In Conclusion, 415
     13.7 Discussion Questions and Cases, 416
     13.8 References, 422

14 The Engineering Professional   **425**
     14.1 Professional Registration, 427
     14.2 Continuing Professional Development, 431
     14.3 Professional Engineering Societies, 434
     14.4 The Engineer as Expert Witness, 439
     14.5 In Conclusion, 443
     14.6 Discussion Questions, 448
     14.7 References, 449

Appendix "Summaries of Reader Responses to Ethics Questions"   **452**

Index   **459**

_____1
The Engineer As Manager

The purpose of this book is to introduce several topics related to the management of technical persons, organizations, and activities. As the graduate engineer assumes increasing responsibility in the engineering organization, it is likely that he or she will be required to take on management as well as technical obligations. Thus we survey a broad range of areas that are generally considered to lie within the scope of engineering management.

We begin with a look at *engineering and management* and the roles the engineering manager is likely to play. Then we consider the *engineering organization,* its structure, and the several intra- and extraorganizational entities with which the engineer is likely to deal. Two chapters are devoted to *total quality management,* a new and important approach to managing people and organizations; we discuss basic concepts as well as quantitative and qualitative techniques and some cases of successful implementation in engineering organizations. We follow the discussions of organization and quality with some of the *human aspects of engineering management,* addressing such topics as motivation, leadership, personnel development, employee recognition, and communication.

A chapter on the *management of technical projects* discusses project organization and leadership as well as planning, scheduling, and control techniques. The engineering manager is likely to confront *legal matters* as well, so we provide information on contract law and its applications to engineering and construction contracts, plus legal aspects of business organizations and a host of other law-related topics.

We end the book with three chapters devoted to *professional aspects of engineering practice,* including professional liability, engineering ethics, licensing, and continuing education.

The topics herein are intended to complement, rather than compete with, the technical and liberal studies that are the heart of the undergraduate curriculum. They are presented from the viewpoint that they may not be immediately applicable at the outset of one's career. Even though this book may be tucked away during the early career years, it is likely to become an important resource as managerial responsibilities increase.

Do not expect to become an expert on these matters by the study of this material! You will not become a lawyer, a project manager, or a quality management expert. But you will become familiar with some of the important terms and concepts with which engineering managers deal. And you will understand the very special professional obligations that engineers and engineering managers have as they serve the public good, as well as the satisfaction that such professional service can bring to the individual engineer and his or her organization.

We begin with a definition of the engineering profession and a look at the broad scope of its members and their employers.

1.1   ENGINEERING AS A PROFESSION AND AS A CAREER

Engineering, according to the Accreditation Board for Engineering and Technology (1993), is "the profession in which a knowledge of the mathematical and natural sciences gained by study, experience, and practice is applied with judgment to develop ways to utilize, economically, the materials and forces of nature for the benefit of mankind."

Engineering, then, includes the application of these mathematical and scientific principles to the planning, design, construction, operation, and maintenance of products, systems, and large fixed works that serve humankind; as such it also includes the management of such activities, research and development related to such outputs, and the education of persons who will be responsible for these myriad forms of activity.

Figure 1.1 shows the distribution of engineering disciplines among the 1,483,014 engineers estimated by the U.S. Bureau of Labor Statistics to be holding engineering positions in 1990. Slightly more than 28% of these were electrical engineers, with electrical, mechanical, and civil accounting for about 56% of the total.

In Figure 1.2 we show the distribution of engineering employment by type of employer. In 1990 almost exactly half of all engineering jobs were in manufacturing industries, mostly in transportation equipment, electrical and electronic equipment, industrial machinery and equipment, instruments and related products, and chemicals and allied products. In the services category, the predominant employers were engineering and architectural services, research and testing services, computer and data processing services, and public and private education. More than half of those employed in government worked for the federal government, while one-quarter were employed by state governments and the balance in local government.

The data reported above are compiled by the Engineering Workforce Commission of the American Association of Engineering Societies from information furnished by the U.S. Bureau of Labor Statistics. The commission cautions that the statistics understate the actual totals slightly, because they are based on surveys of employers and do not include the self-employed (Engineering Workforce Commission, 1991).

Many studies have confirmed that the typical engineer spends at least part of his or her career in some sort of management role. For example, a study by the Engineering Manpower Commission indicated that as many as 82% of all engineers

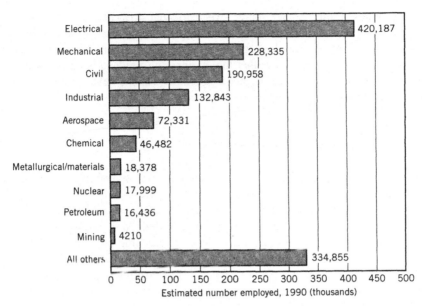

Figure 1.1. Engineering employment by discipline, for the 1,483,014 employed engineers in 1990. (From Engineering Workforce Commission, 1991.)

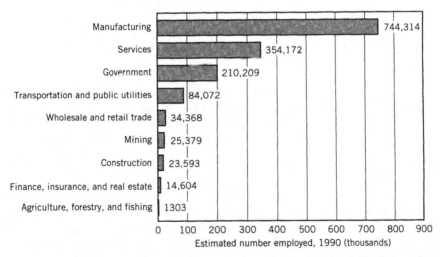

Figure 1.2. Engineering employment by type of employer, for the 1,483,014 employed engineers in 1990. (From Engineering Workforce Commission, 1991.)

in the United States have some form of engineering management responsibility. According to a Carnegie Foundation report, 30 years of surveys revealed that more than 60% of persons who earned engineering degrees either became managers of some kind within 15 years or left the profession to pursue business opportunities of other kinds. From another source, we learn that 40% of industrial executives and 34% of all top corporate managers in the United States have engineering backgrounds (Cleland and Kocaoglu, 1981). Statistics such as these are the real reason for this book. It is clear that the engineering graduate's career is likely to involve some degree of management responsibility. In the next section, we discuss some distinctions between engineering and management.

1.2   ENGINEERING VERSUS MANAGEMENT

Even though we have classified the management of engineering as one facet of the broad spectrum of engineering, it is still helpful to distinguish between the roles and activities of the traditional nonmanager engineer and the manager. First, let us note the ways in which three authors have defined *management*. According to Certo (1985), management is "the process of reaching organizational goals by working with and through people and other organizational resources," while Griffin (1987) calls it "the process of planning and decision-making, organizing, leading and controlling an organization's human, financial, physical and information resources in an efficient and effective manner." Management has also been defined as "the process of acquiring and combining human, financial, informational and physical resources to attain the organization's primary goal of producing a product or service desired by some segment of society" (Pringle, Jennings, and Longnecker, 1988).

Although there are some differences among these definitions, it is not insignificant that each calls management a *process*. A classic description of the functions of this management process was written in 1949 by Henri Fayol in *General and Industrial Management:*

> To manage is to forecast and plan, to organize, to command, to coordinate and to control. To foresee and provide means examining the future and drawing up a plan of action. To organize means building up the dual structure, material and human, of the undertaking. To command means maintaining activity among the personnel. To coordinate means binding together, unifying and harmonizing all activity and effort. To control means seeing that everything occurs in conformity with established rule and expressed command.

Incidentally, Fayol's book, now out of print, is considered by many to be *the* basic work in the area of management process, from which many other authors have drawn. A used copy of this 96-page classic would be an important addition to the library of the well-read engineering manager. Chester Barnard (1960) provides a simpler three-function categorization: (1) formulate and define purpose, (2) provide the system of communication, and (3) promote the securing of the essential efforts of the required members of the organization. This author says that the manager has primarily the roles of planner, facilitator, and communicator. In Section 1.5 we shall develop further some important functions of the engineering manager.

In distinguishing between technical work and management, Kamm (1989) advises young professionals that as engineers they confront a world that is primarily physical: "Managers deal primarily with people, politics, and administration and secondarily with making or approving decisions requiring technical judgment."

A helpful distinction between the roles of the engineer and the manager is given in Table 1.1, based on some research by Morrison (1986). The survey from which this table was developed indicated a belief on the part of many engineering managers that their engineering backgrounds helped them in their roles as managers. In her article, Morrison suggests the following bases for their reasoning:

1. They [engineers] are logical, methodical, objective, and make unemotional decisions based on facts.
2. They use their technical knowledge to check the validity of information.
3. They can analyze problems thoroughly, look beyond the immediate ones, and ask good questions to explore alternative solutions to technical problems.
4. They understand what motivates engineers.
5. They can review and evaluate the work of their subordinates because they understand what they are doing.
6. They can engage in future planning with appropriate consideration for technology and its relationship to cost effectiveness.
7. Engineering backgrounds help in technical discussions with customers.
8. The engineering background increases the manager's credibility with subordinates, customers, and superiors. People attribute qualities, abilities, skills, and knowledge to them, which allows the manager to influence those who have that perception.

Another way to view the distinction between engineering and management is to look at some of the "operational aspects." Here are four aspects of the manager's activities as taken from a book by Gray (1979):

1. The time spent with people, as opposed to time spent on the details of technical work, increases sharply. . . .
2. . . . .The manager becomes a part of the management information system of the firm. . . .
3. The nature of the rewards of the job and the satisfaction to be gained from it change. [Whereas] the engineer takes pride in a tough problem well mastered. . . (a plan, a device, information, a patent, etc.). . . the manager must take satisfaction in the higher plane of personal interactions, namely, the decisions which affect more money, more resources, more accomplishments than were available to him as an engineer. . . .
4. The nature of the reports a manager will read is vastly different from those the engineer will read. . . .

In Section 1.5, we shall look further at some typical roles that constitute the engineering manager's job.

TABLE 1.1 Role Differences Between Engineers and Managers

Position	Engineer	Manager
Focus	More concerned with things technical/scientific	More concerned with people
Decision making	Makes decisions with much information, under conditions of greater certainty	Makes decisions often with inadequate information, under conditions of greater uncertainty
Involvement	Works on tasks and problem solving personally	Directs the work of others to goals
Process outcomes	Work based on facts with quantifiable outcomes	Work based on fewer facts, less measurable outcomes
Effectiveness	Depends on personal technical expertise, attention to detail, mathematical/technical problem solving, and designing	Depends on interpersonal skills in communication, conflict management, getting ideas across, negotiating, and coaching
Dependency	Experiences role as autonomous	Experiences role as interdependent
Responsibility	Individual accomplishment in one project, task, or problem at a time	Many objectives at once, requiring orchestrating a broad range of variables and organizational entities
Creativity	Creative with products, designs, materials	Creative with people and organizations
Bottom line	Will it work?	Will it make/save money for the organization?

SOURCE: P. Morrison, "Making Managers of Engineers," *Journal of Management in Engineering*, Vol. 2, No. 4, © 1986. Reprinted by permission of the American Society of Civil Engineers.

1.3   THE TRANSITION TO ENGINEERING MANAGER

With the distinctions between engineering and management as background, we now turn to some considerations related to the changes, rewards, and challenges that face the engineer who assumes increasing management responsibility. A definition of this transition, suggested by Gray (1979), fits well with the discussion in Section 1.2:

> The transition process may be loosely defined as an exchange of mind sets, an exchange of physical parameters, and an exchange of operational parameters from those operative as an engineer to those operative as a manager. Those operative as a manager include the broader financial, personal, and information/decision aspects.... The new manager begins using measures of success and worth to the firm which are completely different from anything done before as an engineer and which have multiple concerns instead of the technical/budget/time concerns which previously were sole operatives. The manager must, therefore, maintain concern with the *what* and *why* instead of the *how* to become successful and respected.

Notwithstanding the differences between engineering and engineering management, it may be helpful to soften the apparent stark contrast between the two, and thus make the transition a little easier, by describing some ways in which they are similar. Cleland and Kocaoglu (1981) point out that the engineer is primarily concerned with decisions involving the material subsystem, whereas managers' decisions have more often to do with the human subsystem. They remind us, however, that engineering and management have the same basic philosophies:

> Both engineers and managers are trained to be decision makers in a complex environment. They both allocate resources for the operation of existing systems or for the development of new systems. Both have to recognize, identify and evaluate the interfaces among system components.

Hensey (1992) describes four stages of professional development, based on some work by Dalton, Thompson, and Price (1977). Successful engineers and scientists, like other professionals, tend to begin as "apprentices," then move to the "professional" stage, then to the role of "mentor," and finally to the "sponsor" stage (see Table 1.2).

This four-stage breakdown seems especially helpful as we consider the transition from engineer to engineering manager. Stages I and II are most closely connected with the technical aspects of the engineer's job, whereas stages III and IV are more related to the managerial aspects. Stage III is called the mentor stage "because of the increased responsibility individuals in this stage begin to take for influencing, guiding, directing and developing other people" (Dalton, Thompson, and Price, 1977).

It is frequently stated that two-thirds of engineers spend two-thirds of their careers in some sort of management role. There are good surveys that confirm this claim. The results of one such study are shown in Figure 1.3, which is taken

TABLE 1.2 Stages of Professional Development

	Initial Stages of Professional Development		Later Stages of Professional Development	
	Apprentice Stage I	Professional Stage II	Mentor Stage III	Sponsor Stage IV
	• Works under the supervision and direction of a more senior professional in the field.	• Goes into depth in one primary technical area.	• Significant technical contributions while working in several areas.	• Influences future direction of organization through:
	• Is given assignments which are a portion of larger projects or activities; overseen by senior professionals.	• Assumes responsibility for a definable portion of the project, process, or clients.	• Greater breadth of technical skills and application of those skills.	(a) original ideas, leading organization into new areas of work,
	• Lacks experience and status in organization.	• Works independently and produces significant results.	• Stimulates others through ideas and information.	(b) organizational leadership and policy formation,
	• Is expected to exercise "directed" creativity and initiative.	• Relies less on supervisor or mentor for answers; develops own resources to solve problems.	• Involved in developing people in one or more ways:	(c) integrating the work of others.
	• Learns to perform well under pressure and accomplish a task within the time budgeted.	• Develops credibility and a reputation.	(a) acts as an idea person for a small group,	• Influence gained on basis of:
		• Increases in confidence and ability.	(b) serves as mentor to younger professionals,	(a) past ability to assess environmental trends,
		• Still needs guidance on broad aspects.	(c) assumes a formal supervisory position.	(b) ability to deal with the outside effectively,
			• Works outside in relationships with client organizations, developing new business, etc.	(c) ability to affect others inside the organization.
				• Engages and interacts with individuals and groups both inside and outside the organization.
				• Involved in the development of future key people. A sponsor for promising people in other stages. (Sponsor implies the power to move people and assign responsibilities.)

SOURCE: Hensey (1992).

8

from a study performed 20 years ago by the Engineering Manpower Commission, the forerunner of the Engineering Workforce Commission (Engineering Manpower Commission, 1973). Note that from age 36 upward, the percentage of engineers involved in team, project, division, or general management is never less than 66%. It is unlikely that these data have changed significantly since 1973. Thus, the data confirm the transition that is the subject of this section.

"The transition of an engineer to a managerial position is usually one of the most difficult to accomplish within a technical organization" (Koza and Richter, 1988). Amos and Sarchet (1981) use the word "traumatic" to describe most such cases. Why is this so? Two primary reasons are as follows: (1) the differing roles (such as shown in Table 1.1) are neither expected nor well understood, and (2) there is a lack of preparation and training for the new responsibilities. Morrison (1986) quotes some engineering managers from her survey:

"I didn't understand the breadth of the role and the organizational priorities."

"I had no training on company policies and procedures."

"I experienced a lack of rapport with people."

"The fundamentals of management were never taught to me."

A helpful summary of the broader scope of general problems encountered by the engineering manager comes from Koza and Richter (1988), who urge that the new engineering manager be inculcated with an appreciation for these issues:

1. Problems relating to people need more time and attention compared with problems relating to technical matters.
2. Problems relating to information take on a different meaning. Not only does the manager have to deal with a broader set of uncertainties but the problems

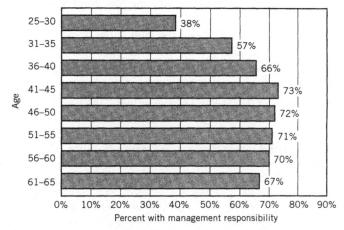

Figure 1.3. Responsibility for team, project, division, or general engineering management as a function of age. (From Engineering Manpower Commission, 1973.)

now deal with unfamiliar subjects such as finance, marketing, organizational structure and politics.

3. Problems relating to engineering's place in the organization's strategic plan are now of critical concern. Previously the issue may have been of just passing interest, but now the matters of budget and personnel allocated to the engineering group may mean the difference between success and failure of the entire organization.

4. Problems relating to the overall profitability of the organization are now of major importance and require immediate, decisive and proper attention as they affect and involve the engineering group.

Thus, the skills required of the engineer (specialist) and the engineering manager (generalist) are considerably different. Koza and Richter (1988) suggest (1) a broad organizational outlook, (2) an understanding of its operations, (3) people-relating skills, (4) communication skills, and (5) some knowledge of specific business areas such as marketing and finance. In later sections of this chapter and others we shall discuss in greater detail the roles, functions, and skills of the engineering manager.

The list of four "problems" just quoted from Koza and Richter should be sufficient warning that engineering management is not all sweetness and light. Three disadvantages of an engineer moving into management that often contribute to failures of many new managers are pointed out by Gray (1979):

First, the promotion removes the specialist from intimate contact with the technical details resulting in the problem of instant obsolescence; this, in turn generates fear and guilt—the engineer no longer has fall-back strength in his specialty. Second, a position in management requires skills different from those which were learned as an engineer (as becomes obvious to the new manager after only a few minutes on the job). Third, and giving rise to most problems, management requires dominance of personality traits and characteristics which are alien to most engineers—dealing with the diffuse, the intangible, the intractable and with insufficient information.

It was suggested earlier that there are similarities between engineering and engineering management, and it has been implied that in many cases the preparation and experience obtained by engineers are important "pluses" in management roles. While the author believes that this is generally true, we should also heed these words by Silverman (1987), which reflect some of the differences discussed above.

Our previous successful technical educational processes might even be a hindrance now since they were based on an orderly and predictable view of the world. It was pragmatic. By understanding and accepting its rules, it was possible to provide an answer to the particular problem being faced. That doesn't equally apply to management. Rules are less obvious since they are based in the emotions of human beings. There are many more of them and they are less predictable. It requires a different kind of thinking to deal with our fellow human beings well. Like the sciences, management is concerned with predicting and controlling change, but that change is almost wholly dependent upon the creativity of individuals, who are much less predictable than nature. There are many more variables.

Whatever our view of the fitness of engineers and scientists to assume increasing management responsibilities, it is clear that the transition will be made easier for the individual and the organization if proper and sufficient planning and training for the transition are provided. Hoffman (1989) identifies the following behavioral traits as important for the technical manager and argues convincingly that virtually all these skills can be learned:

the ability to live with ambiguity and uncertainty along with a strong desire and growing ability to structure the work environment so that these uncertainties and ambiguities become manageable

leadership, or the ability to inspire the confidence of others so that they are successful followers

risk taking, or the willingness a) to look for new and better solutions to a given problem and b) to allow subordinates to make mistakes as a technique for "unfreezing" the search for technical or managerial solutions

delegation, or the ability to let others work their part of the problem

team building and, following on from this, the ability to get others to work with each other constructively with the least amount of guidance possible, given the situation

communication, or the ability to convey ideas to others while understanding what others are telling you

initiative, that is, the willingness to step out front and guide a team effort.

There are many tests and other aids to assist in identifying engineers with particularly strong managerial potential. Also, the young engineer can demonstrate a motivation for managerial responsibilities through success in prior engineering positions, an ability to direct the work of others, and a capacity to manage stress, both in him- or herself and in others. The important role of senior managers in identifying such individuals and then guiding a successful transition program cannot be overemphasized. It is important that a "career roadmap" be developed and followed; here the senior manager's role as mentor and sponsor becomes apparent. There must be a plan, and there must be procedures to guide and track the progress of this plan.

Hoffman (1989) cautions that the process of training competent managers can be long and arduous. Successful managers tend to grow into their jobs over an extended period of time. Furthermore, technical specialists become managerial generalists by gaining experience in their specialized fields and then applying this knowledge to increasingly broader ranges of the organization's activities.

There are many possible approaches to providing engineering management development opportunities. Koza and Richter (1988) urge a comprehensive approach that consists of both formal and informal programs. Formal programs include on-the-job training, in-house classes, outside instruction, and rotation of assignments. Again, the importance of a qualified, interested senior manager as mentor/sponsor of such a program should be stressed.

While formal programs are important, the authors just cited emphasize informal programs, which consist of both "volunteer" and "visibility" opportunities. Examples of volunteer activities include teaching in-house classes and orientation sessions, hosting technical seminars for communication of information, and leading quality classes, quality teams, and the like. Visibility opportunities provide exposure to senior management and others and can include project reports, proposal coordination assignments, interface responsibilities with clients or vendors, and temporary supervisory assignments during vacation periods.

Hoffman (1989) divides management education programs into three elements. The first two are behavioral education (including such "people skills" as motivation, team building, communications, and delegation) and cognitive education (including production, marketing, finance, and control). While these two areas are important, the third, environmental education, is "somewhat newer, somewhat less conventional and perhaps the most critical in today's competitive environment."

By "environmental education," Hoffman means the development of an understanding of the total external environment in which the organization operates— markets, competitors, customers; political, social and economic conditions. The process of acquiring such an understanding is one of envisioning, broadening, and sensitizing. Hoffman says that such a new sensitivity

> manifests itself as a willingness to stay close to the customer, as an ability to develop subordinates' careers, as flexible responses to a broad and oftentimes confusing array of business threats and opportunities, and as a willingness to develop an empathy for diverse forms of cultural and national behaviors. Most CEOs will agree that this environmental sensitivity is the hallmark of the successful executive.

Finally, we present a quotation by Gray (1979) that emphasizes the need for a change in "mind set" and then make a comment. The quotation is as follows:

> The mental aspects of the transition are hidden and make their impact in a more subtle way. The manager whose "mind set" is still in the technical realm will be fascinated by the technical details and give shorter shrift to the financial and people details; this manager will have problems that will get progressively worse without recognizing them until a disaster takes place. The manager whose "mind set" changes from the engineer to that of a person responsible for resources and managerial decisions will become less concerned with retaining familiarity with every technical nuance and will accept the inevitable retreat from the cutting edge of the technical realm. This manager learns to use technical knowledge as an *auditing tool*—a base from which to question, probe and evaluate the technical work done by others. The managerial "mind set" is the one that is opened to new perspectives and makes the mental transition equal, at least, to the physical transition.

The comment is that since often the engineering manager still maintains some degree of responsibility for the technical details, it may not be possible to alter the "mind set" to the extent suggested by Gray. A manager who is stretched in a multitude of directions is pressured to be both generalist and specialist. The brief section that follows considers the need to maintain a degree of technical competence.

1.4 MAINTAINING TECHNICAL COMPETENCE

In a later section, we deal with the responsibility of the engineering manager for the career development, including maintenance of technical competence, of the employees in his or her charge. Here we discuss the importance to the manager of keeping up with personal technical interests.

First, it must be reemphasized that the transition to a management role will necessitate giving up some technical responsibilities. While such a shift may imply a decreased need for technical competence, the role of *engineering* manager brings with it a corresponding need for some expertise in the technical realm. If this sounds like a bit of waffling, it is because the question is not easily confronted and answered. The author feels that even though the need for continuing technical competence varies with the type of management job, it will nearly always be necessary to understand the technical aspects. For example, how can the design manager, legally and ethically, stamp a drawing and state, "This work has been accomplished under my direct supervision" without knowledge of the background that went into that drawing?

Opportunities abound for the engineering manager to keep up with his or her technical field. There is an abundance of literature in every field. Abstract services can assist in identifying and locating publications of particular interest, and computer bibliographic databases provide search services to locate materials worldwide. An example is COMPENDEX PLUS, a computerized database containing over 2.8 million citations, with abstracts, from more than 4500 journals, reports, books, and conference proceedings, in all fields of engineering and technology, including the related areas of applied science, management, and energy (Marcaccio, 1993). Technical society activities, at both local and broader levels, provide one-hour through one-week programs focused on recent research and application. Universities present short courses and long courses, both credit and noncredit. Other organizations offer continuing education opportunities, including seminars, short courses at the worksite, in a more central location and even by television. Examples of this last opportunity are found in the National Technological University, a consortium of more than 40 of the top U.S. schools of engineering, which offers graduate credit courses and master's degrees, as well as a wide range of technical and managerial seminars and short courses, exclusively by satellite television (National Technological University, 1992).

The topic of the engineering manager and continuing technical competence relates not only to the manager as an individual but also to his or her organization. The manager has a major responsibility to assure that engineers and others in the organization keep up to date and that through them, the organization itself resists obsolescence. Thus, a section head can be an important role model by attending, with her engineers, a seminar on the latest trends in VLSI design or a short course on research in earthquake engineering. The engineering manager also can support the maintenance of technical competence of staffers and organization alike, by rewarding individuals for technical competence and by actively helping employees plan for and progress through their careers.

1.5 THE ENGINEERING MANAGER'S FUNCTIONS AND ACTIVITIES

We have already listed some skills likely to be important to the engineering manager, tabulated some important role differences between the engineer and the engineer manager, and reported Fayol's five categories of management functions: planning, organizing, commanding, coordinating, and control. Now it is appropriate to focus on the functions the manager of *technical* activities is likely to undertake and to describe some typical activities.

We adopt here Shannon's (1980) seven functional categories, which are summarized in Table 1.3 and described in a brief paragraph apiece below. In addition to Shannon, we draw from Amos and Sarchet (1981) and Cleland and Kocaoglu (1981), as well from personal experience and observation.

Planning consists of deciding, in advance, what to do, as well as how, where, and when to do it, and who is to do it. It involves setting long-term goals and objectives ("strategic planning") and short-term action plans ("tactical planning"). Technical goals depend on the organization's technical capabilities and on forecasts of future conditions. Technological forecasting helps predict these conditions and their causes and effects. Planning is required for both the engineering organization and individual projects. In both cases, financial planning in the form of budgets is included. Scheduling, especially of project activities, is essential to success, and various project scheduling techniques are available to assist.

TABLE 1.3 Seven Engineering Management Functions

Planning. Anticipating future events. Making preparations to meet those events. Long- and short-range goal setting. Scheduling. Budgeting. Technological forecasting.

Organizing. Establishing communication, authority, and responsibility patterns. Assigning roles, facilities, and equipment. Organizational change. The "informal" organization.

Staffing. Deciding staff needs. Finding, hiring, and training people. Matching organizational needs and employee expectations. Meeting employment regulations.

Motivating. Providing incentives and a productive environment. Balancing "hands off" supervision with a more direct approach. Allowing and encouraging professional development.

Communicating. Writing. Speaking. Reading. Listening. Conveying goals, purposes, information, instructions, and inducements.

Measuring. Monitoring and evaluating individual and group performance. Comparing actual performance with goals and plans.

Correcting. Implementing change, based on the measuring function.

SOURCE: Based on Robert E. Shannon, ENGINEERING MANAGEMENT. Copyright © 1980. Reprinted by permission of John Wiley & Sons, Inc.

Organizing is required to achieve effective utilization of the people, facilities, equipment, and money available to the enterprise. While it does include determining and documenting the patterns of communication, authority, and responsibility ("drawing up the organization chart"), it also involves assigning roles, facilities, and equipment. We devote the next chapter to the engineering organization, its structure, it effectiveness, and the various relationships both within the organization and outside it. The engineering manager may be concerned with establishing and maintaining a temporary, project-type organization or, alternatively, a more permanent organization may be appropriate. Organizing also includes the increasing importance of work teams and clusters, as well as the necessity for organizational change in response to changing needs and technologies. Furthermore, we must recognize the existence of informal organizational structures, in addition to those shown on the organization chart of the project, agency, or company.

Once the organization has been defined and established, management must turn to the matter of *staffing:* deciding what kinds of people are needed, finding them, hiring them, and training them. Staffing also involves matching the organization's needs with each employee's goals and desires, and it requires adherence to equal opportunity regulations. For the technical organization, important staffing challenges include finding persons with (1) a balance of technical, communication, and leadership potential and/or skills, (2) sufficient technical training and background already accomplished, and (3) the potential for maintaining and improving competency during the time of employment.

When the people have been hired and trained, *motivating* them becomes an important engineering management responsibility. We must encourage staff members to meet their own needs as well as those of the organization by providing incentives and an environment that help them to put forth their best efforts. One concern here is the tricky balance between leaving skilled technical people alone to be creative ("their work is sufficient motivation in itself") and actively providing direction, rewards, tools, and information. Another concern is the need to combat technical obsolescence by encouraging employees to keep current in their disciplines. In Chapter 5, we include a major section on employee motivation.

Communication is a multifaceted management responsibility. It has both several purposes and several methods. Engineering managers communicate goals and purposes; they communicate information, instructions, and inducements; they assure that messages are sent and received, in a multitude of directions. Communication is both formal and informal; it is both written and oral. It involves reading and listening as well as writing and speaking. It utilizes paper, telephone, face-to-face, and electronic means. Fayol (1949) complained about his mining engineers' lack of competence in written communication: "Our young engineers are, for the most part, incapable of turning the technical knowledge received to good account because of their inability to set forth their ideas in clear, well-written reports, so compiled as to permit a clear grasp of the results of their research or the conclusions to which their observations have led them." All the actions taken in fulfilling the planning, organizing, staffing, and motivating functions already described must be communicated effectively if they are to be effective. This function is so essential that we devote all of Chapter 6 to it.

Measuring involves monitoring and evaluating performance with respect to technical, scheduling, or financial aspects of a project. This function compares the goals set forth during the planning function with actual performance. Thus, it can include evaluating progress by comparing a project's current schedule status with its baseline schedule. It may involve monitoring and evaluating the performance of a single engineer and comparing that performance with that individual's planned career development to date.

Finally, the engineering manager is responsible for *correcting* aspects of the organization's operations based on the results of the measuring function. If the process that measures costs for a building project indicates that a certain element is over budget, the project manager must attempt to improve that operation. Dysfunctional organizations have to be "corrected" by bringing about change. Individuals who are performing improperly may need to be "corrected," although, as we shall find in Chapter 3, often the system, not the individual, is the element that requires "correcting." The measuring and correcting functions are sometimes called "controlling," which Fayol defined as "seeing that everything occurs in conformity with established rule and expressed command."

The engineer manager carries out Shannon's seven functions, as described above, by means of activities that vary greatly from person to person and from task to task. However, there seem to be some common trends. Research conducted in the United Kingdom (Barclay, 1986) compiled the results of a questionnaire survey of 263 technical managers, who were asked about the kinds of activity in which they frequently engaged and the activities that generated their greatest problems. The results are summarized in Table 1.4, which shows the percentage of respondents who indicated they engaged in the listed activities "frequently," and in Table 1.5, which ranks the seven activities according to percentage of respondents who put them at the top of their list of problem generators.

Table 1.4 shows that three-quarters or more of the technical managers in the survey engaged in people management, projects, interpersonal activities, and innovation frequently, with people management being reported frequently by 93%. With respect to causes of problems, people management was clearly the most "popular" first choice (Table 1.5). This finding further underscores the extreme importance of understanding how communication really works in organizations. The low ranking for interpersonal activities as problem causes (5%) seems to reflect a tendency of engineering managers to view communication faults as lying with the other person ("People are a problem, but my interpersonal relations with them, as exemplified through communication, are not *my* problem.").

1.6   A DEFINITION AND SOME DISTINCTIONS

We conclude this chapter by offering a formal definition of engineering management and describing some distinctions between engineering management and other sorts of management.

TABLE 1.4   Percentage of Engineering Managers Who Report Certain Activities Are Performed "Frequently"

Activity	Percent Who Reported "Frequently"
People management	93
Projects	84
Interpersonal	83
Innovation	75
Profit/efficiency	65
Information processing	62
Money (budgets, etc.)	59
Production services	52
Computing	38
Industrial relations	37
Production systems	28
Corporate	21
Management science	20
Commercial	19
Legal	11

SOURCE: I. Barclay, "A Survey of the Activities, Problems and Training Needs of Technical Managers." ENGINEERING MANAGEMENT INTERNATIONAL, © 1986. Reprinted by permission of Elsevier Science Publishers, B.V.

TABLE 1.5   Percent of Engineering Managers Who Report Certain Activities Give Them the Greatest Problems

Activity	Percent Who Report This as First Choice
People management	34
Projects	10
Industrial relations	10
Money (budgets, etc.)	7
Innovation	5
Profit/efficiency	5
Interpersonal	4
Others	<3

SOURCE: I. Barclay, "A Survey of the Activities, Problems and Training Needs of Technical Managers." ENGINEERING MANAGEMENT INTERNATIONAL, © 1986. Reprinted by permission of Elsevier Science Publishers, B.V.

The late Merritt Williamson (1982) defined engineering management as "the art and science of planning, organizing, allocating resources, directing and controlling activities which have a technological component." He distinguished engineering management from *industrial engineering* by its greater focus on "people problems" rather than system design, and from general management in its requirement that practitioners be competent in some technical field. However, engineering managers "may be found working wherever a blending of managerial and technical knowledge is required, which may be in any organization, whether or not its primary business is technological."

Babcock's paper on distinctions between engineering managers and others (1978) describes the differences as follows:

> The engineering manager is distinguished from other managers because he or she possesses both an ability to apply engineering principles and a skill in organizing and directing people and projects. He is uniquely qualified for two types of jobs: the management of *technical functions* (such as design or production) in almost any enterprise, or the management of broader functions (such as marketing or top management) in a *high-technology enterprise.*

Just what are the significant differences between the technical manager's functions and activities and those associated with other kinds of management? Quoted below are key points excerpted from a helpful description of five of these differences provided by Shannon (1980).

1. Technical functions are associated with creating something new or improving the current method of operation. It [engineering management] is oriented to *innovation* or *change*.... Technical groups are involved with developing the new or changing the old; this includes the activities of machines and people. Operational groups, on the other hand, deal with more predictable, well-defined tasks.... Most operational functions are repetitive....

2. Technical functions usually deal with one-time activities. Once a study is done, a new machine designed, or a new system developed, the work is seldom repeated. Operational work, by contrast, is associated with routines that repeat themselves periodically (hourly, daily, weekly, etc)....

3. The costs of one-time activities are much more difficult to estimate ahead of time. Estimates of operational work can usually be predicted based on historical data....

4. There are differences in the way we can expedite work. We can generally increase the production of an operational activity in proportion to the resources applied.... Too many people on a technical project may be less efficient than not enough people.

5. Our ability to measure performance is different. If 50 percent of the time and resources planned for a technical project are expended, we do not know whether the project is in trouble or not.... Furthermore, most technical functions do not directly bring about increased sales or reduced costs. Their work results in a payoff only when an operational group implements it.... Project performance should be measured on project profitability or return on the invested cost of the technical work, but the profitability is often outside the control of the technical manager.

Shannon concludes:

We can, therefore, sum up the major differences as follows: The technical manager is concerned with one-time, future-oriented tasks directed toward innovation and change. The resources required and end results are highly uncertain and unpredictable. The operational manager, on the other hand, is concerned with present-oriented, periodically repetitive tasks; strict adherence to predetermined policies, procedures and methods is the desired goal. The technical manager must deal with, motivate, and control highly trained, creative people in an uncertain environment that requires flexible planning, policies and procedures. The operational manager must deal with the enforcement of plans, schedules, policies and procedures where the results of decisions are much more predictable.

We have referred extensively to the small volume written by Henri Fayol nearly 50 years ago for its management wisdom that is relevant even today. His "Advice to Future Engineers" seems a fitting conclusion to this chapter.

Your future will rest much on your technical ability, but much more on your managerial ability. Even for a beginner, knowledge of how to plan, organize, and control is the indispensable complement of technical knowledge. You will be judged not on what you know but on what you do, and the engineer accomplishes but little without other people's assistance. (Fayol, 1949)

1.7 DISCUSSION QUESTIONS

1. Find out the number of majors in the various engineering disciplines at your school, and compare the percentages in each discipline with the national statistics shown in Figure 1.1. Why are the proportions different? How do you think the national statistics and the enrollments at your school will change during the next decade? Why?

2. Conduct a brief survey to determine the proportion of time spent in managerial versus technical aspects of the job by various engineers in your community. Each student will contact one or two local engineers to ascertain the estimated time devoted to managerial and to technical work on a typical day. In class, compile the results. Before starting the survey, decide what is meant by "managerial" and "technical," so you can provide a proper framework for the responses of your engineer contacts.

3. Explain what is meant by "The manager becomes a part of the management information system of the firm."

4. Are you convinced by Silverman's suggestion that our technical education is, or could be, a hindrance to success as a technical manager? Explain your position.

5. Hoffman's list of seven behavioral traits of importance in technical management is presented with the claim that each can be learned. Tell how an engineer might learn each skill.

6. Environmental education, according to Hoffman, will be especially important to technical managers in the future. Describe this "environment" and give some examples of how one might acquire such an education.

7. Should continuing professional education be required as a condition for membership in professional engineering organizations and for registration as a professional engineer? Before stating your opinion, give some reasons *for* and *against* mandatory continuing professional development.

8. Suppose you have just been assigned to manage a project for your construction company. Give an example of one task or series of tasks for which you might be responsible in each of the seven functional areas set forth in Section 1.5.

9. If Shannon is correct in his distinctions between the functions and activities of technical managers and those of other managers, then many organizations require vastly varied management functions and activities. Furthermore, it is quite possible that one individual may have responsibility for both technical and operational management. (a) Describe briefly an organization you know about that calls for both types of management. (b) What sort of individual position might require both technical and operational management?

1.8 REFERENCES

Accreditation Board for Engineering and Technology. 1993. *1993 Annual Report.* New York.

Amos, J. M., and B. R. Sarchet. 1981. *Management for Engineers.* Englewood Cliffs, NJ: Prentice-Hall.

Babcock, D. L. 1978. "Is the Engineering Manager Different?" *Machine Design,* Vol. 50, No. 5, March 9, 82–85.

Barclay, I. 1986. "A Survey of the Activities, Problems and Training Needs of Technical Managers." *Engineering Management International,* Vol. 3, No. 4, 1986, 253–260.

Barnard, C. I. 1960. *The Functions of the Executive.* Cambridge, MA: Harvard University Press.

Certo, S. C. 1985. *Principles of Modern Management: Functions and Systems,* 3rd ed. Dubuque, IA: Wm. C. Brown.

Cleland, D. I., and D. F. Kocaoglu. 1981. *Engineering Management.* New York: McGraw-Hill.

Dalton, G. W., P. H. Thompson, and R. L. Price. 1977. "The Four Stages of Professional Careers—A New Look at Performance by Professionals." *Organizational Dynamics,* Vol. 6, No. 1, Summer, 19–42.

Engineering Manpower Commission. 1973. *The Engineer as Manager.* Engineering Manpower Bulletin 25, September.

Engineering Workforce Commission. 1991. "Special Announcement." AAES Manpower Studies Department, American Association of Engineering Societies, December 20.

Fayol, H. 1949. *General and Industrial Management.* London: Sir Isaac Pitman & Sons.

Gray, I. 1979. *The Engineer in Transition to Management.* New York: IEEE Press of the Institute of Electrical and Electronics Engineers.

Griffin, R. W. 1987. *Management,* 2nd ed. Boston: Houghton Mifflin.

Hensey, M. 1992. Personal communication.

Hoffman, G. C. 1989. "Prescription for Transitioning Engineers into Managers." *Engineering Management Journal,* Vol. 1, No. 3, September, 3–7.

Kamm, L. J. 1989. *Successful Engineering: A Guide to Achieving Your Career Goals.* New York: McGraw-Hill.

Koza, H., and P. J. Richter. 1988. "Development of Engineering Managers: An Alternative Approach." *Engineering Management International.* Vol. 4, No. 4, January, 299–306.

Marcaccio, K. Y., Ed. 1993. *Gale Directory of Databases.* Detroit, MI: Gale Research.

Morrison, P. 1986. "Making Managers of Engineers." *Journal of Management in Engineering,* American Society of Civil Engineers, Vol. 2, No. 4, October, 259–264.

National Technological University. 1992. *Bulletin 1992–93: Academic Programs,* Fort Collins, CO: NTU.

Pringle, C. D., D. F. Jennings, and J. G. Longnecker. 1988. *Managing Organizations: Functions and Behaviors.* Columbus, OH: Merrill.

Shannon, R. E. 1980. *Engineering Management.* New York: Wiley.

Silverman, M. 1987. *The Art of Managing Technical Projects.* Englewood Cliffs, NJ: Prentice-Hall.

Williamson, M. A. 1982. "Engineering Management." In Heyel, Carl, Ed., *The Encyclopedia of Management,* 3rd ed. New York: Van Nostrand Reinhold.

2
The Engineering Organization

Nearly all engineers work in some kind of organization. Unless you have the rare one-person office, where you do your own word processing and supply every other internal support service, you are organized in some fashion and have formal and informal relationships with others in the partnership, corporation, or agency. In this chapter, we highlight the creation, maintenance, and evaluation of effective organizations. Several types of formal organizational structure are described, and the significance of the unofficial, informal organization is stressed. Within and outside the organization, the engineer often interacts with nonengineering groups and individuals; several such entities are listed. Finally, we speculate briefly about the future of technical organizations.

2.1 THE EFFECTIVE ORGANIZATION

2.1.1 Teamwork in Practice

Before we discuss organizational design, it is important to recognize that organization structure alone cannot assure a smooth, effective organization. Hensey (1990b) pleads that the degree of effectiveness of a work team is "much more a function of leadership, training, core organizational values, and culture, rather than organizational structure.... Teamwork in practice can assure the needed communication, creativity, and productivity in spite of poor structure. A good structure does not assure teamwork; leadership does."

A team can be defined as a group of people who have different talents and often are assigned different tasks but work together toward a common goal through a meshing of functions and mutual support. (In this context, a *team* is any group that meets this definition, whereas a *self-directed team* is a special category of team that is discussed later: see Section 2.2.6). Hensey (1990a) suggests that the essentials of developing effective teams are confined to three elements: (1) working on common tasks together (mission, strategic plan, budget organization, staffing, etc.), (2) having disciplined, effective working meetings at frequent enough intervals to get those tasks done and (3) using a simple yardstick for assessing the current level of team spirit, tangible teamwork, and/or team development.

Based on work by Francis and Young (1979), Hensey developed a useful chart that outlines the characteristics of work teams at four stages of development: collection, group, developing team, and high-performing team. We show this chart as Table 2.1 (Hensey 1990b). Think of a team of which you are a member, and try to assess the stage at which you would place that team, remembering that an organization's point on this spectrum "is much more a function of leadership and training...than organizational structure."

To utilize this chart to assist an organization, whether it be a project team, a volunteer board of directors, or any other team, Hensey (1990a) proposes four steps, which may be paraphrased as follows:

1. Ask each member to review the stages and characteristics silently and select a number (on a scale of, say, 0 to 12) that represents the development level of this team on a representative day.
2. Ask members to volunteer their numbers, which are written on a pad for all to see, with no further comments until all have reported.
3. Hold a brief discussion of the numbers and their meaning.
4. Then discuss the question, "What might we do, collectively and/or individually, to enable our group to move further along in its development toward being a high performing team?"

TABLE 2.1 Stages of Work Team Development[a]

Collection	Group	Developing Team	High-Performing Team
Cautious	Individual performers	Sorting out roles	Have high group
Testing	Sparks of conflict	and skills	productivity
Wondering	(unresolved)	Setting goals and	Sense of belonging
Playing it safe	Lack group	procedures	Very visible support
Not committed	productivity	Asking for and	by team members
Polite, small talk	People feel stuck	giving assistance	Cover one another's
Differences are	Not supportive of	Feeling like a team,	work during
hidden or	one another	at times	absence
played down	Protecting turf and	Still have conflict;	Have respect and
	resources	usually resolved	regard for one
	Communication not	Communication open	another
	very open	but not very	Clear, direct
	Differences are	efficient	communication
	recognized and	Differences are	Lots of conflict;
	avoided	recognized and	quickly resolved
		acceptable	Differences are
			highly valued

[a]Team development is not a function of time so much as effort and desire; the stage of development of a team can go up or down very quickly as a function of effort/lack of, changes in leader/members/resources/purpose/skills.

SOURCE: M. Hensey, "Organizational Design: Some Helpful Notions," *Journal of Management in Engineering*, Vol. 6, No. 3, © 1990. Reprinted by permission of the American Society of Civil Engineers.

It is suggested that such a review and discussion be conducted regularly, perhaps quarterly, and also upon the occurrence of a significant event (e.g., a change in leadership, the completion of a plan, a revision in the organization).

2.1.2 A Descriptive Model

In Chapter 5, we shall present an interesting "leadership grid" that depicts in a simple, graphical form the managerial style of any particular manager. Developed by Robert Blake and Jane Mouton and originally published a generation ago (Blake and Mouton, 1964), this grid approach has been widely used to assess managerial/ leadership style and assist in making improvements thereto.

More recently, the same authors, with Anne McCanse, have applied this concept to the effectiveness of organizations instead of individuals, in a book entitled *Change by Design* (Blake, Mouton, and McCanse, 1989). In Figure 2.1, we show such a Grid (the basic diagram is copyrighted). In this case, the application was to the organization of airplane cockpits (of all things!). Note that the two axes in Figure 2.1 represent concern for performance and concern for people. For cell 1,1, adjacent to the origin, the authors use the label "impoverished management"; in such a work environment, minimum effort is exerted to perform required work and

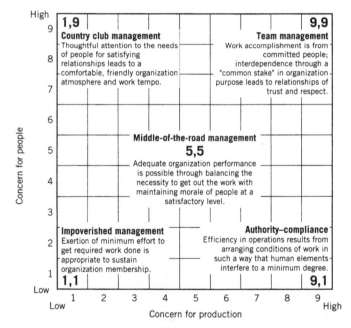

Figure 2.1. The Organizational Effectiveness Grid of Blake and McCanse (1991). [The Leadership Grid Figure from *Leadership Dilemmas—Grid Solutions,* by Robert R. Blake and Anne Adams McCanse (formerly the Management Grid Figure by Robert R. Blake and Jane S. Mouton), Houston: Gulf Publishing Company, p. 29. Copyright © 1991, by Scientific Methods, Inc. Reproduced by permission of the owners.]

to sustain membership in the organization. At the other extreme, in cell 9,9, we have "team management," where committed people accomplish work in relationships of trust and respect. These two extremes seem to correspond quite nicely to Hensey's notions of "collection" and "high-performing team" in Table 2.1. The grid is useful because it makes clear the importance to a high-performing organization of both performance and people.

While the leadership, or, in this case, organizational, Grid® is a descriptive model, it can be used to assist an organization not only in understanding its current stage of team development but also in moving toward a preferable cell. Thus, the title *Change by Design* suggests an orderly, structured approach to organizational development that leads to desired behavioral changes. Where does your organization fit on this grid, and what changes might be implemented to move you closer to cell 9,9? Of course, a part of this organizational self-study involves deciding what it means to "be at 9,9" for our particular organization.

We return to this grid model in Chapter 5 when we apply it to individual leadership styles.

2.1.3 Principles of Organizational Structure

What is a "good" or "appropriate" organizational structure? There is no single answer—every organization is unique with respect to goals, environment, activities, and culture. (Otherwise, we would long ago have found the "perfect" organizational form, and all would look alike!)

But we can offer some generally accepted principles for helping us decide the best form for a given set of circumstances. Badawy (1982) provides a helpful 25-page review of some of the "conventional and fundamental principles of organization," which we shall summarize briefly. They are offered with the caution that they are only guidelines; therefore, they need to be, and are, violated occasionally in every organization. But awareness of these principles is a good starting point as we design our organizational structures.

Badawy divides organizational structure principles into three major categories: (1) authority and power, (2) division of labor and specialization, and (3) coordination and communication. Each contains several principles which may be summarized as follows.

Authority and Power Principles

1. *Scalar Principle.* Authority and responsibility, communication, and decision making flow through a hierarchy of formal relationships from top to bottom.
2. *Unity of Command.* No subordinate reports to more than one supervisor on any given aspect of a job.
3. *Authority and Responsibility.* Authority (the right to act or to direct others to act in order to attain organizational goals) and responsibility (the duty to perform assigned tasks) are defined and balanced; the authority granted to an employee must suffice to permit the assigned responsibility to be fulfilled.

4. *Influence, Power, and Accountability.* The manager's ability to control others by suggestion or example rather than by direct command (influence), the ability to control someone else's behavior (power), and the liability created for the use of authority (accountability) are defined and understood. Accountability cannot be delegated.

Division of Labor and Specialization Principles

1. *The Objectives.* Clearly stated organizational objectives are articulated before the organizational structure is developed.
2. *Job Design and Task Specialization.* The extent to which the overall task is to be divided into specialties is considered.
3. *Departmentalization.* Decisions about how the organization is broken down into administrative units and how employees and activities are arranged by specialty and task are required.
4. *Span of Control.* The structure reflects decisions regarding how many people report to one supervisor.
5. *Decentralization Versus Centralization.* The level at which decisions are actually made (different from the extent of departmentalization, which may not reflect actual decision making) is defined.
6. *Delegation.* The process by which authority is transferred to subordinates is reflected in the structure.
7. *Line–Staff Relationships.* The line organization describes the superior–subordinate relationship in the chain of command from top to bottom; the staff organization provides advice and service outside this formal chain of command.

Coordination and Communication Principles

1. *Formal Structure or Hierarchy.* The formal structure is the "skeleton and framework of official relationships," which is the basis of coordination and communication.
2. *Policies and Procedures.* Guidelines (policies) and prescribed specific behavior (procedures) complement the organizational structure and assist with coordination and communication.
3. *Informal Communications.* The formal organization is developed with the understanding that other patterns of communication will be a real part of the organization.
4. *Committees and Task Forces.* The formal organization is also developed with the understanding that various temporary groups, not shown on the chart, will be necessary.

As we describe the several types of organizational structure, it will become evident that these 15 principles are not always adhered to! Especially in more recent inventions, including matrix organizations and self-directed teams, the organizational design purposely violates one or more of these guidelines. That's okay, as long as the reasons are clear.

2.1.4   Organizational Design: An Approach

Here is technical organizational consultant Mel Hensey's 12-step process for an organizational design or redesign by a management group. He reports that it has been developed during 20 years and has proven effective with technical organizations from 15 to 3000 people. The 12 steps can be accomplished in a one- or two-day intensive workshop or spread out over several weeks (Hensey, 1990b).

Step 1: Do a management function inventory for the key managers, working toward consensus in the group format. Identify each key manager's greatest functional strength, second greatest, and so on.

Step 2: Identify any problems you are trying to solve, improve upon, or address (e.g., weak client responsiveness, need for manager succession, turnover of senior personnel).

Step 3: Develop a list of criteria to be met by any changes you make in the organization. Brainstorm, with no criticism or debate at this step. Some criteria: responsibilities and accountabilities must be clear; use the best leaders where they are most needed; ditto those possessing other management skills; utilize a team approach to work wherever useful; do not fire or lay off anyone simply because of organizational changes; enhance service to clients/users.

Step 4: Prioritize the criteria from step 3 by having each member vote for his or her preferred one-third of the items listed. Use only the top priority criteria for the next steps, based on the top vote getters.

Step 5: Have each member of the planning group develop and sketch on chart paper at least two, preferably three, graphic ideas to improve part or all of the organization. If stuck for ideas, wander around the room and observe others' work. The purpose is to enhance divergent thinking in order to develop a creativity data bank. The keys are two or three ideas per person, graphic presentation, and individual work.

Step 6: Have each member show the group his or her ideas, emphasizing what's useful about each one. The group asks clarifying questions and makes comments on all the ideas, but there are no debates. The purpose here is to understand every idea while rejecting none.

Step 7: Form two-person subgroups, each of which will develop two complete organizational scenarios on chart sheets. Organizational reporting lines must be made clear. This step recognizes viable organizational alternatives; criteria from steps 3 and 4 should be met if possible. New or unusual roles or positions should be explained.

Step 8: Have each subgroup show the total group the scenarios, identifying useful features of each. The group then asks clarifying questions and comments on each scenario, but, again, no debate is allowed. At this step, the group should notice similarities and differences, pros and cons.

Step 9: The entire group participates in a discussion to pull the best from all alternative scenarios to make one final design or proposal to the person or group who will approve the changes. New ideas may show up and can be incorporated if there is general agreement.

Step 10: Test the design with names of actual managers. Who will fill various roles? This can be the most difficult step, but it is well worth the effort if the group can be candid.

Step 11: Turn over the final design recommendations, including the choices of people to fill key new or changed roles, to the person responsible for the organization for approval or modifications.

Step 12: Reconvene the group to hear the decisions, discuss the implications for changes, and plan for the step-by-step implementation of changes.

2.2 ORGANIZATIONAL STRUCTURES

2.2.1 The Organizational Chart Can't Say It All!

Organizational charts, though very useful, as we shall see, do not tell the whole story. They are excellent tools for identifying basic relationships among divisions of the organization and communication pathways, and for providing a convenient framework for budgeting, cost control, scheduling and resource allocation. However, these charts cannot depict organizational functions, cannot indicate how individuals relate to those functions, and cannot specify the varying degrees of authority, influence, and responsibility among persons holding the positions. Furthermore, they cannot show the informal organization structure and its informal lines of communication.

How much should you expect from the organization chart? Some answers are provided in Table 2.2, taken from work by Badawy (1982).

In the sections that follow, we describe and illustrate several types of formal organizational structure. Any such structure is a framework of the relationships that have been officially established between types of work, management levels, and/or geographical locations. While "official," these relationships may be either permanent or temporary. We also discuss the informal organization as a real and integral part of almost any work group.

Our discussion of formal organizations draws heavily on Badawy (1982) and Cable and Adams (1982), although most engineering management references offer helpful insights into and examples of organizations in practice (Babcock, 1991; Shannon, 1980).

2.2.2 Functional

In a functional organization, the major division of the structure is by function or discipline, with key activities grouped into such units as electrical design, finance,

TABLE 2.2   How Much Should You Expect from the Organization Chart?

What the Chart Can Tell You	What the Chart Cannot Tell You
1. The formal structure	1. Nature of various managerial roles and how they relate to different functions.
2. Departments, divisions, and their components	
3. Basic levels and layers of management and the chain of command	2. How the organization is functioning in reality
4. Major reporting relationships including who is (supposed to be) whose superior	3. Variations in people's authority, influence, and responsibility
5. Framework for budgeting and resource allocation	4. The informal lines of communication within the organization
6. Procedures, directives, and other related matters	5. Vast differences between line and staff concepts
7. Nature of activities to be performed by different organizational units	6. The informal organization structure
8. Bases for grouping activities (i.e., functional, regional, project, or product basis)	

SOURCE: M. K. Badawy, *Devloping Managerial Skills in Engineers and Scientists,* © 1982. Reprinted by permission of Van Nostrand Reinhold.

marketing, and organic chemistry. Such a hierarchy is most effective to manage specialists if they are grouped together and supervised by a manager who is similar in background and skills. Under such an arrangement, each individual reports to one supervisor and has a clearly defined career path, and the unit utilizes common resources and a common location.

The functional organization is probably the most prevalent structure in use; it is certainly one of the oldest. Figure 2.2 is an organization chart for an engineering design and research organization that has such a structure. A comparable structure could be used for a portion of a larger organization; for example, the engineering division of a manufacturing company might be subdivided by function. Note that additional details in some or all of the units would be typical of most organizations.

Advantages of the functional organization include the opportunity for technical specialization, minimization of duplication, and emphasis on career development and acquisition of new knowledge based on technical specialty. Drawbacks include the potential for slow flow of work, the difficulty of shifting personnel between departments, and the downplaying of a given project vis à vis departmental discipline. Technological innovation may be hampered by this last drawback, since many such advances come through the cross-fertilization afforded by multidisciplinary teams.

2.2.3   Product

Many organizations choose a type of structure that makes the primary division by product. A major car or soap manufacturer may be divided by brand—Chevrolet

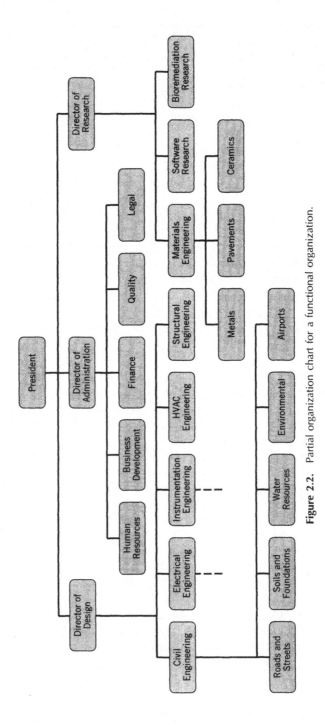

Figure 2.2. Partial organization chart for a functional organization.

Division, Ivory Soap Division, and the like. Within each of these primary divisions, which can be quite autonomous, there would typically be a functional breakdown for the various specialties—engineering, production, research and development, marketing, and so on.

This approach seems to be mainly confined to manufacturing organizations, which turn out "products"; engineering and research and development operations that support such activities—"Engineering and Research for Product Y"—may be placed within the product lines or within an overall engineering division, but with some connection to the actual manufacturing division.

Figure 2.3, a chart for a product organization, illustrates the duplication of functions within various product lines. Such duplication must be listed as a potential disadvantage, along with the possibility that the focus on "product" will diminish the emphasis on overall organizational goals and lead to "empire building" and reduced direction and control from top management. Advantages include a clarity of authority and responsibility, with each product manager in charge of his or her whole show, as well as the potential for matching customer requirements with the product produced through close relationships and teamwork involving marketing and field engineering personnel.

2.2.4 Project

In a pure project, or "projectized" organization, the structure is arranged with projects as the major subdivisions. It may retain functional elements, but the emphasis is clearly on the organization and management of projects. The chart in Figure 2.4, in which several project leaders report to a technical director, may represent only a portion of the company's total organization. In its purest form, each project is independent: its only connection to the permanent organization is through the project manager; it provides its own resources, plans its own work, and generally "does its own thing."

Advantages of this type of organizational structure include the clear emphasis on the project, which is possible because everyone is dedicated to a single goal or purpose, as well as better control over its direction, better customer relations (through the project manager as a single point of contact), clear accountability and responsibility of the project manager, and the potential for fast product development. Drawbacks to this approach include the potential for costly duplication of effort and resources (does each project really need its own data processing center?), the tendency for members of the project team to lose touch with technical developments (since they are connected to the project, not to their discipline), the insecurity associated with project completion, and the difficulty of transferring experience and knowledge from one project to another.

The similarities between the product and the project organization are apparent. We consider this project structure again in Chapter 7 in our discussion of project management.

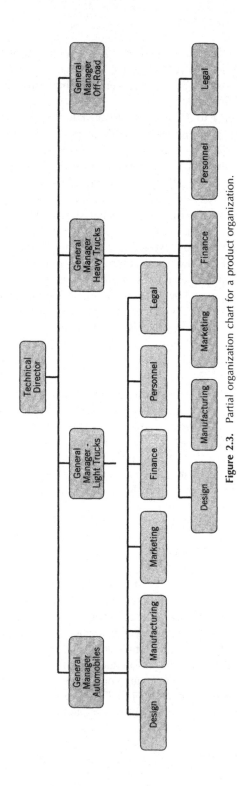

Figure 2.3. Partial organization chart for a product organization.

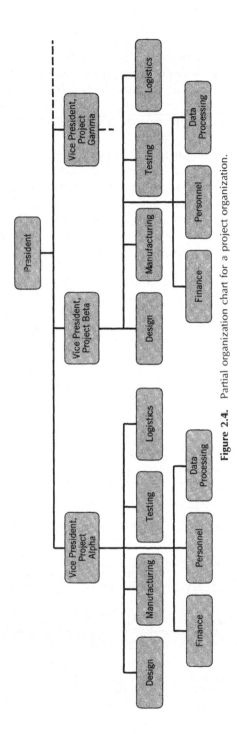

Figure 2.4. Partial organization chart for a project organization.

2.2.5 Matrix

The matrix organization combines, in a very intentional way, features of the functional organization, with its emphasis on functional and technical specialty, and the project organization, with its emphasis on achieving project goals. Originally developed in the U.S. aerospace industry to deal with unprecedented technical and managerial complexity, this structure was diagrammed as a two-dimensional table and thus designated as a "matrix."

Essentially, a project organization is superimposed on a functional organization, as illustrated in Figure 2.5. Staffers from a functional organization who are assigned to a project are shown at the intersection of those respective lines on the chart. They may be assigned full-time or part-time to that project and may work on more than one project during one day or week. They maintain a tie to their technical discipline as well as a temporary tie to their project or projects. As explained by Shannon (1980):

> The object of this superimposed configuration is to allow projects to be born, mature and terminate in an orderly manner. Each project manager has direct access to the specific personnel and expertise required for these projects for a planned time duration and to the rapid release of surplus manpower as project needs decrease.... The project organization can expand and contract as dictated by budget and technical needs without the need for layoffs or new hires.

A certain degree of difference of opinion has arisen over the effectiveness of the matrix as a management structure. Peters and Waterman (1982) describe

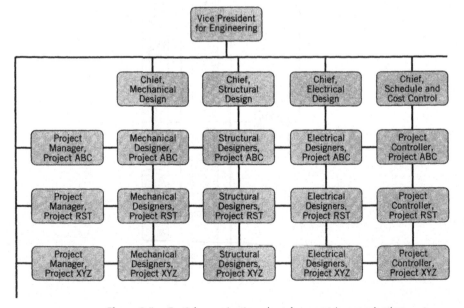

Figure 2.5. Partial organization chart for a matrix organization.

the matrix as "our favorite candidate for the wrong kind of complex response" to large complex systems and structures. They say that it "virtually always ceases to be innovative...has particular difficulty in executing the basics...and regularly degenerates into anarchy and rapidly becomes bureaucratic and noncreative." On the other hand, many companies have found the matrix to be particularly well suited to their needs. CH2M-Hill, an international consulting engineering firm, reports project success and company growth under this system (Poirot, 1991). In fact, the CH2M-Hill structure can be described as a three-dimensional matrix, the third dimension being by location. An employee "in the trenches" is likely to have three bosses—a technical discipline director, a project manager, and a regional manager through department and division managers. Table 2.3 lists Poirot's suggestions for making the matrix work.

Since the matrix organization, or some variant of it, is often used for technical projects, we deal with it further in Chapter 7. Here let us report some research results and then summarize advantages and disadvantages. Lawson (1986) conducted an extensive literature search and reported that the most prevalent matrix values were increased market responsiveness, better coordination and policy decisions and more effective cost management. Shortcomings identified most often were personal dissatisfaction, difficulties in adjusting to multiple managerial styles, and difficulties in setting priorities. Gobeli and Larson's (1986) study of more than a hundred project management professionals determined that for overall project effectiveness, the pure project organization was most preferred, but various types of matrix organization were effective for at least three-quarters of their projects.

Among the advantages of the matrix organization are high visibility of project objectives, a high degree of coordination by the project manager, flexibility in utilizing limited resources, rapid response to contingencies, potential

TABLE 2.3   Key Elements for Matrix Success

1. Allow time to define responsibilities/authorities.
2. Commit senior management time.
3. Develop people who want to make the matrix work.
4. Make decisions based on what is good for the client and the firm.
5. Promote open communication with no secrets.
6. Eliminate politics at high levels.
7. Commit energy to evaluate and compensate on a common basis.
8. Use consensus management.
9. Hire top people.
10. Consolidate net income at corporate level and reward everyone in the firm.

SOURCE: J. W. Poirot, "Organizing for Quality: Matrix Organizations," *Journal of Management in Engineering,* Vol. 7, No. 2, © 1991. Reprinted by permission of the American Society of Civil Engineers.

minimization of conflicts, and the security afforded personnel, who can return to their functional organizations when the project ends. Disadvantages include the confusion inherent in reporting to two bosses, the need for more overhead (e.g., project managers plus functional managers) than a simpler organization requires, possible disagreements over allocation of resources, and complexities of communication and control.

2.2.6 Self-Directed Teams

A new concept in work organization has emerged in which small groups of interdependent individuals share responsibility for outcomes of their organizations in a very active way (Sundstrom, DeMeuse, and Futrell, 1990). These "self-directed teams" have been introduced with considerable success, and some resistance, in many manufacturing and service organizations. Also known as empowered teams, empowered work groups, egoless groups (Bennett, 1993) or "ensemble groups, acting as intellectual commandos" (Toffler, 1980), these groupings have several characteristics that may be quite attractive to technical organizations.

Wellins, Byham, and Wilson (1991) give a quick working definition—a small group of people empowered to manage themselves and the work they do on a day-to-day basis. They explain further that such a self-directed team is:

> an intact group of employees who are responsible for the "whole" work process or segment that delivers a product or service to an internal or external customer. To varying degrees, team members work together to improve their operations, handle day-to-day problems, and plan and control their work. In other words, they are responsible not only for getting work done but also for managing themselves.

This rather revolutionary concept has been applied to such diverse enterprises as a Brazilian manufacturer of marine pumps, digital scanners and commercial dishwashers (Semler 1989), the Philadelphia Zoo, an AT&T subsidiary that provides financing for lease customers (Hoerr, 1989), Federal Express, Aetna Life and Casualty, and a Wisconsin sausage maker (Dumaine, 1990). Although their structure and activities vary greatly, in general they consist of eight to fifteen members responsible for producing a well-defined product or service. Team members tend to learn all the tasks their team must perform, and thus they rotate from job to job. With increasing maturity, experience, and confidence, the team takes over such responsibilities as scheduling, hiring, training, firing, troubleshooting, maintenance, and material ordering. Typically, the number of management layers is minimal, since the team itself is doing supervisory/managerial tasks (Lee, 1990). The Brazilian manufacturing company, for example, has counselors, partners, coordinators, and associates—four titles and three layers of management (Semler, 1989).

Diagramming such an organization is difficult, at least with our traditional mind-set, since the team itself has no hierarchy. Figure 2.6 presents the STAR

Figure 2.6. The STAR concept for a self-directed team. (Reprinted courtesy of Hannaford Bros. Co.)

(Situation or Task, Action, Result) structure as used in the warehouse operations of a grocery chain, the Hannaford Brothers Company. The points of the star emphasize that each member of the team is responsible for managing a function for that team. Another important aspect of self-directed teams is their fit into the structure of the overall organization. Typically, some or all members of a team have responsibilities outside the team, as depicted in Figure 2.7. Such assignments, which help decide company policy, set overall goals, and the like are periodically rotated among team members (Wellins, Byham, and Wilson, 1991).

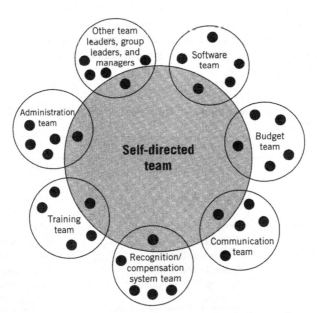

Figure 2.7. Job assignments within a team. (From Richard S. Wellins, William C. Byham, and Jeanne Wilson, *Empowered Teams: Creating Self-Directed Work Groups That Improve Quality, Productivity and Participation.* Copyright © 1991. Reprinted by permission of Jossey-Bass, Inc.)

Table 2.4, based on work by Wellins, Byham, and Wilson (1991), cites some of the key differences between self-directed teams and the more traditional structures discussed in Sections 2.2.2 to 2.2.5.

2.2.7 Clusters

Quite similar to the self-directed work teams described in Section 2.2.6, clusters are collections of work teams. People come from different disciplines, and they work together on a semipermanent basis. Clusters handle most of their administrative functions, including most decision making, they are accountable for business results, and they share information broadly (Sarchet, 1991). Born of a new emphasis on thinking more about "customers, relationships and information" than about organization (Mills, 1991), clusters are made up of five- to seven-person work teams that feature open communication and easy accessibility. The total membership of a cluster is normally between 30 and 50, which means that the cluster is large enough to provide not only the numerical strength but also the multidisciplinary expertise necessary to operate virtually independently.

The management levels and reporting relationships are reduced in number but not eliminated! Within a division, the cluster is answerable to division management. That means there is still a need for a cluster manager who provides leadership for the cluster and also relates to upper management. An organization chart might look like Figure 2.8, where we indicate, perhaps optimistically, just four levels: team, cluster, division, organization leader.

Sarchet (1991), who calls this new arrangement a "communication-driven approach," says that it "thrives where communications are free and open. It is designed to augment and replace the limited communication channels based on a chain of command. A communications system should be available that includes a

TABLE 2.4 Key Differences Between Traditional Organizations and Empowered Team Organizations

Element	Traditional Organization	Self-Directed Teams
Organizational structure	Layered/individual	Flat/team
Job design	Narrow single-task	Whole process/ multiple-task
Management role	Direct/control	Coach/facilitate
Leadership	Top-down	Shared with team
Information flow	Controlled/limited	Open/shared
Rewards	Individual/seniority	Team-based/skills-based
Job process	Managers plan, control, improve	Teams plan, control, improve

SOURCE: Richard S. Wellins, William C. Byham, and Jeanne Wilson, *Empowered Teams: Creating Self-Directed Work Groups That Improve Quality, Productivity and Participation.* Copyright © 1991. Reprinted by permission of Jossey-Bass, Inc.

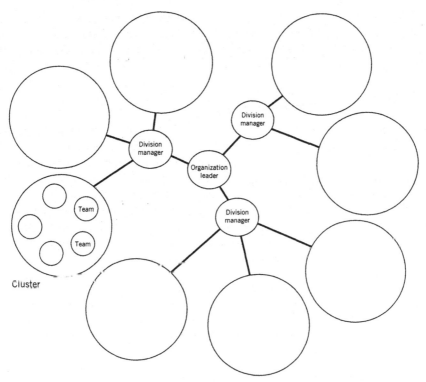

Figure 2.8. Schematic diagram of a cluster management structure.

complete data base about the business and its activities, including both financial and non-financial elements."

2.2.8 Geographical

Since many engineering design and consulting organizations have multiple offices that are disbursed geographically, we include the "geographical" organization as a separate category. Such a setup has been alluded to in our discussion of the matrix organization, where one dimension of the matrix may be locational. Such organizations have special challenges surrounding a number of issues, including the extent of services the "satellite" offices receive from headquarters, the methods and extent of communications between home base and branch, consistency of policies throughout the organization, allocation of corporate overhead, and special traits required of the branch office manager.

A 1992 survey of U.S. architecture, engineering, planning, and environmental consulting firms that used satellite offices (White, 1992) revealed that the number of such offices per firm ranged from one to one hundred, with a median of six. The median office had twenty employees and was located 400 miles from headquarters. All but 3% of the office managers held at least one degree in architecture, engineering, or science. Required leadership qualities include such entrepreneurial skills as the abilities to work independently and market successfully, but the extent to which the satellite office manager should be a "maverick" or "boat rocker" has not been established.

2.2.9   The Throwaway Organization

While the organizations discussed so far are all permanent or semipermanent to some degree, we must not neglect the temporary organization, established to go out of business when its task is completed. Certainly the projectized and matrix organization structures have such features, but employees of those groups still tend to have some official tie to the parent organization after project completion.

Babcock (1991) reminds us of many examples of "disposable organizations"; an example is the team assembled from several cooperating organizations to develop a proposal and then disband. As many as a thousand persons may work in this way for periods up to several months. In *Future Shock,* Alvin Toffler (1970) wrote:

> We are, in fact, witnessing the arrival of a new organizational system that will increasingly challenge, and ultimately supplant, bureaucracy. This is the organization of the future. I call it "Ad-hocracy".... Teams are assembled to solve specific short-term problems.... What we see here is nothing less than the creation of a disposable division—the organizational equivalent of paper dresses or throwaway tissues.

Interestingly, 10 years later, Toffler observed in *The Third Wave* (1980) that many large companies had moved to incorporate temporary groupings in "a radically new formal organization called the matrix organization."

2.2.10   The Informal Organization

Thus far we have presented formal organizations—"officially established relationships" among the parts of an organization. In addition, we must recognize the existence of the informal organization. Badawy (1982) defines the informal structure of an organization as "a set of voluntary and unplanned (as opposed to officially and consciously developed) networks of interpersonal relationships, methods and procedures that emerge and are developed among members of an organization." It is loosely developed, maintained unconsciously, and spontaneous; it is not shown on the organization chart. Rather, it is based on such social interactions as group gossip, coffee break conversations, friendships, cliques and grapevines.

The informal organization tends to satisfy social, security, and identification needs that the formal structure does not, and is not expected to, satisfy. The engineering manager will be wise to recognize the existence of such a structure, understand it, and even use it to advantage. Table 2.5 lists Badawy's (1982) set of characteristics of formal and informal organizations that should be useful in understanding the important differences.

Shannon (1980) identifies three types of informal organization: (1) the social organization, which arises from social interactions and shared values and fulfills the human needs of association, belonging, and importance, (2) the power structure, based on the ability of an individual to induce others to produce by giving or withholding rewards (but not through the formal organization structure), and (3) the status or prestige structure, especially important in technical organizations because it is based on technical skills, abilities, accomplishments, and experience.

Shannon suggests that an ideal situation arises when the engineering manager holds a high position in the formal organization and in all three types of informal organization. He cites as an example Wernher von Braun, who was (1) director of the George C. Marshall Space Flight Center (formal authority), (2) renowned for his personal engineering accomplishments (status/prestige), (3) a charismatic leader with a great capacity to inspire workers to commitment and enthusiasm (power), and (4) a close personal friend and colleague of many of his managers, based on previous associations and those at the Marshall Center (social) (Shannon, 1980).

2.3   RELATIONSHIPS WITHIN THE ORGANIZATION

The engineer, whether employed within a design engineering group or elsewhere, can expect to interface with a variety of other groups. In this section and the next, we list and briefly describe several of the entities within and outside the organization that the engineer is likely to encounter. We present this discussion under the simplifying assumption that the engineer is a part of the design engineering group, even though engineers are frequently found in manufacturing, sales, quality, cost control, and other parts of the organization. The listing of categories is not intended to be all-inclusive; many possibilities have been omitted. On the other hand, not all organizations contain all the categories described; a state highway department, for example, is unlikely to have manufacturing and marketing groups. The intent here is to make the reader aware of the several constituencies beyond his or her immediate group that may be relied on for information or work products; alternatively, such entities may rely on the individual engineer to perform their tasks effectively. Figure 2.9 shows the various components that may interact with design engineering. No attempt is made to show overlapping functions or to depict the relations of the non–design engineering entities with other parts of the organization or outside groups. The 11 internal entities appearing in Figure 2.9 are defined briefly as follows.

- *Marketing and Sales.* This part of the organization is responsible for selling its products and services. Activities include advertising, proposal preparation, estimating and bidding, product demonstration, and negotiation with potential buyers. There may be many engineers in this group, providing direct one-on-one contact with prospective customers. In a professional services organization, this function may be termed "business development."
- *Purchasing and Expediting.* This group purchases and assures the delivery of the raw materials, components, equipment, and systems that will be assembled or otherwise used in the organization's operations. Purchasing includes soliciting price quotations and other proposal information for acquisitions,

TABLE 2.5 Characteristics of Formal and Informal Organizations

Characteristic	Formal Organization	Informal Organization
1. Nature of system	Static—organizations without people.	Dynamic—organizations in action, with people.
2. Objective	Achieving overall organizational goals.	Possible combination, such as knowing what is going on, getting even with the boss, resisting changes, and protecting group members.
3. Leadership	Authority and responsibility are given or assigned to managers as a function of their formal positions.	The informal leader emerges as the representative or the spokesman for the group.
4. Authority	Based on principles of sound organization including specialization and division of labor; basically bureaucratic in nature	Peer interaction and respect are important: "If you really want to know what's happening see John, he always knows"; "Max is the expert on that subject, he'll be glad to help."
5. Power and control	Directly related to one's position on the organization chart, and job description	A function of one's clout, influence, connections, and informal impressions made on individuals "upstairs."
6. Reward system	Prescribed packages for sharing or withholding organizational "goodies" depending on loyalty and contribution—or lack of it—to the organization.	Reflected in terms of more or less status in the group.

evaluating them and placing orders. Expediting consists of assuring that the correct items, as ordered, arrive at the proper place at the proper time.

- *Manufacturing.* The manufacturing group is responsible for making the product, from the assembly of raw materials and/or components to final testing and packaging. For an ongoing operation, engineers will provide support through process maintenance. For new products, engineers provide process as well as product design. In an organization that provides construction services, field operations "manufacture" the final "product."

- *Research and Development.* This part of the organization can entail activities ranging from basic research involving materials, systems, components, or structures to the early stages of product development, sometimes called proof of concept. After this stage, a new product is generally turned over to an engineering design team. Engineers play a major role in many research and development efforts.

TABLE 2.5   continued

Characteristic	Formal Organization	Informal Organization
7. Status and Symbols	Prescribed to fit the organization's "image"; sharply differentiate between employees at different levels	Behavior and dress codes are sharply enforced; higher status individuals are informally recognized in various ways by the group.
8. Communications	A formal communication system is designed to follow the chain of command.	The grapevine is a powerful source of information which might reinforce or detract from formal communications.
9. Organizational norms	Formal rules, policies, and procedures strictly guide individual group behavior; a system of formal standards of behavior and expectations is developed.	A strong desire to be accepted by the group, to be "in," to "fit," to be "one of the boys"; a strong tendency to conform to group expectations.
10. Discipline	Enforced based on organizational formal policies and procedures.	Imposed by the group on individuals who do not conform to the group's norms and standards; group acceptance and satisfaction of the social need are more valued than material rewards earned by the "rate buster."

SOURCE: M. K. Badawy, *Developing Managerial Skills in Engineers and Scientists,* © 1982. Reprinted by permission of Van Nostrand Reinhold.

- *Quality.* This operation is responsible for assuring that the product meets technical specifications as well as safety and reliability requirements. Inspection and testing are traditional activities. This technology-based function often involves engineering personnel. We devote Chapters 3 and 4 to total quality management.

- *Logistics.* The logistics organization provides such support services as transportation before and after manufacture, material movement inside and outside the factory, and housing, food service, utilities and other support for remote operations.

- *Budgeting, Accounting, and Cost Control.* Budgeting complements the strategic plan by casting that plan in financial terms; budgets may be developed on a product, area, project, and/or overall basis. Accounting accumulates and summarizes cost and profit data, while cost control is responsible for monitoring progress by comparing actual cost experience with the budget, as well as controlling costs by suggesting or implementing corrective measures as indicated. Cost control engineers are often employed on projects of various kinds.

- *Labor Relations.* The company's labor, or industrial relations, office is often a separate department, responsible for negotiating collective bargaining

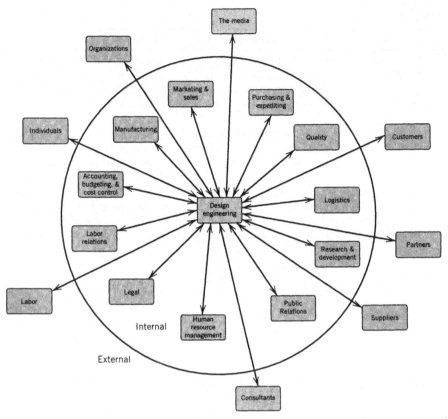

Figure 2.9. Relationships between design engineering and other entities, both internal and external.

agreements for nonprofessional employees and administering the resulting contracts. There may also be a separate entity that includes collective bargaining unit representatives (shop stewards), who are company employees whose primary function in the unit is to represent union members' interests.

- *Legal.* The organization's legal affairs department negotiates contracts with customers and suppliers. It represents the organization in litigation involving breach of contract and liability. Other areas of concern include real property matters, intellectual property rights such as patents, and the legal aspects of collective bargaining agreements. Engineers often become involved in litigation involving product and services liability. Chapters 8 through 12 deal with legal and liability matters of concern to engineers.

- *Human Resource Management.* The human resource management, or personnel, office handles a variety of activities ranging from recruiting of new employees to maintaining and expanding skills. Career planning, team building, and dissemination of information about benefits programs are among its responsibilities. Chapter 5 discusses some human resource issues in engineering management.

- *Public Relations.* This group is responsible for representing the organization to the public. Its activities complement those of the marketing and sales departments. The purposes are to make the organization more widely known and better understood. Engineers often assist in this effort as speakers at public gatherings and as liaison agents with community organizations.

2.4   RELATIONSHIPS OUTSIDE THE ORGANIZATION

Just as important entities within the organization relate to the engineer's activities and responsibilities, so too the engineer will probably deal with groups and individuals "on the outside" of the employing organization. We elaborate here on the eight external entities shown in Figure 2.9.

- *Customers.* This group is the reason for our organization's existence! Customers are the people and groups to whom we provide services and products. Often the engineer provides direct field support to these clients. In a construction setting, the engineer usually interfaces with the project's owner directly and often.
- *Suppliers.* Suppliers, or vendors, furnish items ranging from raw materials to equipment and systems, interfacing primarily with the organization's purchasing and expediting group. Suppliers often provide engineering design information, including specifications on materials and equipment. An organization's engineers may spend considerable time in suppliers' facilities during design, fabrication, and testing.
- *Consultants.* These outside specialists support the organization's activities on a temporary basis. They are utilized when hiring full-time permanent employees is not efficient. Consultants provide an outside, independent viewpoint that can offer a needed and healthy balance to the sometimes ingrained attitudes and approaches of the permanent organization.
- *Partners.* The "partnering" concept provides for alliances of two or more organizations for specific projects or on an ongoing basis. Typically, the partners join together with a view to offering a fuller scope of services than either could provide by itself. The engineer may establish important relationships with representatives of these colleague organizations. We discuss partnering in Section 11.2.3.
- *Labor.* The offices and representatives of external collective bargaining units represent a very real constituency to the company. They are involved in bargaining for the contract and then watchdogging the implementation of it. The engineer is less likely to interface with labor unions than with other groups outside the organization.
- *Public Organizations.* Often the engineer deals with members of the general public through a variety of community, regional, or national organizations,

frequently representing special interests. Examples include environmental lobbies, taxpayer associations, regulatory boards, and professional societies.

* *The Media.* The engineer may be called on to explain a concept, new product, or proposed project to the general public through the broadcast or print media. These opportunities to write, to be interviewed for a newspaper article, or to appear on a radio or television talk show can frighten some engineers, but they are an important professional responsibility.

* *Individuals.* We must not forget the individual who is not a customer or client. The engineer's work can impact the individual property owner, road user, recreationalist or man on the street, despite the absence of any official relationship.

2.5   THE FUTURE OF TECHNICAL ORGANIZATIONS

Speculating about organizations is risky, but it may be helpful to suggest some probable trends. First, we can expect an increase in the importance of *temporary, task force–type organizations.* We have discussed, in this chapter, the "throwaway organization" that is developed for a particular purpose and then disbanded when that purpose has been fulfilled. Chapter 7 is devoted exclusively to structures and techniques for management of project organizations.

Second, organizations of the future are likely to be much "flatter," with fewer levels in the hierarchy, and they will embrace the *team concept* to a greater extent, thus migrating even farther from the notion of "hierarchy." Finally, we can expect to see a greater use of *partnering,* with permanent or semipermanent relationships developed between the technical organization and outside groups, especially suppliers and joint venture partners.

The team concept and the notion of a limited number of suppliers as partners are important elements of total quality management, which we consider in Chapters 3 and 4.

2.6   DISCUSSION QUESTIONS

1. Try Hensey's design approach: identify the present development level of an organization you know about (Section 2.1.1), and suggest ways for the firm to move toward being a high-performing team. Be prepared to defend your answer.

2. Consider the organization you work for or another one you are familiar with. If this team tried to define "being at cell 9,9" on the Grid, the organizational model of Blake, Mouton, and McCanse, what would some of the characteristics be?

3. Where does your organization fit on the model of Blake, Mouton, and McCanse? What might be done to move you closer to cell 9,9?

4. Pick two organizational structures described in Section 2.2. For each, tell which of Badawy's principles of organizational structure (Section 2.1.3) seem to be violated. Comment on your analysis.

5. In our discussion of the importance of teams to successful total quality management (Section 3.3.3), we will be quoting Scott as follows: "I know of no new business operating beyond its shakedown period under the old or pyramidal organization." Is this observation accurate? If so, why?

6. Select an organization with which you are familiar and draw its organizational chart. Pick a somewhat complex organization—at least three levels of hierarchy, a matrix organization, and/or a combination of line and staff. If it is a very large or very complex organization, show a skeleton of the chart. Describe briefly some aspects of the organizational structure that are especially effective and other aspects that are less so.

7. What role might the informal organization play in each of the following?
   (a) A group of employees, concerned about their department head's leadership, express their displeasure to the divisional vice president.
   (b) A competitor is about to announce the introduction of a new breakthrough product.
   (c) Charlie got fired.
   (d) The corporate vice president for engineering will visit next Tuesday.

8. Is the informal organization more important in an organization that is functioning smoothly or in one that is in crisis? Why?

9. Describe a temporary, "throwaway" organization you have been a part of. Discuss how it was formed and organized, what role the leader played, its relationship with outside groups (if any), the means by which it was terminated, and the extent to which you believe it succeeded in accomplishing its objectives.

10. For your organization in Question 6, identify some of the outside persons and groups with which the organization must deal. Describe briefly the nature of those relationships.

11. How is the "information superhighway" likely to affect the structure and functioning of engineering organizations of the future?

2.7   REFERENCES

Babcock, D.L. 1991. *Managing Engineering and Technology.* Englewood Cliffs, NJ: Prentice-Hall.

Badawy, M. K. 1982. *Developing Managerial Skills in Engineers and Scientists.* New York: Van Nostrand Reinhold.

Bennett, A. L. 1993. Personal communication.

Blake, R. R., and A. A. McCanse. 1991. *Leadership Dilemmas—Grid Solutions.* Houston, TX: Gulf Publishing.

Blake, R. R., and J. S. Mouton. 1964. *The Managerial Grid.* Houston, TX: Gulf Publishing.

Blake, R. R., J. S. Mouton, and A. A. McCanse. 1989. *Change by Design*. Reading, MA: Addison-Wesley.

Cable, D., and J. R. Adams. 1982. *Organizing for Project Management*. Drexel Hill, PA: Project Management Institute.

Dumaine, B. 1990. "Who Needs a Boss?" *Fortune,* Vol. 121, No. 10, May 7, 52–60.

Francis, D., and D. Young. 1979. *Improving Work Groups*. San Diego, CA: University Associates.

Gobeli, D. H., and E. W. Larson. 1986. "Matrix Management: More than a Fad." *Engineering Management International,* Vol. 4, No. 1, October, 71–76.

Hensey, M. 1990a. "Editor's Letter: Team Spirit and Teamwork." *Journal of Management in Engineering,* American Society of Civil Engineers. Vol. 6, No. 3, July, 235–237.

Hensey, M. 1990b. "Organizational Design: Some Helpful Notions." *Journal of Management in Engineering,* American Society of Civil Engineers, Vol. 6, No. 3, July, 262–269.

Hoerr, J. 1989. "The Payoff from Teamwork." *Business Week,* Issue No. 3114, July 10, 56–62.

Lawson, J. W. 1986. "A Quick Look at Matrix Organization from the Perspective of the Practicing Manager." *Engineering Management International,* Vol. 4, No. 1, October, 61–70.

Lee, C. 1990. "Beyond Teamwork." *Training,* Vol. 27, No. 6, June, 25–32.

Mills, D. Q. 1991. *Rebirth of the Corporation*. New York: Wiley.

Peters, T. J., and R. H. Waterman, Jr. 1982. *In Search of Excellence: Lessons from America's Best-run Companies*. New York: Harper & Row.

Poirot, J. W. 1991. "Organizing for Quality: Matrix Organization." *Journal of Management in Engineering,* American Society of Civil Engineers, Vol. 7, No. 2, April, 178–186.

Sarchet, B. R. 1991. "Cluster Management—An Organization for the '90s." *Engineering Management Journal,* Vol. 3, No. 4, December, 27–32.

Semler, R. 1989. "Managing Without Managers." *Harvard Business Review,* Vol. 67, No. 5, September–October, 76–84.

Shannon, R. E. 1980. *Engineering Management*. New York: Wiley.

Sundstrom, E., K. P. DeMeuse, and D. Futrell. 1990. "Work Teams: Applications and Effectiveness." *American Psychologist,* Vol. 45, No. 2, February, 120–133.

Toffler, A. 1970. *Future Shock.* New York: Random House.

Toffler, A. 1980. *The Third Wave*. New York: Morrow.

Wellins, R. S., W. C. Byham, and J. M. Wilson. 1991. *Empowered Teams: Creating Self-Directed Work Groups That Improve Quality, Productivity and Participation*. San Francisco: Jossey-Bass.

White, F. D. Ed. 1992. *Satellite Office Survey of A/E/P & Environmental Service Firms*. Natick, MA: Mark Zweig & Associates, Inc.

___3
Total Quality Management: Principles and Approaches*

In this chapter and the next, we present an overview of a new approach to management that has literally swept across large segments of management practice in the past 10 years. Since total quality management (TQM), under any of the names by which this emphasis on quality is known, is considerably more than a passing fancy, it is important that the engineering manager be acquainted with some of its basic notions.

It is not important whether these ideas are really new, or whether they are old ideas packaged in a different way. The importance of quality products and services, the emphasis on defining and satisfying the customer, and the involvement of teams to accomplish the organization's goals are receiving so much newfound interest that without some understanding of the principles, approaches, and techniques of quality management, the manager of technical activities will be unprepared to function in today's commercial environment.

The challenge here is to pick from literally hundreds of sources and distill the available information into a meaningful overview. We begin this chapter with a brief history, followed by some definitions of quality and quality management. Some primary principles and some of the gurus and their contributions are then presented. Since much of the early quality management applications were in product-producing industries, it is important to recount the importance of quality and quality management in such service-producing activities as engineering design and construction.

We then describe benchmarking, a technique for comparing the quality status of one organization with that of another, and follow that material with a presentation on the "cost of quality." Several standards and awards in the quality area are described, and the chapter concludes with a consideration of how total quality management can be implemented in the engineering organization.

*I am deeply indebted to H. Andrew Scott, P.E., for introducing me to the principles and practices of total quality management, for providing many good suggestions for Chapters 3 and 4, and for reviewing the manuscript in its many stages of completion.

3.1 SOME HISTORY

Exactly where and when the so-called quality revolution began is both unknown and unimportant. In the United States, the years following World War II were marked by great strides in manufacturing practices that placed special emphasis on large quantity production at low cost, in response to a recovering world economy. At the same time, other nations, particularly Japan, began their postwar recovery in poor condition but with an increasing recognition of the importance of quality.

From the U.S. perspective, two significant events helped to push the interest in quality that is a keynote of management practice in the 1990s (Hunt, 1992). First, the shrinking world marketplace, with vastly improved communications and intricate webs of trading networks, has led to stiff foreign competition of a type never before experienced. In the mid-1980s U.S. companies "discovered" this competition and realized that the only effective way to compete was to provide quality goods and services.

Second, the early 1990s brought the collapse of communism, the end of 40 years of Cold War, and the development of the European Economic Community. For a period lasting at least until the beginning of the twenty-first century, it is likely there will be three economic superpowers: Japan, Europe, and the United States. But to compete in this interconnected global economy, all competitors will be expected to emphasize quality as well as quantity and price. Furthermore, this interconnectedness means that sources of supply, as well as markets, will become increasingly global. Figure 3.1 illustrates the dramatic shift in the international marketplace. According to Hunt (1992):

> The message it delivers loud and clear is this: Because of these intricate global webs in which American companies are now full-fledged partners, they have little or no choice but to keep up with the other global competitors in the web if they want to remain competitive. Whether or not to adopt improved quality practices is no longer a real option for most American companies. The only option, really, is when to shift and whether your company will do it soon enough to remain competitive.

Between 1950 and 1980, the U.S. share of worldwide automobile sales declined from 76% to less than 21% (Bowles and Hammond, 1991). In 1974, 80% of the automobiles leaving a Ford assembly line went immediately to a rework facility. In 1978, Ford sold steel to small European countries from one of its mills that it considered "ultramodern" but bought steel for its products from Japan in order to obtain the required quality. The exporting mill closed in 1980 (Jablonski, 1992).

In 1955, 96% of all radios sold in the United States were manufactured in this country; in 1975, the proportion was nearly zero. In semiconductor production, the U.S. worldwide market share declined from 60% to 40% in the 1980s. In 1980, Hewlett-Packard reported testing 300,000 16k RAM chips from one Japanese and three U.S. manufacturers. When the chips were inspected upon receipt, the Japanese product had a failure rate of zero, while the American-made chips had between 11 and 19 failures per thousand. After 1000 hours of use, the U.S. chips failed at

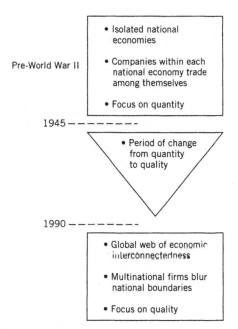

Figure 3.1. Shift toward global integration and focus on quality. (From V. Daniel Hunt, *Quality in America,* Business One Irwin, copyright © 1992. Reprinted with the permission of the author.)

a rate of 27 per thousand, while the Japanese product had a failure rate between 1 and 2 per thousand (Bowles and Hammond, 1991).

Bowles and Hammond (1991) suggest that the short answer to the question of why such dramatic changes occurred in the U.S. position abroad is that "American industry grew fat, lazy and arrogant."

Elsewhere, Japan was recovering from military and economic defeat through the help of its own experts and some from the United States. Taiichi Ohno, Kaoru Ishikawa, W. Edwards Deming, and Joseph Juran were among the leaders of this significant "revolution." Soon after the end of World War II, the Japanese demonstrated their uncanny ability to copy by creating a city named Usa (pronounced oo-sa), where products were labeled "Made in USA," to take advantage of the reputation of U.S. products, which at that time was much higher than that of Japanese goods (Jablonski, 1992). In addition, they copied various statistical process control techniques suggested to them by Deming and Juran in the early 1950s.

Ishikawa's ideas were a vital key to the transformation in Japan. His two fundamental concepts have to do with focus on the customer and a quality emphasis throughout the organization and its processes: "quality means quality of work, quality of service, quality of information, quality of process, quality of division, quality of people, including workers, engineers, managers and executives, quality of company, quality of objectives, etc." Bowles and Hammond (1991) summarize Ishikawa's fundamental message as follows:

Commit to continuous improvement throughout the entire organization.

Fix the problem, not the blame.

Strip down the work process—whether it is the manufacture of a product or the performance of a service—to find and eliminate problems that prevent quality.

Identify the customer, internal or external, and satisfy that customer's requirements in the work process or the final product.

Eliminate all waste.

Instill pride in performance, encourage teamwork, and create an atmosphere of innovation for continuous and permanent quality improvement.

In the balance of this chapter and the next, we shall have an opportunity to confront many of these fundamental ideas again.

So, we find an increasing emphasis on quality in Japan during the years after World War II and a corresponding concern for things other than quality in the United States during that same time, until, perhaps, the early to mid 1980s. As might be expected, there have always been pockets of interest in quality in the United States. Table 3.1 contains the well-known "Penney idea," on which J.C. Penney built his retail sales operation beginning in 1913.

Among the early advocates of quality management in the United States was Philip B. Crosby. His book, *Quality Is Free,* published in 1979, is among the first of a large number of publications that are helping to transform the way organizations around the world do business. Crosby's writing grows out of his experience at ITT Corporation beginning in 1965. Early in its quality program, ITT Corporation gave itself four objectives: (1) establish a competent quality management program in every operation, both manufacturing and service, (2) eliminate surprise nonconformance problems, (3) reduce the cost of quality, and

TABLE 3.1   J.C. Penney's Seven Ideas

1. To serve the public, as nearly as we can, to its complete satisfaction.
2. To expect for the service we render a fair remuneration and not all the profit the traffic will bear.
3. To do all in our power to pack the customer's dollar full of value, quality, and satisfaction.
4. To continue to train ourselves and our associates so that the services we give will be more and more intelligently performed.
5. To improve constantly the human factor in our business.
6. To reward men and women in our organization through participation in what the business produces.
7. To test our every policy, method and act in this way: "Does it square with what is just and right?"

(4) make ITT the standard for quality worldwide. Crosby lists the four "legs" of the ITT quality program as management participation, professional quality management, original programs, and recognition. His Quality Management Maturity Grid is a means of assessing whether an organization is in the management stage of uncertainty, awakening, enlightenment, wisdom, or certainty (Crosby, 1979).

The recognition of the importance of quality grew rapidly in the late 1980s. Armand V. Feigenbaum, author of a highly acclaimed book entitled *Total Quality Control* (1991), whose first edition was published in 1951, reported that more than 80% of American consumers placed quality ahead of price in 1989, whereas fewer than 40% were so concerned with quality in 1978. Chapple (1990a) cites Feigenbaum's perception that "[the] doubling of buyer emphasis on quality may be the single most significant marketplace change in the U.S. since World War II. ... Demand for quality is now the driver behind everything." The economic incentive favoring the current quality emphasis is evident in Feigenbaum's Quality Improvement Potential (QIP) index. In 1990 the index predicted that widespread implementation of TQM in American organizations would increase the gross national product by approximately 7% (Chapple, 1990b).

Increasing competition, a more global economy, demonstration that a concern for quality is economically advantageous—these and other reasons bring us to the high level of current interest in quality. The chief executive officer at Corning Glass, Jamie Houghton, is supposed to have described the current atmosphere by quoting the old saw, "Nothing focuses your attention like the hangman's noose!" (Bowles and Hammond, 1991).

Worldwide the interest parallels that in the United States. The Building Production Laboratory at VTT, Finland's national research institute, conducted two projects related to the construction industry, completed in 1994. "TQM the Nordic Way" identified "the best Nordic methods for developing and implementing contractors' Total Quality Management," while the "Quality Management Systems" project was intended to support contractors in developing their own quality systems while also producing a Finnish model for both contractor and subcontractor quality systems (VTT, 1992).

A survey of top European managers, reported in 1990, revealed that 91% believed that quality was absolutely critical. Reasons cited were satisfying the ultimate customer (89%), reducing costs (66%), improving flexibility and responsiveness (58%), and reducing throughput times (40%). Fifteen percent said quality was THE top company priority, while 72% said it was one of their top priorities (Aune, 1990).

The first National Quality Forum, sponsored by the American Society for Quality Control, was held in 1985. About 150 people participated, and the keynote speaker described quality as the key business issue of the 1980s, indeed, a matter of survival. Quality Forum VIII, the 1992 conference, involved 200,000 persons (many attending by means of video hookup). Edwin L. Artzt, Procter & Gamble's chairman and chief executive officer, said in his keynote address:

> But even with the progress each of us has made in the last several years, quality remains as critical to American business in the 90s as it was in the 80s—because it

is the only way we'll manage what I believe is one of the key business challenges of the 1990s: meeting customers' increasing demand for value. (Artzt, 1992)

The popularity and importance of quality management may well be demonstrated by the large number of publications and publishers, organizations, meetings, degree programs, and both credit and noncredit courses in evidence. In the realm of publications there are such periodicals as *Quality Progress, Quality Digest, Journal for Quality and Participation, TQM Magazine, The Quality Observer, The Quality Executive, Total Quality Newsletter, Quality by Design,* and *Quality,* as well as a multitude of in-house organs such as *Q Plus: The Greiner Quality Connection* of Greiner Engineering, Inc., and *Quality Abstracts* (Advanced Personnel Systems, 1992), which publishes "100–120 carefully selected abstracts gleaned from over 200 journals" every two months. Publishers such as Quality Resources Press, Quality Press of the American Society for Quality Control, and Technical Management Consortium provide many new titles each month, only a few of which can be cited here.

One can attend an increasing number of meetings, such as the annual Quality Forum (forums I and VIII were cited above), the International Service and Quality Forum (see, e.g., *International Service and Quality Forum Update,* 1992), and the Total Quality Management Workshop TQM-IV, sponsored by the American Society for Quality Control and Iowa State University Workshop.

When the National Society of Professional Engineers established its National Institute for Engineering Management and Systems in 1990, three of its four stated missions were directly related to total quality management (NIEMS, 1990).

Most colleges and universities offer credit and/or noncredit courses in some aspect of quality. Examples include an MS in quality at Eastern Michigan University, the Quality First program at Tennessee's Northeast State Technical Community College, and a graduate course in quality management at the National Technological University (NTU). In addition, there are such NTU Advanced Management and Technology Program noncredit courses as "Applying Total Quality Principles to Software Development" and "Novice-Friendly Tools for Total Quality Management," an American Society of Civil Engineers home study course ("An Effective A/E Quality Management Program"), and a tape-and-workbook offering by the National Society of Professional Engineers ("Constant Improvement: Improving the Quality of Engineering Management").

3.2   SOME DEFINITIONS

Thus far we have not attempted to define quality or quality management. It is time to face that task. At least six different approaches have been taken to the definition of *quality.* The *customer-* or *user-based* definition is that "Quality is fitness for use" (Juran, 1989) or "goods that best satisfy [customers'] preferences are those that they regard as having the highest quality" (Garvin, 1984). The so-called *manufacturing-based* definition says simply that "Quality means conformance to requirements" (Crosby, 1979). The American Society of Civil Engineers used this basis for defining quality in *Quality in the Constructed Project* (1990), which says:

> Quality in the constructed project is achieved if the completed project conforms to the stated requirements of the principal participants (owner, design professional, constructor) while conforming to applicable codes, safety requirements and regulations.

Note that in the view of ASCE, the concept applies to constructed projects (the kinds of activity in which civil engineers are primarily engaged), as well as to manufactured products.

Quality is sometimes defined in other ways, as well:

Product-based: "A precise and measurable variable, relative to the amount of some ingredient or attribute possessed by a product" (Garvin, 1984).

Value-based: "...best for certain customer conditions. These conditions are (a) the actual use and (b) the selling price of the product" (Feigenbaum, 1991).

Transcendent: "Quality is synonymous with 'innate excellence'" (Garvin, 1984). "Quality is neither mind or matter, but a third entity independent of the other two...even though Quality cannot be defined, you know what it is" (Pirsig, 1981).

Systems: "The ability of a product to function as intended, as part of the systems it belongs to or is associated with" (Sarantopoulos and Liebesman, 1992).

Hunt (1992) asks that we consider this question: "Is quality resident only in a firm's product or service, or is quality also found in the process that produces the product or service?" He concludes that "quality *is* actually present in, or missing from, every aspect of a firm's operation from top to bottom and side to side." He concludes further that "the days of limiting the definition of quality to the 'soundness' of the product—its hardness or durability, for example—are gone. The new kind of quality American firms rediscovered in the 1980s is far more cultural than physical, far more the way things are done than the nature of the things themselves."

In Chapter 4, we shall present several techniques for assisting the engineering manager to achieve quality products and services. It is important, however, to understand that total quality management is much more than these techniques. It is "not quality control nor statistics, nor inspection in the conventional sense even though certain of these functions may be employed in achieving customer satisfaction and eventually perhaps customer delight!" (Scott, 1992). In fact, as Deming has written, "statistics are essential but relatively unimportant" (quoted in Sproles, 1990).

A useful definition of *total quality management* is that offered by Jablonski (1992). We have enhanced his words slightly and suggest the following as a working definition:

TQM is a cooperative form of doing business that relies on the talents and capabilities of both labor and management to continually improve quality and productivity using teams with the ultimate goal of delighting both internal and external customers.

The section that follows offers some fundamental total quality management principles derived from this basic definition.

3.3 SOME PRIMARY PRINCIPLES

We shall rely on the definition of total quality management suggested above to present four primary principles: (1) continuous improvement, (2) focus on the customer, (3) the importance of teams, and (4) management leadership, support and involvement. Bob Bosshart of Construction Quality Management says that "At first pass, TQM can seem complex and cumbersome." But, he continues, after the "Ph.Deeze" statistical formulas and the consultants' black magic have been stripped away, it is really quite simple (*Quality by Design,* 1992). An understanding of these four principles will provide a good basis for implementing quality management in the engineering organization.

3.3.1 Continuous Improvement

Consider the following well-known adage: "If it ain't broke, don't fix it!" Now consider W. Edwards Deming's alternate suggestion: "Improve constantly and forever the system of production and service" (Deming, 1986). Total quality management boldly advocates the second quotation!

There are few processes or products that cannot be improved! Is a 99% performance level acceptable? At such a level, we would have unsafe drinking water "only" three or four days per year, doctors in the United States would write just 200,000 incorrect prescriptions per year, you would be without electricity, water, heat, telephone, and television for only 15 minutes per day (Bowles and Hammond, 1991), and about a hundred commercial airline flights would not reach their intended destinations each day worldwide.

Motorola Corporation instituted a defect reduction program which it called Six Sigma, based on the goal of having its manufacturing processes produce defects less than 6 standard deviations (6 SD; 6σ) to the right of the mean in a normal statistical distribution. In contrast, the typical U.S. manufacturing company operates at about three sigma. Motorola has not yet achieved this six sigma goal, but the company's efforts toward constant improvement reportedly yielded a $500 million savings in 1990 (Bowles and Hammond, 1991).

IBM Corporation was operating at about 3 sigma when it decided to adopt the Motorola six sigma approach to quality improvement. An officer was quoted as saying, "A virtually defect-free operation sounds impossible but it's not. Many Japanese companies and a few American companies like Motorola are well on their way to being there. We intend to do it through scrupulous gathering of data, applying measurements, identifying root causes of defects and then eliminating them in a systematic way" (Bowles and Hammond, 1991).

Continuous improvements may be of several types, including improved product and/or service that gives the customer better value, reduction of errors and defects, improved responsiveness and cycle-time, and improved effectiveness and efficiency in the use of all resources (Hunt, 1992). Deming (1986) emphasizes the need to build quality into the design phase:

> It may be too late once plans are on their way. Every product should be regarded as one of a kind; there is only one chance for optimum success.... Downstream, there will be continual reduction of waste and continual improvement of quality in every activity of procurement, transportation, engineering, methods, maintenance, locations of activities, sales, methods of distribution, supervision, retraining, accounting, payroll, service to customers.

With increased emphasis on recycling and the environment, the concept of continuous improvement can be applied throughout the entire product life cycle from supplier to recycler. Consider the potential cost reduction for such products as tires and refrigerants if this philosophy of continuous improvement is widely implemented.

An excellent means of visualizing the process of continuous improvement is the simple Shewhart cycle shown in Figure 3.2. The process is clearly cyclical, inasmuch as the plan–do–study–act sequence is repeated over and over. *Planning* involves defining the problem, the customer and the important quality characteristics, then developing solutions for improvement. *Doing* consists of implementing the improvement. *Checking* means identifying what happened during the doing (by comparing the performance after improvement with that before the changes were installed). *Acting* incorporates successful improvements as work standards, which leads to identification of new opportunities for improvement in the next *plan* step.

Although much of the foregoing discussion has obvious applications in product manufacturing, it can apply equally to such service activities as engineering design and construction. In materials prepared for application to architecture and

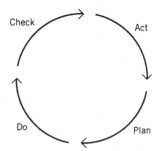

Figure 3.2. The Shewhart cycle.

engineering firms, Markert, Simon, and Miller (1992) state the goal succinctly: "Quality must be a vision we pursue—forever."

3.3.2 Focus on the Customer

At Disney World in Orlando, Florida, the world's most popular tourist destination, safety is first priority, courtesy is second, show is third, and efficiency is last. Disney President Frank Wells says that the heart of the Disney dream is "making the intangible tangible. There must be a daily commitment to excellence that people take with them to work each morning. We see that there is, so that our guests and customers can take home a feeling of quality each night" (Bowles and Hammond, 1991).

McDonald's opened a restaurant on Pushkin Square in Moscow in 1989. The fast-food facility was designed for 15,000 customers a day, but on some days it handles upwards of 50,000. On a typical morning, 500 are queued up outside the building by 9:30, awaiting the 10:00 A.M. opening. "When the doors open at 10, the surging crowds are treated to what is almost certainly the [former] USSR's first attempt at customer satisfaction. The young workers, donning beatific smiles, applaud for the customers as they come through the doors and motion them to come forward" (Bowles and Hammond, 1991).

If it can happen in the tourist and fast-food industries, a *focus on the customer* can surely happen in the management of engineering! First, we must recognize who our customers are. The National Institute for Engineering Management and Systems defines the customer as follows:

> The recipient or beneficiary of the outputs of your work efforts or the purchaser of your products or services. May be either internal or external to the company. The recipient who must be satisfied with the output. (*Quality Management Portfolio,* 1990b).

Note well that our customers may be both inside and outside our organizations. Jablonski (1992) refers to the customer outside the organization as the big "C," whereas the little "c" is the customer inside the company, the one with whom we work on a daily basis. Consider your own organization—who are those internal customers? They might include the copy pool, the filing clerk, your direct supervisor, the CADD operator, and anyone else who receives directly the results of your operations. Jablonski suggests that we tend to relate well to the big "C," but our support and enthusiasm often wane when it comes to the little "c's." Because funds are not exchanged for the internal services, we forget to draw a connection between those efforts and the revenue they generate. But if a primary objective of total quality management is to *delight the customer,* then we must be aware of *all* those persons and groups who receive outputs. Jablonski says, "As we implement TQM, we shift to a heightened awareness of all our customers, both the big 'C' and the little 'c.' "

Hunt (1992) explains customer focus and involvement as follows:

Attracting, serving and retaining customers is the ultimate purpose of any company, and those customers help the organization frame its quality consciousness and guide its improvement effort. A process, product or service has no relevance without customers; everything done in an organization is done for the customer. The quality of a product or service is defined by customer behavior and response. Process improvement must be guided by a clear understanding of customer needs and expectations.

Increased customer satisfaction is the ultimate result of customer focus and involvement. The responsibility for assuring customer focus and involvement starts with a top management focus on the organization's external customer and extends down through every level of activity to involve all customers in the improvement process. "Quality First" emphasizes satisfying all customers, internal and external. . . . To serve its customers adequately, the organization must continually reassess its customers' needs and requirements and factor them into its improvement efforts.

Thus, for any member of an organization, the *supplier* is the employee whose output he or she receives, and the *customer* is the next person in the sequence (Scott, 1992). Hayden (1992) suggested that the process in an engineering design and construction environment might be depicted as shown in Figure 3.3; he calls this a client-driven process and comments that "each of us is a customer (client) for work done by our fellow workers. We each have a right to expect (and contribute) work acceptable for its intended purpose."

3.3.3 The Importance of Teams

One significant characteristic of successful quality management programs is heavy reliance on self-directed work teams like those described in Section 2.2.6, as

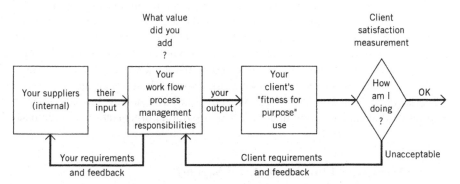

Figure 3.3. Supplier–customer relations in an engineering design and construction environment. (From W. M. Hayden, Jr., "Management's Fatal Flaw: TQM Obstacle," *Journal of Management in Engineering,* copyright © 1992. Reprinted by permission of the American Society of Civil Engineers.)

well as partnerships of other kinds. The important concept is that members of the team must represent a cross section of the process. Thus, an internal team (as in Section 2.2.6) is one that meets regularly to manage and improve its part of the operations. Supervisors serve as team leaders and as key links to other teams and to higher management; their managerial roles are those of coach, leader, and facilitator. It is essential that the team have linkages to organizational goals and major improvement opportunities.

Sam Walton (Walton and Huey, 1992) vividly describes the application of the team concept to retail merchandising. Warehousing and "middle men" have all been eliminated from the distribution chain. The result has been a reduction in both delivery time and inventory costs, and the customer enjoys better quality, sooner and at a lower cost.

Batson offers the following cautions with respect to team operations ("Basic Steps in the Total Quality Process" 1992):

- Teams are much easier to form than problems are to solve.
- There is always the possibility for team formation wherever any organizational interface occurs.
- Team consensus is not always the correct way to do business.
- The team must not degenerate into the 1990s version of a *committee*.
- Tight structure, purpose and planning for team meetings are necessary elements.

Beyond these internal teams, most quality management leaders have found that other kinds of partnership also are valuable. "Quality partnerships, simply defined, mean a cooperative effort between customers and suppliers to improve supplier performance which focuses on customer expectations and improvement, is data based, requires sharing of information and the solving of problems, and insists on improved communications" ("TQM Implementation and Quality Partnerships," 1992). Although this description was intended to apply solely to "teams" comprised of persons from within AND outside the organization, it could just as easily describe the successful coordination of the suppliers and clients as shown in Figure 3.3.

The ideal team, it would seem, is that which includes those who work within the process, those who are its suppliers, and those who are its customers (Jablonski, 1992). Quality improvement can be realized by the work of such a team, regardless of the source from which its members are drawn. These teams have proven so successful that Scott (1993a) has announced, "I know of no new business operating beyond its shakedown period under the old or pyramid organization." Enough said!

3.3.4 Management Involvement and Support

One of the key elements put forth by Deming (1986) and many other writers on quality management is the importance of top management involvement in and support of the atmosphere and the process. Indeed, Deming's experience indicated that 94% of troubles and possibilities for improvement are due to "common causes"—that is, conditions that belong to the system rather than to the individual employee, and are thus the responsibility of management.

Tribus (1988) makes the following bold statements (among others!):

1. The workers work *in* the system. The manager should work *on* the system to improve it with [the workers'] help.
2. The involvement of top management is essential to the success of the program. Assent is not enough.
3. When there is a problem, 85% of the time it will be with the system, not with the workers.

In connection with point 3, both Juran (1989) and Deming (1986) argue strongly for the elimination of quotas, targets, and slogans. Tribus says, "The more the management is diverted into blaming the workers, the longer it will take to improve the system."

Bowles and Hammond (1991), in introducing their descriptions of several winners of the Baldrige Award (see Section 3.7.2), report that "One of the great lessons of the quality revolution is that leadership is the most important ingredient for launching and sustaining a quality improvement process. If the top dog and his immediate poundmates don't believe in the process and behave accordingly, it won't happen."

At a TQM seminar sponsored by the National Society of Professional Engineers, the importance of managerial leadership was described as follows:

Managerial leadership means consistency directed toward achieving a vision. You are a leader if your people are lining up to help you achieve your vision of the organization in the future.

Management's challenge is to make everybody feel like they are somebody.... The role of supervisors becomes one of a coach, leader and facilitator, and all employees become involved in managing part of the business. ("Key Elements of Total Quality Management," 1992)

Those words sound strangely like those we used in connection with the importance of teams! In Chapter 5, we shall have further opportunity to consider that sometimes murky concept of leadership. For now, we offer this perception of the chief executive officer of Corning Glass, one of the top leaders in TQM in the United States: Jamie Houghton says that, if he blinks, TQM will flounder; the whole company is watching and relying on his leadership. Top management can do many things to make TQM successful; its actions influence the whole culture of the organization, and the reward systems to which people respond are the responsibility of top management (Jablonski, 1992).

3.4   THE GURUS AND THEIR APPROACHES

In its current, early stages of development, total quality management has been urged along by several "gurus." As would be expected, their ideas and approaches are far from identical, but it is important that we catalog briefly their primary contributions. From there, the reader is in charge of formulating a personal TQM philosophy! The material that follows draws heavily from Bowles and Hammond (1991), *Qplus: The Greiner Quality Connection* (n.d.), *Quality Management Portfolio* (1990b), and Samson (1992).

3.4.1   Philip B. Crosby

Crosby's approach was developed from his work at both the Martin Corporation and at ITT, where he was vice president for quality from 1965 to 1979. In 1962 Martin delivered on time and with no defects a Pershing missile that was fully operational within 24 hours. From then on, "zero defects" became a standard for many American industries.

Crosby's "four absolutes of quality management" are the following:

1. Quality is defined as conformance to requirements.
2. Quality results from prevention.
3. The performance standard is zero defects.
4. The measure of quality is the price of nonconformance.

Crosby describes a 14–step quality improvement process for structuring and positioning the organization for improved communications and operational improvements. The steps are as follows:

1. Management commitment
2. Quality improvement team
3. Measurement
4. Cost of quality
5. Awareness
6. Corrective action
7. Zero defects planning
8. Employee education
9. Zero defects day
10. Goal setting
11. Error cause removal
12. Recognition
13. Quality councils
14. Do it all again

3.4.2   W. Edwards Deming

The best-known quality guru was born in 1900 and was an active proponent of quality until his death in late 1993. W. Edwards Deming earned a Ph.D. in physics, learned statistical quality control from Walter Shewhart at the U.S. Department of Agriculture in the 1930s, and was recruited by General Douglas MacArthur to help Japan prepare for the 1951 census. His work with many industries in Japan is credited with making a large impact on that nation's recovery after World War II. Deming was virtually ignored in the United States until he was 80 years old, when his work became known to a broad audience through a television documentary, "If Japan Can...Why Can't We?" One of his foremost points is the essential role of management in quality. "Question Deming on what the work force can do to improve quality and productivity, and he'll admonish the poor soul who dares inquire that the problem is not with the workers but rather with management" (*Quality Management Portfolio*, 1990b).

Deming's 14 points "for transformation of American industry" are well known; he says they apply to small as well as large organizations and to service activities as well as manufacturing. The National Society of Professional Engineers has produced a helpful interpretation of these points in "Dr. W. Edwards Deming's 14 Points Adapted for the A/E Firm" (*Quality Management Portfolio*, 1992). NSPE's paraphrase is as follows:

1. Create constancy of purpose toward improvement of product and service, with the aim to become competitive and to stay in business and to provide jobs.
2. Adopt the new philosophy. We are in a new economic age. Western management must awaken to the challenge, must learn their responsibilities, and take on leadership for change.
3. Cease the dependence on inspection to achieve quality. Eliminate the need for inspection on a mass basis by building quality into the product in the first place.
4. End the business of awarding business on the basis of price tag. Instead, minimize total cost. Move toward a single supplier for any one item, on a long-term relationship of loyalty and trust.
5. Improve constantly and forever the system of production and service, to improve quality and productivity, and thus constantly decrease costs.
6. Institute training on the job.
7. Institute leadership. The aim of supervision should be to help people and machines and gadgets to do a better job. Supervision of management is in need of overhaul, as well as supervision of production workers.
8. Drive out fear so that everyone may work effectively for the company.
9. Break down barriers between departments. People in research, design, sales and production must work as a team, to foresee problems of production and in use that may be encountered with the product or service.

10. Eliminate slogans, exhortations and targets for the work force asking for zero defects and new levels of productivity. Such exhortations only create adversarial relationships, as the bulk of the causes of low quality and low productivity belong to the system and thus lie beyond the power of the work force.

11. Eliminate work standards (quotas) in the work place. Substitute leadership. Eliminate management by objectives. Eliminate management by numbers, numerical goals. Substitute leadership.

12. Remove barriers that rob the workers of their right to pride of workmanship. The responsibility of supervisors must be changed from sheer numbers to quality. Remove barriers that rob people in management and in other departments of their right to pride of workmanship. This means abolishing the annual or merit rating and management by objective.

13. Institute a vigorous program of education and self-improvement for everyone.

14. Put everybody in the company to work to accomplish the transformation. The transformation is everybody's job.

3.4.3 Armand V. Feigenbaum

Feigenbaum has already been cited for his well known book, *Total Quality Control* (1991). He coined the term "total quality control" in a 1956 *Harvard Business Review* article. Like many others, he stressed that quality relates to an organization's every function and activity, not simply manufacturing but also such areas as marketing and finance. At age 24, Feigenbaum became the top quality expert at General Electric's headquarters in Schenectady, New York. Later he was GE's worldwide manager of manufacturing operations, and in 1968 he formed General Systems Company, where he still is actively sought as a consultant.

Feigenbaum is well known for his 10 benchmarks of total quality control (Chapple 1990a), as follows:

1. Quality is a companywide process.
2. Quality is what the customer says it is.
3. Quality and cost are a sum, not a difference.
4. Quality requires both individual and teamwork zealotry.
5. Quality is a way of managing.
6. Quality and innovation are mutually dependent.
7. Quality is an ethic.
8. Quality requires continuous improvement.
9. Quality is the most cost-effective, least capital intensive route to productivity.
10. Quality is implemented with a total system connected with customers and suppliers.

3.4.4 Kaoru Ishikawa

Professor Ishikawa taught engineering at the Science University of Tokyo and at the University of Tokyo. At the time of his death in 1988, he was president of Musashi Institute of Technology in Tokyo. We described his impact on the Japanese quality revolution in Section 3.1. The cause-and-effect, or fishbone, diagram, which we shall present in Chapter 4, is also known as the Ishikawa diagram. Ishikawa's name is often associated with the term quality function deployment (QFD), an approach to new product development used by engineers to determine true customer requirements and translate them into engineering specifications. His philosophy of company-wide quality assurance emphasizes the customer and societal issues through the following issues:

1. Quality first—not short-term profit first.
2. Customer orientation—not producer orientation; think from the standpoint of the other party.
3. The next process is your customer—breaking down the barrier of sectionalism.
4. Using facts and data to make presentations—utilization of statistical methods.
5. Respect for humanity as a management philosophy—full participatory management.
6. Cross-function management—the approach used to solve problems.

3.4.5 Joseph M. Juran

Born only a few years after Deming, Juran has also had a major influence on the transformation of quality in Japan. He worked for the Western Electric Hawthorne Works in the 1920s, at a time when one-eighth of the employees—5000 out of 40,000—were assigned to inspection duties. His *Quality Control Handbook* (Juran and Gryna, 1988), first published in 1951, has been called the bible of the quality improvement movement in Japan and the United States. Ishikawa wrote that Juran's visit to Japan in 1956 was a major factor in helping the Japanese to change from an effort "dealing primarily with technology, based on factories, to an overall concern for the entire management" (quoted in Bowles and Hammond, 1991).

Juran's "quality trilogy" focuses on planning, control, and improvement:

Quality Planning

* Determine who the customers are.
* Determine the needs of the customers.
* Develop product features that respond to customer needs.
* Develop processes able to produce the product features.
* Transfer the plans to the operating forces.

Quality Control

• Evaluate actual performance.
• Compare actual performance to product goals.
• Act on the difference.

Quality Improvement

• Establish the infrastructure.
• Identify the improvement projects.
• Establish project teams.
• Provide teams with resources, training, and motivation to diagnose the causes, stimulate remedies, and establish controls to hold the gains.

3.4.6 Genichi Taguchi

Since the 1960s Taguchi, a Japanese engineer, has introduced many statistical tools and quality improvement concepts that rely heavily on design of experiments theory. His two-part thesis is that engineering specifications with acceptable upper and lower limits hamper quality and that *any* departure from optimum will increase costs and reduce customer satisfaction. His approach to parameter design provides a systematic and efficient method for conducting experimentation that results in optimum manufacturing processes and optimum product characteristics (Magowan, 1991; Unal and Bush, 1992). Taguchi's method is being paired with Ishikawa's QFD approach at the design stage, to enhance product reliability and to facilitate process simplification, with a view to minimizing life cycle and production costs.

3.5 APPLICATIONS TO SERVICE INDUSTRIES

Despite claims to the contrary, it may seem that total quality management really applies much more to product-producing activities than those that provide services. Because so much of engineering involves activities other than manufacturing, this section tries to convince the reader otherwise!

First, we acknowledge that service-producing activities are different from product-producing activities in several respects, which we paraphrase after Bowles and Hammond (1991) as follows:

1. Most services are intangible because they are performances, rather than objects.
2. Services tend to be heterogeneous, with quality of the performance varying from provider to provider, from customer to customer, and from day to day.
3. Production is often inseparable from its consumption. Consider a haircut, for example.
4. Services are perishable and thus cannot be inventoried, saved and later resold. Consider the services of restaurants, hotels, banks, health care providers,

all government agencies, transportation and communication providers, and educational institutions (although some do not meet all four characteristics noted!).

Coming closer to engineering management, Kline and Coleman (1992) have provided two useful comparisons of the manufacturing and construction processes. Table 3.2 shows differences between design for manufacturing and design for construction, while Table 3.3 presents the steps toward achieving production, satisfied customers, and quality.

In the design and construction of facilities, each operation is unique and, it is suggested, therefore unmeasurable. However, the processes that contribute to those one-time designs are usually fairly common, being repeated daily, weekly, or monthly. Consider activities such as permit application, proposal preparation, drawing preparation, specification preparation, cost estimation, shop drawing review, invoicing, project change orders, project control, and information transmittal. As one example, design firms commonly find that about 40% of all permit applications contain errors that must be corrected before approval can be obtained (Quality Management Portfolio, 1990a). Some of the tools to be discussed in Chapter 4 could be applied in a systematic way to this problem, to improve quality, lower cost, and increase timeliness.

When Texas Instruments started a construction quality process in 1985, the cost of nonconformance was about 12%. That means that for every million dollars of work, $120,000 was for rework or warranty work. In the years since 1985, that percentage has been reduced to less than 1%, and the goal is to eliminate the cost of nonconformance. Design errors added 1.8% to costs, and changes requested by TI during the construction process added 4%. After applying quality management

TABLE 3.2   Differences Between Design for Manufacturing and for Construction

Design for Manufacturing	Design for Construction
Usually medium to huge organization	Usually small to medium organization
Make many units of product	Make one unit to design
Customer unknown	Customer known
Has relatively little contact with ultimate consumer	Has much contact with ultimate customer
Customer contact: mostly after sale	Customer contact: always before sale
Sale made after product made	Services done after they are sold
Product is mostly result of a machine process	Product is mostly result of mental effort
End result is product	End result is a service, with documents
Craftsmanship very important	Craftsmanship less an issue
Personnel mix centered on "skilled"	Personnel mix centered on "professional"

SOURCE: D. H. Kline and G. B. Coleman; "Four Propositions in Quality Management," *Journal of Management in Engineering,* Vol. 8, No. 1, © 1992. Reprinted by permission of the American Society of Civil Engineers.

TABLE 3.3   Steps in Production, Satisfying Customers, and Achieving Quality in Manufacturing and in Construction

Design for Manufacturing	Design for Construction
Estimate customer wants	Locate potential client
Design product to meet those wants	Learn client wants: designer pursues or bows out
Make products to meet design	Two parties agree to get together
Promote products to public	Negotiate customer's and designer's wants
Distribute products to public	Design project, meet client's wants
Do final "sell" job	Help client build project
Get customer feedback	Changes are expected or should be
	Direct communication is important for entire length of project
Start cycle over	Start cycle over

SOURCE: D. H. Kline and G. B. Coleman, "Four Propositions for Quality Management of Design Organizations," *Journal of Management in Engineering,* Vol. 8, No. 1, © 1992. Reprinted by permission of the American Society of Civil Engineers.

principles, the company was able to reduce by 85% the costs due to design errors (Lawson, 1992).

Kline and Coleman (1992) have put forth four propositions for the quality management of design organizations. They are as follows:

1. The best quality management program for a design organization is one that is participative to the maximum extent compatible with the organization's culture and management style—both those that are historical and those programmed for the future.
2. The best quality management for any organization is one that recognizes and consciously builds on the managerial philosophy that really exists or is really going to exist.
3. The approach to quality on a project is highly dependent on the attitude of the owner/client/superior; the design organization is greatly influenced by the quality philosophies of the client, customer, or other superior; attempts by the design organization to depart from those precepts and procedures may be a source of conflict that should be acknowledged; and an organization's long-term quality improvement is limited to the narrow perception of quality of its collective clients/customers/superiors.
4. Management styles may need to be varied to meet internal or external needs, but quality management must, at least, meet the test of what could be called the seven C's: consistent, conscious, continual, cost-effective, concrete, claim-reducing, and client-driven.

Scott (1993b), who comments that the approach of Kline and Coleman may cast the design consultant in an inappropriately passive role, suggests that "the engineering firm must demonstrate to the customer/client that their quality management system is the *best* way to do the job." That is, the owner/client and the organization itself may *need* an attitude and culture change!

Chapter 4 includes three examples of successful implementation of quality management in engineering organizations.

3.6 COST OF QUALITY

Crosby (1979), Feigenbaum (1991), Hunt (1992), and others provide helpful guidelines for identifying the various components of quality costs—or, more accurately, the costs of lack of quality. Overall, the *cost of quality* measures the cost of not making the product or service that conforms to the customer's requirements "right the first time" (Merino, 1991).

Figure 3.4, taken from Hunt (1992), identifies these costs as those for conformance and nonconformance. The costs will never be reduced to zero, but it is clear that the introduction of a successful quality management program will reduce significantly the cost of nonconformance.

The following outline, suggested by Feigenbaum (1991), provides further breakdown of the various elements of quality costs.

1. Conformance
    (a) *Prevention*
- Quality planning
- Process control
- Design and development of quality information equipment
- Quality training and work force development
- Product design verification
- Systems development
- Other prevention costs

    (b) *Appraisal*
- Test and inspection of purchased materials
- Laboratory acceptance testing
- Laboratory and other measurement services

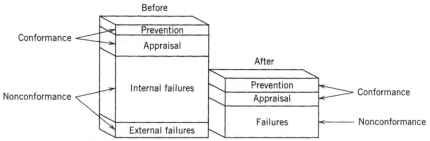

Figure 3.4. "Cost of quality" categories, before and after institution of quality management. (From V. Daniel Hunt, *Quality in America,* Business One Irwin, copyright © 1992. Reprinted with the permission of the author.)

- Inspection
- Testing
- Checking labor
- Setup for test and inspection
- Test and inspection equipment and material and minor quality equipment
- Quality audits
- Outside endorsements
- Maintenance and calibration of quality information test and inspection equipment
- Product engineering review and shipping release
- Field testing

2. Nonconformance

(a) *Internal Failures*
- Scrap
- Rework
- Material procurement costs
- Factory contact engineering

(b) *External Failures*
- Complaints in warranty
- Complaints out of warranty
- Product service
- Product liability
- Product recall

With respect to the relative importance of these cost categories, it has been suggested (Quality by Design, 1992) that most managers believe that conformance costs ("acceptable" costs of quality) are about 75% of the total, whereas in fact nonconformance ("unacceptable") costs are typically about 75% of the total. Feigenbaum's firm, during more than 20 years of tracking quality costs using their Quality Improvement Potential (QIP) index, found that the total costs of quality—"an aggregate of product and service costs for the improvement and assurance of quality, as well as those costs caused by the failure or inadequacy of the improvement or assurance"—range from 5% to 25% of the value of sales (Chapple, 1990b).

In an often-quoted passage, Crosby (1979) claimed:

Quality is free. It's not a gift, but it is free. What costs money are the unquality things—all the actions that involve not doing jobs right the first time. Quality is not only free, it is an honest-to-everything profit maker. Every penny you don't spend on doing things wrong, over, or instead becomes half a penny right on the bottom line.

With the implementation of total quality management and the associated new relationships of trust with suppliers and their own improved procedures, many of the appraisal costs on Feigenbaum's list will be markedly reduced. Thus, we agree that overall costs will surely be reduced as defect rates (or other measures of "unquality") are reduced by introducing quality management principles. However, it is likely that some additional costs for the conformance activities will be required. Juran (1988) provides an excellent conceptual graph to indicate the tradeoffs between internal and external failure costs (nonconformance) and appraisal and prevention costs (conformance). Figure 3.5 indicates the optimal conformance level, at which the total quality costs are minimized.

3.7   STANDARDS OF EXCELLENCE

Now we review some mechanisms for judging quality management systems. Section 3.7.1 introduces a series of international standards, while Sections 3.7.2 to 3.7.5 describe various awards for quality excellence.

3.7.1   ISO 9000

The series of standards referred to as ISO 9000 was developed by the International Organization for Standardization, a body comprising 91 countries which sets standards for use in the global marketplace. ISO 9000 is not an award system per se, but, instead, qualifies a company's quality system to sell directly to countries in the European Common Market without having an independent inspector review the quality of the product or service. The series is a set of five documents specifying quality management systems standards for use in external quality assurance. Note well that the standards specify requirements for a quality management

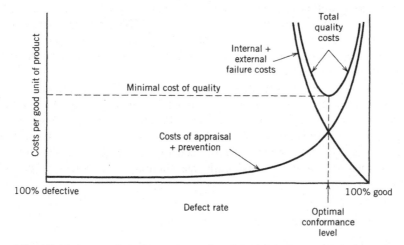

Figure 3.5.   Minimal cost of quality curve. (Adapted with permission of The Free Press, A Division of Macmillan, Inc., from PLANNING FOR QUALITY by J. M. Juran. Copyright © 1988 by Juran Institute, Inc.)

system, *not* technical standards for products or services ("International Standards to Become Paramount in 90's," 1993; Jablonski, 1992).

The five documents are as follows:

- *ISO 9000:* The generic document outlining the series and its fundamental concepts
- *ISO 9001:* A comprehensive standard providing a model of quality management and assurance for companies involved in the design and development, production, installation, and servicing of products or services.
- *ISO 9002:* A model for quality assurance in production and installation.
- *ISO 9003:* A model for quality assurance in final inspection and testing.
- *ISO 9004:* Guidelines on the essential elements and implementation of a quality system. (Basconi, 1992; Chapple, 1992)

More than 50 countries, including the member states of the European Community, Japan, and the United States, had adopted the ISO 9000 series as a national standard by the end of 1992. An individual company desiring to be registered, however, must submit to a lengthy and rigorous review and audit process; total costs may be as high as $100,000 for a review period lasting well over a year (Chapple, 1992). A registered organization that provides design services in the international marketplace must have policies and procedures related to quality policy, contract review, document control, corrective action, training, statistical techniques, quality systems, design control, purchasing, quality records, and internal audits ("International Standards to Become Paramount in 90's" 1992).

While the emphasis is on international activities, the supplier of domestic engineering services may find that sophisticated clients require adherence to ISO 9000 standards. Registration is proving to be a distinct competitive advantage; Mohitpour (1992) says that "ISO 9000 tunes your engine." Hayden (1993) concludes that "ISO 9000 is rapidly evolving as a client-driven guide for consistency, especially among service organizations where the end product is information and the transfer of knowledge—not hardware."

The Dow Corning Corporation began the ISO 9000 registration process in 1987, even though it was no stranger to quality management. In addition to continuously improving the quality of its customers' products and services, the company lists many other benefits, including greater interest in quality by customers, a greater interest in customers' needs, reduction in customer audits, clear and well-documented procedures, and improved training in a variety of disciplines (Schnoll, 1993).

3.7.2 Baldrige Award

The Malcolm Baldrige National Quality Improvement Act of 1987, Public Law 100-107, signed by President Ronald Reagan on August 20, 1987, establishes the annual United States National Quality Award, now known as the Baldrige Award.

Its purposes are to promote quality awareness, to recognize quality achievements of U.S. companies, and to publicize successful quality strategies (Jablonski, 1992). No more than two awards may be given each year in each of three categories: manufacturing companies or subsidiaries, service companies or subsidiaries, and small businesses.

Table 3.4 lists the winners since the award was first presented in 1988. There are seven primary evaluation categories: leadership, information and analysis, strategic quality planning, human resources development and management, management of process quality, quality and operational results, and customer focus and satisfaction. Table 3.5 gives the categories, items, and point values used in establishing the 1994 award. Note that "customer focus and satisfaction" is weighted much more heavily than other categories.

The Baldrige Award has sparked some controversy and debate, with critics suggesting that the award is not necessarily an indication of good product or service quality, that it does not gauge competitiveness or profit potential, and that applying for the award is too costly. Defenders reply that the award reflects judgment of process, not results, and that the expense of preparing the award application should be considered an investment in quality. A fascinating, and sometimes rather sarcastic, part of this debate is found in two *Harvard Business Review* articles (Garvin, 1991; Sims et al., 1992).

3.7.3   Deming Prize (Japan)

The Union of Japanese Scientists and Engineers (JUSE) established the Deming Prize in 1951 to recognize companies that have successfully applied total quality control, based on statistical methods. The prize is named after Deming because the Japanese wanted to honor the American statistician for his friendship and achievements in industrial quality control. The medal bears the words: "The right quality and uniformity are foundations of commerce, prosperity and peace." The prize can be awarded to Japanese companies, overseas companies, small enterprises, divisions, and factories; it makes no distinction between private and public institutions or between manufacturing and service organizations.

Applicants are judged on the basis of 10 criteria:

1. Company policy and planning
2. Organization and its management
3. Education and dissemination
4. Collection, dissemination, and use of information on quality
5. Analysis
6. Standardization
7. Control
8. Quality assurance
9. Results
10. Planning for the future (Dooley et al., 1990; Hunt, 1992)

TABLE 3.4 Winners of the U.S. National Quality Award (the Baldrige Award), 1988–1993

Year	Small Business	Manufacturing	Service
1988	Globe Metallurgical, Inc. Cleveland, Ohio	Motorola, Inc. Schaumburg, Illinois Westinghouse Commercial Nuclear Fuel Division Pittsburgh, Pennsylvania	
1989		Milliken & Company Spartanburg, South Carolina Xerox Business Products and Systems Stamford, Connecticut	
1990	Wallace Co., Inc. Houston, Texas	Cadillac Motor Car Company Detroit, Michigan IBM Rochester Rochester, Minnesota	Federal Express Corporation Memphis, Tennessee
1991	Marlow Industries, Inc. Dallas, Texas	Solectron Corporation San Jose, California Zytec Corporation Eden Prairie, Minnesota	
1992	Granite Rock Company Watsonville, California	AT&T Network Systems Group Transmission Systems Business Unit Morristown, New Jersey Texas Instruments Incorporated Defense Systems & Electronics Group Dallas, Texas	AT&T Universal Card Services Jacksonville, Florida The Ritz-Carlton Hotel Company Atlanta, Georgia
1993	Ames Rubber Corporation Hamburg, New Jersey	Eastman Chemical Company Kingsport, Tennessee	

TABLE 3.5   Baldrige Award: Examination Items and Point Values for 1994

1994 Examination Categories/Items	Point Values
1.0 Leadership	95
1.1 Senior executive leadership 45	
1.2 Management for quality 25	
1.3 Public responsibility and corporate citizenship 25	
2.0 Information and Analysis	75
2.1 Scope and management of quality and performance data and information..15	
2.2 Competitive comparisons and benchmarking................... 20	
2.3 Analysis and uses of company-level data...................... 40	
3.0 Strategic Quality Planning	60
3.1 Strategic quality and company performance planning process..... 35	
3.2 Quality and performance plans 25	
4.0 Human Resource Development and Management	150
4.1 Human resource planning and management.................... 20	
4.2 Employee involvement...................................... 40	
4.3 Employee education and training 40	
4.4 Employee performance and recognition....................... 25	
4.5 Employee well-being and satisfaction 25	
5.0 Management of Process Quality	140
5.1 Design and introduction of quality products and services........ 40	
5.2 Process management: Product and service production and delivery processes..35	
5.3 Process management: Business and support service processes 30	
5.4 Supplier quality... 20	
5.5 Quality assessment... 15	
6.0 Quality and Operational Results	180
6.1 Product and service quality results........................... 70	
6.2 Company operational results 50	
6.3 Business and support service results..........................25	
6.4 Supplier quality results 35	
7.0 Customer Focus and Satisfaction	300
7.1 Customer expectations: Current and future.....................35	
7.2 Customer relationship management 65	
7.3 Commitment to customers 15	
7.4 Customer satisfaction determination..........................30	
7.5 Customer satisfaction results 85	
7.6 Customer satisfaction comparison............................70	
Total points	1000

SOURCE: United States Department of Commerce. *1994 Award Criteria, Malcolm Baldrige National Quality Award.*

3.7.4   State Awards

In 1993, 70% of the states in the United States were involved in some sort of quality award program to honor outstanding quality efforts by private organizations, by state government agencies, or both. Individuals may also be recognized. In Arkansas, exemplary state employees and quality management project teams are given the Governor's Quality Award. The Arizona Celebration of Excellence Award is presented to outstanding state employees and teams and to especially worthy partnerships between the public and private sector. Many states use the criteria for the U.S. National Quality Award (the Baldrige Award) as the basis for their selections (Bemowski, 1993).

3.7.5 President's Award

Modeled after the Baldrige Award, the President's Award for Quality is administered by the Federal Quality Institute and recognizes federal agencies with outstanding programs. Examiners from both the public and the private sectors conduct on-site assessments based on scoring guidelines that include the following: top management leadership and support, strategic planning, focus on the customer, employee training and recognition, employee empowerment and teamwork, measurement and analysis, quality assurance, and quality and productivity improvement results (Bowles and Hammond, 1991; Jablonski, 1992).

3.8 IMPLEMENTING TOTAL QUALITY MANAGEMENT IN THE ENGINEERING ORGANIZATION

Before we leave this chapter, with its principles, gurus, and awards, we need to consider the implementation of total quality management. We can declare our intentions, but what should happen then? We begin by reminding ourselves, once again, that implementing TQM demands a fundamental culture change; then we consider some approaches to implementation.

3.8.1 Culture Change

We have already cited W. Edwards Deming's challenge to Western management to adopt the new philosophy, learn the accompanying responsibilities, and "take on leadership for change."

Miller (1990) said it this way: "The TQM philosophy calls for a major transformation in perception. The way we saw things in the past must be abolished." The American Institute of Architects told its members in 1992, "TQM is far more than just the latest management-of-the-month fad; it's a paradigm shift, a global movement, yet our profession has never addressed it until now" (Quality by Design, 1992). And Hensey reports a conversation with Bill Hayden that included the following observations: "TQM is a culture change, not a 'program.' It's a cause, not a result" (Hensey, 1992).

Thus, the first word of guidance for those considering implementing total quality management is to be prepared to impact the *entire* way your organization does its business. Texas Instruments even borrows the word "culture" in its statement: "The Total Quality Culture at Texas Instruments demands a continuously improving quality process which will be achieved only through the personal leadership of TI managers at all levels and the complete involvement of every TIer" (Heilmeier, 1990).

3.8.2 Approaches to Implementation

Of all the ways that have been suggested to implement total quality in engineering organizations, perhaps the most sensible is that provided by Sproles (1990). Based on his work at Tennessee Eastman Company and considerable activity with the National Society of Professional Engineers, he has outlined a process consisting of four steps, which he says are basic and doable. The steps can be briefly given as follows:

1. Assess Your Organization

Define each group's or team's mission.

Conduct a supplier–input–process–output–customer analysis in order to identify key processes.

Establish a vision for your organization.

Determine key performance areas and establish criteria to measure performance.

2. Manage Normal Operations

Prepare control charts for each key performance area.

Conduct performance reviews.

Identify problems.

3. Plan Your Improvement Efforts

Gather information from (a) your team's vision of where it wants to be in the future, (b) your measures of performance, and (c) customer needs.

Develop an action plan for your major improvement opportunities (MIOs).

4. Execute Improvement Projects

Carry out the action plan.

To this, the author would only add, "do it all over again," in the true spirit of continuous improvement!

Many other references may be helpful in implementing total quality management, including assuring top management support, establishing various quality councils and committees, assessing needs, conducting training, and determining performance measures. See especially Hunt (1992), Jablonski (1992), Miller (1990), Quality Management Portfolio (1990a, 1992), and "TQM Implementation and Quality Partnerships" (1992).

3.9 CONCLUDING REMARKS

How does one conclude a chapter on continuous improvement? Let us simply emphasize some of the points already made by reporting what three writers have said. According to the writer of a letter to a publication of the American Society of Civil Engineers, *ASCE News,* TQM "is a leadership philosophy that fosters and nourishes pride of ownership, responsibility for quality and a quest for continuous improvement and customer satisfaction. It is not a program. I believe exceptional benefits in effectiveness and efficiency can accrue to civil engineering business areas through awareness and implementation of TQM principles" (King, 1993).

On the same theme, Hayden (1993) wrote, "Quality is an attitude, not a program. It requires a deep and widespread commitment to continuous improvement by persistent examination of what is going on. It emphasizes perpetual improvement in products, services, information, processes and performance standards. This approach builds in quality; it does not inspect out defects."

And finally, Barker (1992) called total quality "the most important paradigm shift of the twentieth century."

3.10 DISCUSSION QUESTIONS

1. The four primary principles of total quality management, it is claimed, can be applied to any organization.
   (a) Pick a nonengineering organization that you know about or are involved with (PTA, university, house of worship, youth group, etc.) and suggest specific ways that each of the four principles might be applied.
   (b) Do the same thing with an engineering organization.
2. The Shewhart cycle shown in Figure 3.2 is a continuous process. If the organization has not yet practiced this approach, at which step should it start?
3. Explain how each step of the Shewhart cycle provides feedback to the next step.
4. Who are the customers at your university?
5. "The ideal team, it would seem, is that which includes those who work in the process, those who are its suppliers, and those who are its customers." Identify the players who might be part of a real or hypothetical team of this type, and describe the advantages of such a team concept.
6. Section 3.4 provides brief descriptions of the philosophies of six quality "gurus." Try to identify some common themes among their philosophies.
7. If the implementation of total quality management represents a change in an organization's "culture," what can the following personnel do to assist in an orderly and effective culture change: top-level leadership, midlevel management, entry-level employee?
8. Try to identify some characteristics of *engineering* organizations that make total quality management especially important, effective, easy, and/ or difficult.
9. What impact has quality management had on the worker? The company? The economy?

10. Suppose that total quality management fulfills the claim of some proponents that it can eliminate the cost of between 10 and 30% of the labor in the American economy. How can we afford this loss of job opportunities for the population?

3.11 REFERENCES

Advanced Personnel Systems. 1992. *Quality Abstracts,* Vol. 1, No. 2, July.

American Society of Civil Engineers. See *Quality in the Constructed Project.*

Artzt, E. L. 1992. "Redefining Quality: The New Relationship Between Quality, Price and Value." Keynote address to Quality Forum VIII, New York, October 1.

Aune, A. 1990. "Total Management to Universities: A Challenge and a Necessity." American Society for Quality Control Annual Conference, San Francisco, May.

Barker, J. A. 1992. *Future Edge.* New York: William Morrow.

Basconi, M. A. 1992. "What Is ISO 9000?" *Quality at Work,* the Business Journal of Upper East Tennessee and Southwest Virginia, October 1, 19.

"Basic Steps in the Total Quality Process." 1992. *Industry Engineer,* National Society of Professional Engineers, Vol. 9, No. 4, April/May, 4.

Bemowski, K. 1993. "The State of the States." *Quality Progress,* Vol. 26, No. 5, May, 27–36.

Bowles, J., and J. Hammond. 1991. *Beyond Quality: New Standards of Total Performance That Can Change the Future of Corporate America.* New York: Berkley Books.

Chapple, A. 1990a. "Quality Guru Engineer Pursues Quiet Revolution." *Engineering Times,* National Society of Professional Engineers, Vol. 12, No. 11, November, 10.

Chapple, A. 1990b. "Quality Guru Feigenbaum Estimates GNP Gains if U.S. Installs TQM." *Engineering Times,* National Society of Professional Engineers. Vol. 12, No. 11, November, 1, 11.

Chapple, A. 1992. "Quality Standard Key for Work in Europe." *Engineering Times,* National Society of Professional Engineers, Vol. 14, No. 12, December, 1, 3.

Crosby, P. B. 1979. *Quality Is Free: The Art of Making Quality Certain.* New York: McGraw-Hill.

Deming, W. E. 1986. *Out of the Crisis.* Cambridge, MA: Massachusetts Institute of Technology Center for Advanced Engineering Study.

Dooley, K. J., D. Bush, J. C. Anderson, and M. Rungtusanatham. 1990. "The United States' Baldrige Award and Japan's Deming Prize: Two Guidelines for Total Quality Control." *Engineering Management Journal,* Vol. 2, No. 3, September, 9–16.

Feigenbaum, A. V. 1991. *Total Quality Control,* 3rd ed., rev. New York: McGraw-Hill.

Garvin, D. A. 1984. "What Does 'Product Quality' Really Mean?" *Sloan Management Review,* Vol. 26, No. 1, Fall, 34.

Garvin, D. A. 1991. "How the Baldrige Award Really Works." *Harvard Business Review,* Vol. 69, No. 6, November–December, 80–93.

Hayden, W. M., Jr. 1992. "Management's Fatal Flaw: TQM Obstacle." *Journal of Management in Engineering,* American Society of Civil Engineers, Vol. 8, No. 2, April, 122–129.

Hayden, W. M., Jr. 1993. "ISO 9000: Is the Day of Decision Near?" *The Military Engineer,* No. 554, January–February, 12–13.

Heilmeier, G. H. 1990. "Quality—A Personal Responsibility of Managers." *Engineering Management Journal,* Vol. 2, No. 3, September, 5–8.

Hensey, M. 1992. "Conversation on TQM with Bill Hayden." *Journal of Management in Engineering,* American Society of Civil Engineers, Vol. 8, No. 3, July, 221–223.

Hunt, V. D. 1992. *Quality in America: How to Implement a Competitive Quality Program.* Homewood, IL: Business Irwin One.

International Service and Quality Forum Update. 1992. Vol. 1, No. 5, October/November.

"International Standards to Become Paramount in 90's." 1993. *Industry Engineer,* National Society of Professional Engineers, Vol. 10, No. 2, December 1992/January 1993, 2.

Jablonski, J. R. 1992. *Implementing TQM: Competing in the Nineties Through Total Quality Management,* 2nd ed. Albuquerque, NM: Technical Management Consortium.

Juran, J. M. 1988. *Planning for Quality.* New York: Free Press.

Juran, J. M. 1989. *Juran on Leadership for Quality: An Executive Handbook.* New York: Free Press.

Juran, J. M., and F. M. Gryna. 1988. *Juran's Quality Control Handbook,* 4th ed. New York: McGraw-Hill.

"Key Elements of Total Quality Management." 1992. *Industry Forum,* National Society of Professional Engineers, Vol. 9, No. 3, February/March, 4.

King, C. H. 1993. "Total Quality Management: A Philosophy, Not a Program." Letter to *ASCE News,* American Society of Civil Engineers, Vol. 18, No. 2, February, 4.

Kline, D. H., and G. B. Coleman. 1992. "Four Propositions for Quality Management of Design Organizations." *Journal of Management in Engineering,* American Society of Civil Engineers, Vol. 8, No. 1, January, 15–26.

Lawson, W. H. 1992. "Why a Quality Process for Design/Construction?" *Quality Management Portfolio,* The National Institute for Engineering Management and Systems, Vol. 3, Issue 1, Winter.

Magowan, R. E. 1991. "Using the Taguchi Method to Enhance the Quality of Products and Processes." *Engineering Management Journal,* Vol. 3, No. 4, December, 33–42.

Markert, C. D., R. C. Simon, and R. M. Miller. 1992. "Dr. W. Edwards Deming's 14 Points Adapted for the A/E Firm." *Quality Management Portfolio,* The National Institute for Engineering Management and Systems, Vol. 3, Issue 2, Summer.

Merino, D. N. 1991. "Cost of Quality." *Engineering Management Journal,* Vol. 3, No. 3, September, 8–12.

Miller, R. E. 1990. "Quality: Are We Up to It?" *Engineering Management Journal,* Vol. 2, No. 3, September, 45–50.

Mohitpour, M. 1992. "ISO 9000 Tunes Your Engine." *Quality by Design.* William M. Hayden, Jr. Consultants, Vol. 5, No. 4, October, 3, 6.

NIEMS. 1990. National Institute for Engineering Management and Systems *Mission Statement.* Alexandria, Virginia: National Society of Professional Engineers.

Pirsig, R. M. 1981. *Zen and the Art of Motorcycle Maintenance.* New York: Bantam Books.

QPlus: The Greiner Quality Connection, Vol. 1, No. 3. n.d. Irving, TX: Greiner Engineering, Inc.

Quality by Design. 1992. William M. Hayden, Jr., Consultants, Inc., Vol. 5, No. 4, October 1992.

Quality in the Constructed Project: A Guide for Owners, Designers and Constructors. 1990. New York: American Society of Civil Engineers.

Quality Management Portfolio. 1990a. Alexandria, VA: The National Institute for Engineering Management and Systems, Vol. 1, Issue 1, Spring.

Quality Management Portfolio. 1990b. Alexandria, VA: The National Institute for Engineering Management and Systems, Vol. 1, Issue 2, Fall.

Quality Management Portfolio. 1992. Alexandria, VA: The National Institute for Engineering Management and Systems, Vol. 3, Issue 2, Summer.

Samson, C.H. 1992. "NSPE's 'Quality Journey.' " *The Professional Engineer,* Professional Engineers of North Carolina, January–February, 12–14.

Sarantopoulos, P.D., and B.S. Liebesman. 1992. "The Systems Approach Toward a Comprehensive Approach to Product Quality." Working Paper #92-103, Rutgers, the State University of New Jersey, Department of Industrial Engineering.

Schnoll, T. 1993. "One World, One Standard." *Quality Progress,* Vol. 26, No. 4, April, 35–39.

Scott, H.A. 1992. Personal communication, June 16.

Scott, H.A. 1993a. Personal communication, January 16.

Scott, H.A. 1993b. Personal communication, May 4.

Sims, A.C., et al. 1992. "Does the Baldrige Award Really Work?" *Harvard Business Review,* Vol. 70, No. 1, January–February, 126–147.

Sproles, G.W. 1990. "Total Quality Management Applied to Engineering and Construction." *Engineering Management Journal,* Vol. 2, No. 4, December, 33–38.

"TQM Implementation and Quality Partnerships." 1992. *Industry Engineer,* National Society of Professional Engineers, Vol. 9, No. 6, August/September, 4.

Tribus, M. 1988. "Managing to Survive in a Competitive World." *Quality First,* National Society of Professional Engineers, Publication 1459, August.

Unal, R., and L.B. Bush. 1992. "Engineering Design for Quality Using the Taguchi Approach." *Engineering Management Journal,* Vol. 4, No. 1, March, 37–47.

VTT Building Production Laboratory. 1992. "Ongoing and Lately Concluded Projects in the Section of Construction Management in the Building Production Laboratory." Tampere, Finland.

Walton, S., and J. Huey. 1992. *Sam Walton, Made in America: My Story.* New York: Doubleday.

4

Total Quality Management: Techniques and Applications

Against the background of Chapter 3, this chapter describes several tools that may be useful to the engineering manager who is using total quality management. We consider a number of numerical and nonnumerical tools. We then describe the benchmarking technique that is used to compare one organization's productivity and performance against the best practices of others. To demonstrate the real-world use of total quality management, we present descriptions of implementation in three engineering organizations—a design consultancy, a professional construction organization, and a process industry. Finally, we speculate about the future of TQM in the management of engineering organizations.

4.1 A SAMPLING OF NUMERIC TOOLS

4.1.1 Control Charts*

A control chart is "a graphical method for evaluating whether a process is or is not in a 'state of statistical control'" (Feigenbaum, 1991). Dr. Deming makes a major point of distinguishing between *common causes* and *special causes* of trouble. He defines common causes as faults of the system that are inherent or "uncontrollable," whereas special causes are faults from fleeting events (operator error, an out-of-adjustment machine, etc.) and are "controllable." In the state of statistical control, all special causes so far detected have been removed. The remaining variation must be left to chance—to common causes inherent in the system.

Control charts help us to determine whether special causes are present in the system (i.e., whether the system is "out of statistical control") by pinpointing any occurrences that are outside established control limits. Some variation is inevitable, even in a state of statistical control; but the process is stable in such a state, even though it is random. Special causes—those that cause occurrences outside the control limits—can be improved by an individual employee. Common causes—those that cause occurrences within the control

---

*I am grateful to Dr. E. R. Baker for assistance in preparing this section on control charts.

limits—cannot be improved by individual employees. An organization in a state of statistical control is not within the control of individual employees because once this state has been achieved and maintained, improvements can be brought about only by improving the process (Deming, 1986).

A control chart, then, is a device for tracking the variation in a particular measure of process performance (e.g., the diameter of the rotor shafts produced in a lathing operation or the number of visible defects in the paint job on a new automobile) to determine whether the variations are attributable to individual performances or to the random operation of the system. Several types of control chart are in use, and each is designed to report on a different measure of process performance. Table 4.1 presents a complete list of the standard statistical process control charts, which are categorized as control charts for variables, such as "part size," and for attributes, such as "fraction defective." Note that we would seldom use more than one or two of these tools with a given process.

We shall demonstrate one such application, with the warning that much important theory has been omitted. Works by Feigenbaum (1991), Ishikawa (1982), and Juran and Gryna (1988) provide details that must be understood before the theories can be applied to practical cases.

The most commonly used control charts are the so-called x-bar (for $\bar{x}$) chart and the R chart (the symbol is $\bar{R}$). The x-bar chart is designed to report on the mean or average value of some measure of system performance. For example, consider a lathing operation, in which we take periodic measurements of shaft diameter after the units have been produced. By grouping five randomly selected observations each hour, say, we may calculate an average diameter for those five shafts. That average diameter varies as a function of chance (inherent) variation (Deming's "common causes") and, if the process is going out of control, assignable causes (changes in controllable variables, or Deming's "special causes").

Conceptually, the application of a control chart, regardless of type, is simple. Consider again our control chart for the mean value of the shaft diameters. Since the inherent variation of a process is the result of the variation in many different variables, the central limit theorem tells us that we should expect the variation to be normally distributed about the mean. The mean sample values would vary about the overall process mean in a normally distributed fashion.

To determine whether a process is "in control," we must establish upper and lower control limits. Whenever a sample mean is calculated and plotted outside the "control limits," we say that the process is out of control and look for assignable

TABLE 4.1   Types of Control Charts

Control Charts for Variables	Control Charts for Attributes
$\bar{x}$ = mean value	p = proportion defective per batch
$\bar{R}$ = range of the variable	np = number of defective per batch
s = standard deviation	c = number of defects per batch
M = median	u = number of defects per unit

causes—that is, controllable variables whose values have changed. Since we want to be notified when the process is out of control, we need to determine the control limits such that most of the area under the normal curve is within the control limits. For the normal distribution, it can be shown that 99.7% of the area under the curve is contained within limits of plus or minus 3 standard deviations of the mean. This "three sigma" range, which represents 3 parts per thousand, is commonly used to establish control limits, although the "six sigma" limit, which achieves a maximum of 10 parts out of control per million, is a goal of many companies, as explained in Section 3.3.1.

Figure 4.1 shows three conceptual control charts to illustrate the use of the upper and lower control limits (UCL and LCL). In Figure 4.1a, the process is in control, since all sample means are contained between the upper and lower control limits. In Figure 4.1b, the process is out of control, and it is evident that the mean value is increasing, whereas in Figure 4.1c, the process is also out of control, but it is the standard deviation or range that is increasing.

Determination of the upper and lower control limit values is accomplished by developing a confidence limit for the mean value of the control variable. For the x-bar chart, a common approach is to define this confidence limit as a function of the mean, the range and a factor that depends on the number of samples used to calculate the process mean. This method avoids having to calculate the value of

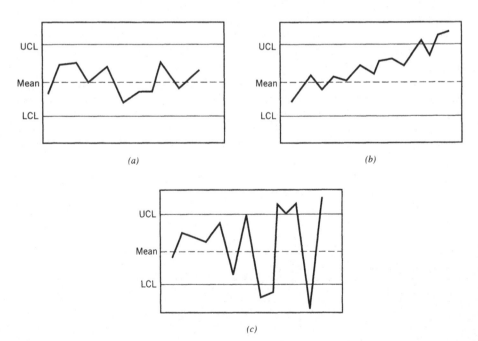

Figure 4.1. Three conceptual control charts. (a) Process in control. (b) Process out of control, mean increasing. (c) Process out of control, range increasing.

the standard deviation. Thus,

$$\text{UCL} = \bar{x} + A_2\bar{R}$$

$$\text{LCL} = \bar{x} - A_2\bar{R}$$

where $\bar{x}$ = average of the sample means
$\bar{R}$ = average of the sample ranges
A_2 = a factor from Table 4.2 based on the number of samples

To develop the control chart, we sample our process output for n samples, and then calculate $\bar{x}$, the UCL, and the LCL. Once the chart has been constructed, each new sample mean is plotted on the graph, and when behavior similar to that illustrated in Figure 4.1b or 4.1c occurs, we shut down the process and look for the problem. Consider the example in Table 4 3 We have measured the shaft diameter after the lathing operation and recorded its diameter in centimeters.

Using the first 15 samples we can develop a control chart for $\bar{x}$.

$$\bar{x} = 4.064$$

$$\bar{R} = 0.3933$$

$$A_2 = 0.223, \text{ from Table 4.2, based on 15 samples}$$

$$\text{UCL} = 4.064 + (0.223)(0.3933) = 4.152$$

$$\text{LCL} = 4.064 - (0.223)(0.3933) = 3.976$$

Figure 4.2 shows the control chart developed from the preceding calculations. To control the mean of this lathing process, we plot the sample mean values for observations subsequent to the first 15 samples. In Figure 4.2, we note that the mean values for samples 16, 17, 20, 22, and 23 fall outside the control limits for this example.

To control the variance (or standard deviation) of the process, we would normally use an R chart, which uses control limits for the range of values for each sample. In fact, we would seldom use one chart without the other.

In a fashion similar to that of the x-bar chart, we may develop an estimate of the mean range, and appropriate upper and lower control limits to the process. Thus,

$$\text{UCL} = D_4\bar{R}$$

$$\text{LCL} = D_3\bar{R}$$

$$\bar{R} = \text{ average range}$$

and D_4 and D_3 are taken from Table 4.2. For our example, we again use $\bar{R} = 0.3933$, taking D_4 (0.1652) and D_3 (0.348) from Table 4.2. Then we can write:

TABLE 4.2   Factors for Quality Control Charts

| | $\bar{x}$ Chart | | R Chart | | | |
| | Factors for Control Limits | | Factors for Central Line | Factors for Control Limits | | |
n^a	A_1	A_2	d_2	D_3	D_4	n
2	3.760	1.880	1.128	0	3.267	2
3	2.394	1.023	1.693	0	2.575	3
4	1.880	0.729	2.059	0	2.282	4
5	1.596	0.577	2.326	0	2.115	5
6	1.410	0.483	2.534	0	2.004	6
7	1.277	0.419	2.704	0.076	1.924	7
8	1.175	0.373	2.847	0.136	1.864	8
9	1.094	0.337	2.970	0.184	1.816	9
10	1.028	0.308	3.078	0.223	1.777	10
11	0.973	0.285	3.173	0.256	1.744	11
12	0.925	0.266	3.258	0.284	1.716	12
13	0.884	0.249	3.336	0.308	1.692	13
14	0.848	0.235	3.407	0.329	1.671	14
15	0.816	0.223	3.472	0.348	1.652	15
16	0.788	0.212	3.532	0.364	1.636	16
17	0.762	0.203	3.588	0.379	1.621	17
18	0.738	0.194	3.640	0.392	1.608	18
19	0.717	0.187	3.689	0.404	1.596	19
20	0.697	0.180	3.735	0.414	1.586	20
21	0.679	0.173	3.778	0.425	1.575	21
22	0.662	0.167	3.819	0.434	1.566	22
23	0.647	0.162	3.858	0.443	1.557	23
24	0.632	0.157	3.895	0.452	1.548	24
25	0.619	0.153	3.931	0.459	1.541	25

$^a n > 25$: $A_1 = 3/\sqrt{n}$. n = number of observations in sample.

SOURCE: W. W. Hines and D. G. Montgomery, *Probability and Statistics in Engineering and Management Science*, 3rd ed. Copyright © 1990. Reprinted by permission of John Wiley & Sons, Inc.

$$UCL = (1.652)(0.3933) = 0.6497$$

$$LCL = (0.348)(0.3933) = 0.1869$$

Figure 4.3 plots the mean range and the upper and lower control limits, plus samples 16 through 25. Note that the range for sample 24 is out of control.

4.1.2   Histograms

A histogram is a visual representation of the spread or dispersion of variable data, such as number of defects per lot. Often these diagrams show a tendency for many

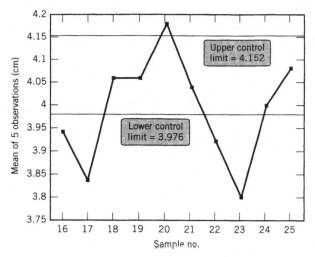

Figure 4.2. Control chart for mean of shaft diameter.

items to congregate toward the center of the distribution, with progressively fewer items as the distance from this central tendency increases (Hunt, 1992).

The information is represented by a series of bars proportional in height to the frequency of the group being represented, with the heights of the rectangles indicating the proportions of data points in each group. If the data have been reduced to frequencies that are fractions of 1.0, the total of all frequencies will equal 1.0.

A histogram assists in identifying changes in processes as changes are made. It indicates how much variability there can be in a process or product, and it can help in setting standards. Figures 4.4 and 4.5 are histograms based on an illustration given by Tribus (1988). Figure 4.4 shows variations in a part dimension when an automatic machine tool was running freely. Figure 4.5 shows the distribution when the machine was placed under the control of an automatic controller. Comparison of these two histograms readily indicates that the automatic controller was overcontrolling, and the resulting statistical distribution is broader than the distribution reported before the control was used.

4.1.3   Pareto Charts

An Italian economist, Vilfredo Pareto, observed in the late 1800s that typically 80% of the wealth in a region was concentrated in about 20% of the population. Later, Joseph Juran drew on those observations to put forth a generalized idea that he called the "Pareto principle," according to which, in any system, only a few elements, say about 20%, account for a majority (say, 80%) of the system's problems (Ross, 1991). Simply stated, this could be called the principle of the "vital few and the trivial many" (Brocka and Brocka, 1992).

TABLE 4.3 Shaft Diameter Measurements and Mean and Range Calculations for Lathing Operations

Sample Number	Observation Number					Mean, $\bar{x}_i$	Range, R_i
	1	2	3	4	5		
1	4.3	4.5	4.3	3.8	4.4	4.26	0.7
2	3.9	3.8	4.2	4.1	4.0	4.00	0.4
3	3.9	3.9	4.1	4.2	4.1	4.04	0.2
4	3.8	3.8	3.9	4.0	4.0	3.90	0.2
5	4.0	4.1	4.1	4.0	4.2	4.08	0.2
6	4.3	4.2	4.1	4.3	4.2	4.22	0.2
7	4.1	4.2	4.5	3.9	4.2	4.18	0.6
8	4.3	4.5	4.1	3.8	4.5	4.24	0.7
9	4.1	4.0	4.0	4.2	4.1	4.08	0.2
10	3.8	3.7	3.9	4.0	3.8	3.84	0.3
11	3.8	4.1	4.2	4.0	4.0	4.02	0.4
12	4.6	4.1	4.0	4.2	4.2	4.22	0.6
13	4.1	3.7	3.8	3.7	3.9	3.84	0.3
14	4.0	3.6	3.8	3.9	3.9	3.84	0.4
15	4.2	4.2	4.5	4.1	4.0	4.20	0.5
16	4.2	4.1	3.8	3.9	3.7	3.94	0.5
17	3.6	3.8	3.9	3.9	4.0	3.84	0.4
18	4.1	4.0	3.9	4.0	4.2	4.06	0.3
19	4.0	4.0	4.2	4.3	3.8	4.06	0.5
20	4.1	4.1	4.3	4.5	3.9	4.18	0.6
21	4.1	4.1	4.2	4.0	3.8	4.04	0.4
22	3.8	3.9	4.0	4.0	3.9	3.92	0.2
23	3.8	3.9	3.9	3.8	3.6	3.80	0.3
24	4.0	4.0	4.5	3.7	3.8	4.00	0.8
25	4.4	4.3	3.9	4.0	3.8	4.08	0.6

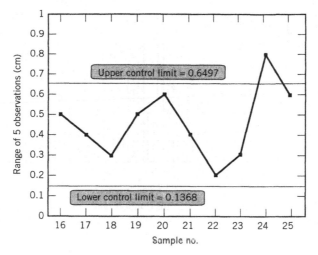

Figure 4.3. Control chart for range of shaft diameter.

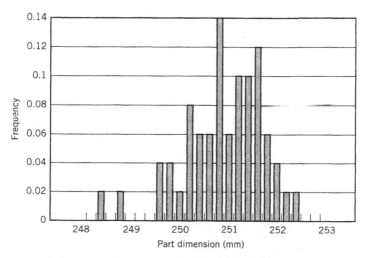

Figure 4.4. Sample histogram for part size distribution without automatic control system. (From Tribus, 1988, by permission of Myron Tribus through the National Society of Professional Engineers.)

Examples of the rather universal application of this principle include diverse phenomena: perhaps 20% of your firm's employees account for 80% of its tardiness, 10% of your accounts are responsible for 80% of the revenue, 10 to 20% of inventory items account for 80 to 90% of the total inventory value, or 15% of the firm's engineers hold 90% of its patents.

A Pareto chart is a bar chart in which the bars are arranged in descending order of their importance. Each bar represents one type of activity (typing, telephoning,

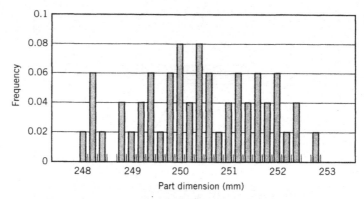

Figure 4.5. Sample histogram for part size distribution with automatic control system. (From Tribus, 1988, by permission of Myron Tribus through the National Society of Professional Engineers.)

copying, mailing, etc. for a clerical staff), one group of employees, one area, or some other grouping. The length of the bar is proportional to the importance of that item. The arrangement of the chart allows the user to identify easily those "vital few" areas that, if improved, would have the greatest overall impact on quality and productivity.

Figure 4.6 shows a type of Pareto chart for the results of an interesting study performed by VTT, the Technical Research Center of Finland, to identify the primary causes of quality problems in construction. The chart makes it clear that more than 60% of the quality problems are due to project and site management and the designer. If contractors address only local authorities and company management in attempting to remedy quality problems, it is unlikely that substantial improvement

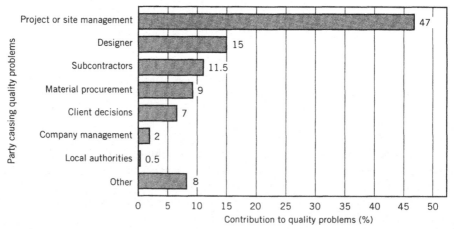

Figure 4.6. Sample Pareto chart: quality problems in Finnish construction. (From VTT, the Technical Research Center of Finland, 1992.)

will be forthcoming! While this example is from an entire industry, the approach can be used just as easily and effectively for a single firm or one element thereof. The case study in Section 4.5.1 shows an example of a Pareto chart used by a design consultancy in Florida.

4.1.4 Run Charts (Timeline Charts)

A run chart, or timeline chart, is a plot of how an attribute varies with time. Successive observations are plotted on a graph as a function of time. Tribus (1988) suggests that the advantage of run charts is that the eye is good at noticing patterns, whereas extracting trends from tabular data is more difficult.

Figure 4.7 shows a run chart for the elongation of 50 springs tested in the order of their manufacture, as published by Deming (1986). The chart also shows the distribution of those 50 readings. Dr. Deming commented as follows: "The distribution is fairly symmetrical, and both tails fall well within the specifications. One might therefore be tempted to conclude that the process is satisfactory. However, the elongations plotted one by one in the order of manufacture show a downward trend. Something is wrong with the process of manufacture, or the measuring instrument...." Thus an attempt to use the distribution and its standard deviation would not help us to understand the process because the distribution is not stable.

4.1.5 Scatter Diagrams

Scatter diagrams, or *XY* plots, are "raw" data displays that plot one variable against another. They permit the examination of the two factors and the determination of the relationship that may exist between them. The pattern of the points describes the strength of the relationship between the variables. There may be a positive correlation, a negative correlation, or no correlation between the variables. Statistical studies, such as regression analysis, can assist in determining the kind of correlation that exists.

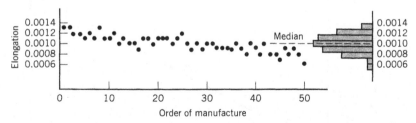

Figure 4.7. Sample run chart for 50 springs tested in order of manufacture. (Reprinted from *Out of the Crisis*, by W. Edwards Deming, by permission of MIT and W. Edwards Deming. Published by MIT Center for Advanced Engineering Study, Cambridge, MA 02139. Copyright © 1986 by W. Edwards Deming.)

The scatter diagram in Figure 4.8 shows the relationship between the cost of preparing proposals and the lengths of those proposals, in pages, for 30 proposals. A least-squares linear regression analysis determined that there is a positive correlation between the two variables (costs tend to increase as number of pages increases). The line on the diagram is the best linear regression line for the data. The analysis showed a relatively strong positive correlation, with an R^2 regression coefficient equal to $+0.71$.

4.2 A SAMPLING OF NONNUMERIC TOOLS

4.2.1 Brainstorming

Brainstorming is a group technique in which members of the group *storm* or generate ideas in a way that is free of criticism and second thoughts. The idea is to come up quickly with many ideas on a focused problem; the emphasis is on quantity rather than quality. Brainstorming can be an excellent way to tap the creative thinking of a team. Rather than generating a single "solution," brainstorming proposes many possible solutions. Other techniques, such as the nominal group technique described in Section 4.2.5, may be used to select the best solution.

The "rules" for conducting a brainstorming session include the following:

- Be sure the purpose is stated and understood.
- Urge each participant to suggest ideas without fear of being criticized.
- Allow no criticism and no discussion of suggestions.
- Encourage participants to build on the ideas of others.
- Record the ideas on flip charts or other easily visible media.

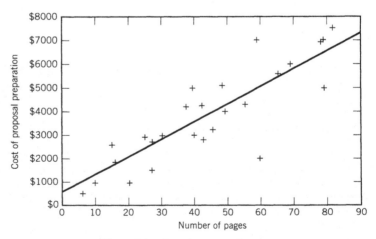

Figure 4.8. Sample scatter diagram.

After the ideas have been expressed, they should be clarified to make sure everyone understands them. Then, the ideas may be evaluated to eliminate duplications, irrelevancies, or off-limit issues. Note once again that full-fledged discussion of the alternatives and selection of *the* solution are not part of the brainstorming process (Hunt, 1992).

Brocka and Brocka (1992) describe a special type of brainstorming called storyboarding, in which individual participants write ideas on cards, which are immediately attached in a random pattern to a board that is visible to the whole group. Again, no comments or criticisms are allowed, and members are encouraged to use the ideas of others to develop their own. After the group has run out of ideas, the cards can be placed into categories, with duplicates and irrelevant ideas discarded. The remaining, organized cards can then be taped together for later use in a selection process.

4.2.2 Cause-and-Effect Diagrams (Ishikawa or Fishbone Diagrams)

A cause-and-effect diagram provides a graphical representation of the relationship between phenomenon and its potential causes. These diagrams are also called fishbone diagrams, because they look like the head and bones of a fish, or Ishikawa diagrams, because Kaoru Ishikawa was the first to use them, in 1943 (Brocka and Brocka, 1992). Often these graphic devices are helpful in recording and organizing the results of a brainstorming session.

A box on the right (the "fish head") contains the effect, or the problem, whose causes are being identified. The causes are then organized and displayed in categories on the left, as "bones" of the fish. The major categories sometimes used to organize the diagram are money, machines or tools, methods or processes, materials, people, management, and environment.

The cause-and-effect diagram of Figure 4.9 shows what happens to the impact strength (effect) of PVC pipe as a result of several aspects of its manufacture. Note that the diagram organizes the causes into five categories: process, raw material, blending, environment, and human care (Yazici, 1990). A diagram such as this helps quality analysis teams to examine all possible causes, and it provides a tangible process for such groups to follow.

4.2.3 Flowcharting

Tribus (1988) points out that a cause-and-effect diagram helps to relate the elements of a process, but it does not really show how the process works. A flowchart helps us to understand how people and the organization interact with the process by depicting the steps in a process, thus indicating how the work "flows" from beginning to end. Such charts can also model the "ifs, ands, and buts" of a process, or its logic, much as a flowchart is often used to model and plan a computer program.

Flowcharts can be used to depict (and thus potentially improve) such diverse activities as the flow of materials, the steps in making a sale, and the actions needed

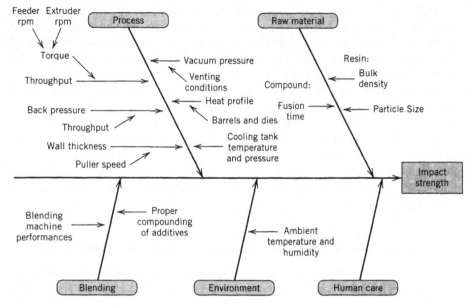

Figure 4.9. Sample Cause and Effect Diagram. (From H. Yazici, "Implementation of SPC Techniques," *Engineering Management Journal,* Vol. 2, No. 3, © 1990. Reprinted by permission of *Engineering Management Journal.*)

to service a piece of equipment. The engineering manager seeking to improve a process might employ a simple flowchart—called by Hunt (1992) a top-down flowchart—where only the most fundamental steps are shown, or a more detailed chart that gives specific information about process flow by representing every decision point, feedback loop, and process step. Often, detailed flowcharts are used for selected critical portions of overall processes or systems.

Figure 4.10 is a flowchart for the preparation of an engineering drawing. It includes several steps and a decision point after each review period. If the chart represents a process that is believed to be too unwieldy or time-consuming, the various steps and the relationships among the steps could be examined to try to identify potential improvements.

4.2.4 Force Fields

Force field analysis is a method for analyzing the positive and negative forces that impact on a proposed improvement alternative. The user identifies the forces likely to affect a given situation in a positive or negative way and then assigns a value indicating the degree to which each force is positive or negative. These values can be totaled, with a relatively large positive total indicating that the proposal is a good one (Brocka and Brocka, 1992).

In the example shown in Figure 4.11, a company is considering installing a modified launch vehicle guidance and tracking system. A team consisting of an

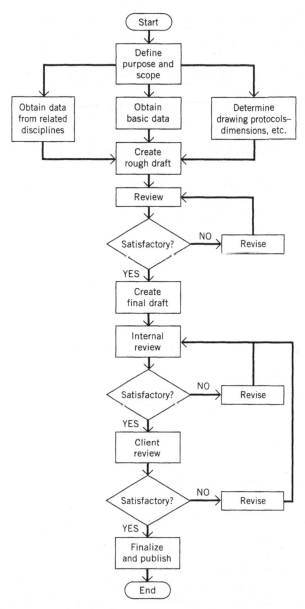

Figure 4.10. Flowchart for preparation of an engineering drawing.

aeronautical engineer, an instrumentation engineer, a cost and schedule analyst, and a project manager identified the several forces that might impact such an implementation. The forces are displayed in no particular order, but the "polarity" of each force (positive or negative) has been determined, and weighting factors

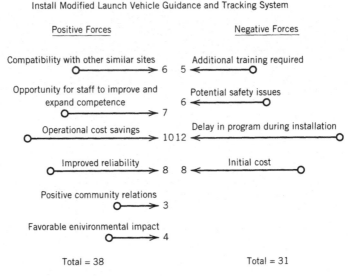

Figure 4.11. Sample force field analysis.

have been assigned to each. In balance, it appears that positive factors outweigh negative factors by a considerable degree.

Force field analysis, which permits comparison of several proposed improvement alternatives, may be an excellent group technique to be used after a brainstorming session and the preparation of a cause-and-effect diagram. This mode of analysis, however, is limited to relatively small proposals and by the somewhat arbitrary nature of the weighting factor values.

4.2.5 Nominal Group Technique

The nominal group technique (NGT) is a method for reaching group consensus that is similar to brainstorming but is more structured, more tolerant of conflicting ideas, more likely to involve all participants, and more decision oriented (Brocka and Brocka, 1992; Hunt, 1992). The first part of the process could be called "silent brainstorming." After the problem or issue has been presented, the group (ideally, 10 to 15 persons) is asked to generate ideas silently for 10 to 15 minutes. Then there is a round-robin session, during which each person is asked to present one of his or her ideas, which is captured by the leader on a flip chart or other device; the ideas are presented without discussion or elaboration, and the leader makes no attempt to editorialize. The process continues until all participants have presented all their ideas.

Next, the ideas are discussed for clarification and to eliminate duplications. Members of the group then evaluate all ideas silently, using a coherent method to establish priorities. Tribus (1988) suggests that with N ideas, each person be allowed $1 + 0.5N$ votes. Thus if there had been 30 ideas, each member would select 16 ideas and rank them from 1 to 16, with 16 representing the most important idea. Individual selections and rankings are tallied with respect to which ideas were selected and how they were ranked. An example is shown in Table 4.4, based on a list of suggestions for improving telephone response to technical questions. A Pareto chart could be used to display the results.

4.3 BENCHMARKING

Benchmarking is a method of comparing the operational processes and procedures of one organization against those of another, with the goal of finding ways of improving those processes and procedures. The term "benchmark" is taken to mean a standard of excellence or achievement that is "world class." Hunt (1992) describes benchmarking as a four-step process that consists of figuring out what to benchmark, finding out what the benchmark is, determining how the benchmark is achieved, and determining how to meet or exceed the benchmark by changing one's own practices.

Benchmarking may be carried out at several levels (Cecil and Ferraro, 1992). It can encompass an entire organization, or a particular department within the

TABLE 4.4 Results of Nominal Group Technique Ranking for Improved Telephone Service for Technical Questions

Idea	Number of Votes	Total Ranking[a]
Provide an "answer book" to phone answerers	8	59
Provide technical training to phone-answering staff	8	56
Dedicate a phone answer line	7	47
Make a list of top 10 problems	6	43
Have different numbers for orders and for technical questions	7	36
Have different numbers for different categories of technical questions	5	22
Make our product easier to use	6	21
Provide better instructions on how to use the products	5	18
Have a guru available in the background	5	16
Hire more people	3	13
Live with the situation	4	11
Get rid of the phones	3	8
Make the caller read the instructions first	2	5
Don't worry about the time spent	1	2
Give us more vacation time	1	2
Find better educated customers	1	1

[a]The higher the number, the higher the rank.

organization. An example might be proper staffing level for CADD operators in a certain department or for all departments in the organization. Internal or external comparisons may be used. We might compare our department's CADD staffing with that of other departments in our company, or we might compare the staffing level for our company with that in other similar companies.

What kinds of activity might be benchmarked? One of the most successful examples is the "competitive benchmarking" program initiated by the Xerox Corporation in 1979. In the area of warehousing and distribution, the giant maker of copiers benchmarked against L. L. Bean, Hershey Foods, and Mary Kay Cosmetics; for billing and collection, the company compared itself to American Express. For automatic inventory control, it was American Hospital Supply, and for manufacturing floor layout, Xerox benchmarked against Ford Motor Company. Since the Cummins Electric Company had a near-perfect record of meeting daily production schedules, Xerox compared its activities in this area with those of Cummins, studied and implemented ways to improve, and succeeded in bettering its schedule performance by 75%. At Xerox, benchmarking is defined as "the continuous process of measuring products, services and practices against the company's toughest competitors and against companies recognized as industry leaders" (Bowles and Hammond, 1991). Such an approach helped Xerox win the Baldrige Award in 1989.

Eastman Chemical Company's Cellulose Esters Division was not certain of the quality of its cellulose acetate, compared to similar products of competitors, until a benchmarking study was conducted. This evaluation revealed the overall level of quality among Eastman and its five major competitors was roughly the same. On the basis of these results, improvements were implemented, and another benchmarking study was carried out. From Eastman's standpoint, the outcome was very satisfactory. Table 4.5 shows the results of that study, which compared six cellulose acetate characteristics from six producers, including Eastman's Cellulose Esters Division (*Eastman News*, 1992).

4.4 A COMMENT ON TOTAL QUALITY MANAGEMENT TOOLS

We have somewhat arbitrarily designated TQM tools as numeric or nonnumeric. For example, the nominal group technique contains elements of both, even though we have called it "nonnumeric." Also, force field analysis involves some elements of numeric analysis, and we could have included benchmarking as a numeric technique.

The tools presented above are but a sampling. For further details on all of them, and information on such other techniques as auditing, check sheets, Delphi technique, multivoting, and problem selection matrix, works by Brocka and Brocka (1992), Hunt (1992), and Walton (1990) are recommended.

We now examine the application of total quality management principles and techniques in three engineering organizations.

TABLE 4.5 Results of Eastman Chemical Benchmarking Study on Quality of Ester Products (from *Eastman News*, 1992)[a]

Characteristic	Producer (Count)					
	Eastman (U.S.A.)	Competitor 1 (U.S.A.)	Competitor 2 (Japan)	Competitor 3 (England)	Competitor 4 (Italy)	Competitor 5 (Japan)
Acetyl	3	1	2	5	6	3
Viscosity	2	3	1	5	4	6
Coulter Count	1	6	4	2	3	5
Filtration	2	5	6	4	3	1
Haze	1	1	4	3	5	6
Fiber	3	6	5	4	1	2
Total	12	22	22	23	22	23

[a]The lower the number, the higher the quality.

4.5 TOTAL QUALITY MANAGEMENT APPLICATIONS IN ENGINEERING

4.5.1 Gee & Jenson: Design Consultants

Gee & Jenson, Engineers-Architects-Planners, Inc., of West Palm Beach, Florida, began its quality improvement efforts in 1985. Discovering a lack of examples of successful implementation of TQM in the consulting engineering arena, the design consultants set out to create their own program, based on ASCE's draft manual of quality in the constructed project, some early work by the National Society of Professional Engineers, and the principles set forth by W. Edwards Deming. Drawing from the successful experience of such organizations as Eastman Kodak and Florida Power & Light (hardly design consultants, but very successful!), Gee & Jenson translated those experiences into a philosophy and technique that would apply in the design arena. All the firm's 360 employees were fully informed about the language, the tools, and the concepts.

Richard M. Miller, P.E., retired president of Gee & Jenson, wrote to this author about the early efforts: "The cultural transformation was a big problem. Getting a meaningful consensus in management was a big problem. Developing the vision was not well done. The mission was unclear" (Miller, 1992).

A fundamental premise of TQM that Gee & Jenson followed closely was the need to create a system in which errors can be detected as they occur, rather than later on. Such a philosophy is at the heart of Deming's approach to quality. Miller has written elsewhere (1990):

> No amount of checking or inspection for defects will reveal what is going on in a defective process, or indicate the capacity of the process for improvement. In almost every case checking or inspection is too late, very costly and, above all, cannot add one iota of quality to a product. Deming's analogy of "scraping burnt toast" is the best metaphor for our detection (inspection or checking) method of production. Everything that has to be corrected or done over is waste, costs the customer excess money and casts doubt on the acceptability of the materials that passed "quality control."

A leader at Gee & Jenson offers the following (quoted in Miller, 1990):

> We understand that we can never get rid of checking completely. However, we can do it at the right location and at the right time, before errors accumulate and compound. This approach allows us to stop wasting time doing the wrong things or doing things wrong, with errors building on errors. We are trying to avoid having to do the job over—or worse, pay a contractor to correct the errors that get through the checking/inspection/detection system.

> So many people talk about quality process—quality assurance/quality control— because they inspect or check their work "carefully and often" before anything goes out. Even assuming that all bad products or errors can be eliminated, nothing can actually improve through such a system. Inspection may cause products to be rejected, reworked or trashed, but it cannot improve quality.

The inspection system manages for failure by creating more new ways to find errors or defects. This has nothing to do with ongoing quality improvement. Are two checkers better than one? Are ten better than two? Might the second or ninth checker think that they are wasting their time or assume that the first checker caught all the errors? Might the first checker think, "If I miss an error, the next checker will catch it."?

Instead, through a Quality Improvement Program, we can manage for success. We can learn to correct our process for production of plans and specifications to be free of errors or defects. We can create a system in which all of a firm's employees work together to find out what causes mistakes and eliminate the causes. Then just one checker can do a final review.

Management at Gee & Jenson decided to engage an outside training firm to begin the learning process. "First and foremost, we must start training and never stop training for everyone, especially top management. Training in management and leadership, as well as training in variation, statistical process control, and tools for improvement, are all required" (Miller, 1990). As well as training employees to work in teams to identify and solve quality problems, the sessions developed trainers who were then able to propagate the program company-wide. During the training sessions themselves, many production problems were worked on and even solved.

Gee & Jenson reached a major milestone in its learning process when its leaders came to realize that everything the firm does is a process in a system. The important concept is that workers in the system can be held responsible for very little improvement until the system is changed, and system changes are solely the responsibility of management. The consultants' guideline was Deming's principle number one: "The workers work in the system. The manager should work on the system—to improve it, with their help." After that realization, the operation began to improve seriously.

A steering committee was formed, and one important task was to rewrite Deming's 14 principles, based on the company's philosophy, needs, and activities, to distribute the rewrite, and to assure that all employees understood it. Table 4.6 reproduces these points, as "Gee & Jenson's 14 Obligations of Management."

The company's quality improvement teams have addressed such challenges as contract deadlines, cost overruns, department coordination, budgets, computer utilization, plan preparation process, letters of interest and proposals, sick time, and recruitment (Miller, 1990). The drafting process at the Coral Springs office became the subject for a study that resulted in significant improvements. First, it was necessary to define three states of completion for a drawing: 30, 60, and 90%. A checklist was developed to help assure that all work expected to be completed at a certain stage is actually completed. Improved uniformity in design, decreased rework and delays, and improved customer relations have all resulted from this one quality improvement (*Quality Management Portfolio,* 1990).

TABLE 4.6   Gee & Jenson's 14 Obligations of Management

1. Create a constancy of purpose towards improvement of our client services and work product, competitiveness, and our ability to provide secure jobs.
2. Adopt the new philosophy of continual improvement throughout our firm and management system.
3. Build quality into our services from the beginning to reduce or eliminate the need for massive checking and revisions of the final deliverables to the client.
4. End the practice of buying supplies and awarding subcontracts based solely on lowest unit costs. Rather, minimize total cost by developing long-term supplier relationships based on loyalty, trust, and performance. Preference should be accorded to suppliers and subcontractors who are committed to the same principles of continual improvement.
5. Improve constantly and forever our systems of service and production to improve quality for our clients and our employees. Perpetual, system-wide improvements in quality will provide the desirable by-products of improved overall productivity and decreasing costs.
6. Institute training on the job throughout all levels of our firm so that each employee knows what to do and how to do it.
7. Institute supervision rather than mandates, to manage and help every employee to work more effectively at what they do.
8. Drive out fear as the control and motivator for employees and replace with knowledge and openness. Fear is an inhibitor and false motivator that prevents us from working effectively.
9. Break down the physical, monetary, and attitudinal barriers between individuals, departments, and regional offices so that we can work effectively as a team.
10. Eliminate slogans, exhortations, and targets for the staff calling for more "care," increased productivity, and "zero defects." The management and the system, not the workers, are responsible for most of these problems. Slogan campaigns only breed resentment.
11. Eliminate prescribed quotas and management by numbers of departments, regional offices, and staff. Focus rather on leadership to improve the overall efficiency of the system while continually improving service to our clients.
12. Remove the barriers that deprive the staff of their right to enjoy their work and be proud of it. Annual merit ratings, management by objectives, and individual profit centers are all barriers that must be removed for us to work as a team.
13. Vigorously educate and train employees throughout the company so that each person knows what to do and better ways to do it. Encourage employees to attend professional training seminars and evening classes to increase their knowledge and expertise at their job to work towards career advancement.
14. Involve everyone in the process of continual improvement from all levels of the firm. Making sure everyone becomes involved is very important. People who fully understand the transformation will be more enthusiastic and feel that their participation is important. The importance of the transformation must be impressed upon every employee.

Based upon Dr. W. Edwards Deming's 14 Points

SOURCE: Richard M. Miller, "Quality: Are We Up to It?" *Journal of the Florida Engineering Society,* 1990.

Some of the TQM tools described earlier in this chapter have been used with success at Gee & Jenson. Figure 4.12 is an example of a control chart applied to in-house printing, while Figure 4.13, a Pareto chart, depicts "hours lost" and identifies vacation and sick leave and personal leave as the "significant few" that accounted for more than 75% of lost time for the week being analyzed (Miller, 1990).

Highlights of the Gee & Jenson program include an emphasis on "doing it right the first time," the essential role of engineering management in improving the system, the significant role of teams, and the use of numerical tools. And it is working! Miller (1990) writes, "Through quality improvement projects, we can manage for success. We can learn to correct our process for production of plans and specifications to be free of errors or defects. We can create a system in which all of a firm's employees work together to find out what causes mistakes and eliminate the causes."

4.5.2 Heartland Professional Construction Services*

Heartland Professional Construction Services Inc., is a construction services firm based in Lansing, Michigan. The company does about $10 million in construction annually, more than half of which is currently for the Michigan Department of Natural Resources Parks Division. The work involves project management services on projects in the $50,000–1,800,000 range at state parks in Michigan's lower peninsula. Typical project size is between $200,000 and $300,000.

The first 22 of these projects were completed in 1990, and President William O. Frank reported considerable frustration at the end of 1990:

> Out of the 22 projects, we had 5 contractors default and 4 nearly default. We had 2 very serious claims which could have put us out of business. Projects were completing late and the cost of change orders was running about $7\frac{1}{2}\%$. The defaulted contractors and claims were absorbing a tremendous amount of our time and energy. The problems with contractor defaults and near defaults centered around their poor performance and poor quality of construction. (Frank, 1993)

The 1990 problems led Frank and his company to consider total quality management. An early question, in addition to general considerations about philosophy, vision, and approach, had to do with measurement. What should be measured? A result of Heartland's adoption of total quality management has been the identification and measurement of a number of parameters; study of these data, in turn, has lead to great success in implementing improvements.

When the company began its quality journey, it tended to blame others for its problems. For example, trade contractors had to be selected by low bid, so

*The contributions of William O. Frank, president of Heartland Professional Construction Services, are acknowledged, with thanks.

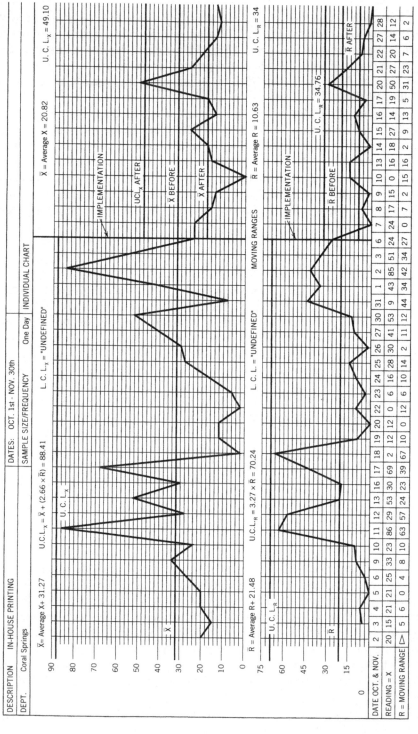

Figure 4.12. Sample control chart, Gee & Jenson. (From R. M. Miller, "Quality: Are We Up To It?" *Engineering Management Journal*, Vol. 2, No. 3, © 1990. Reprinted by permission of *Engineering Management Journal*.)

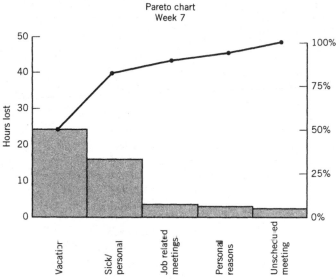

Figure 4.13.   Sample Pareto chart, Gee & Jenson. (From R. M. Miller, "Quality: Are We Up to It?" *Engineering Management Journal,* Vol. 2, No. 3, © 1990. Reprinted by permission of *Engineering Management Journal.*)

the managing contractor (Heartland) was stuck with them regardless of their qualifications or past performance. However, headway began to be gained when it was realized that challenges and situations the company *could* control were very much available.

To begin to control its destiny, the company compiled a list of its most serious problems with trade contractors. The list included the following:

- Claims
- Poor quality
- Failure to complete on schedule
- Incorrect paperwork
- Defaulted contractors

Although the 1990 season had not gone as well as Frank would have liked, Heartland was able to obtain a second phase of 23 state parks projects for 1991; with this new work in front of it, the company decided to apply total quality management to see whether improvements would be forthcoming. One of the first positive steps was the development of vision and mission statements (Table 4.7). Note the special emphasis both on the customer and on continuous improvement.

Several elements in Heartland's TQM program are especially noteworthy. The goals and guidelines, as outlined in "Heartland Cuts Change Orders 77%" (1992), included the following.

TABLE 4.7   Heartland PCS, Inc., Vision and Mission Statement

Heartland PCS, Inc. September 3, 1991

VISION
To revolutionize the contruction industry by achieving new
standards of quality and productivity that are significantly above
current industry standards.

MISSION
Our mission is to amaze our customers by providing
outstanding and ever improving construction services—through
leadership, continuous improvement, anticipating and exceeding
our customers' expectations, maintaining a perfect safety record,
and reducing cost through productivity and technological
improvements.
  Through careful hiring, ongoing training and education,
continuous process improvement, and doing it right the first time,
we will develop an efficient, highly skilled, harmonious work
force to provide exceptional customer service while providing our
employees with job satisfaction, growth opportunities, and
opportunities for financial security and self-actualization.

SOURCE: "Heartland Cuts Change Orders 77% with TQM Concepts," 1992.

- All employees are encouraged to continually look for ways to improve "the system" and to submit suggestions for improvements. Creativity is encouraged. It is suggested that employees spend 10% of their time on the TQM effort.
- Ongoing training and education for all employees.
- Process Action Teams and Corrective Action Teams pursue improvement efforts.
- A weekly, 2-hour TQM meeting is attended by all employees to discuss improvement efforts, improve communications, and facilitate "constancy of purpose" with regard to TQM.
- Prebid conferences are held with all bidders to communicate quality and performance expectations.
- Postbid meetings are held with successful trade contractors to ensure their understanding of quality expectations and contract requirements.
- Trade contractors are required to hold "quality standards meetings" for all their tradespersons who will be working on the job. Heartland project managers also attend these meetings.
- The Heartland Award of Excellence is given to trade contractors for outstanding performance in the areas of safety, on-time completion, minimal punch list, and proper paperwork.

Invoicing was a major problem. A portion of the flowchart developed to show how complicated the process of reviewing trade contractor invoices had become is shown in Figure 4.14. Study of the flowchart led to the establishment of an invoice review board comprising anyone involved with the approval or payment of invoices and meeting at the end of each month. Whereas invoice approval used to take 2 to 3 weeks, an hour or less is now sufficient. The end result is that trade contractors are paid within 20 days, whereas the same contractors have to wait 60 to 90 days on other State of Michigan projects.

Another example of the successful use of a TQM technique involved the preparation of operations and maintenance manuals for the owner. A fishbone diagram was prepared to try to determine why it took so much time to obtain and compile these manuals. The diagram showed, basically, that no one knew what was required in such an O & M manual. This discovery resulted in the development, review, modification and publication of a suggested list of contents, and the institution of a procedure for its use. The time required, from a date of substantial completion until the manuals are forwarded to the owner, was reduced from an average of 336 days to 143 days, a 57% improvement.

Figure 4.15 is an example of a fishbone diagram used by the Heartland organization to study the reasons for its late completion of projects.

Figures 4.16 and 4.17 are bar charts (they could also be called run charts, since they show trends over time) that indicate dramatic results of Heartland's total quality management program. Figure 4.16 shows that the average deviation from scheduled project duration decreased from 170 days late for the 22 projects in 1990 to 3 days early for the 27 projects in 1992. During the same period, the cost of change orders as a percentage of total project cost declined from 7.5% to 0.18%, as shown in Figure 4.17. The company also reports significant decreases in project durations, from 259 days to 108 days, and a reduction in shop drawing review time from an average of 54 days to 21 days between 1990 and 1992 (Frank, 1993).

4.5.3   Texaco: Process Industry*

In 1988 Texaco USA underwent a major restructuring. New business units were formed, layers of management were reduced, and authority levels were increased. The On-Shore Producing Division was formed as a separate business unit, headquartered in New Orleans, with operations primarily in Texas, Louisiana, Mississippi, and Alabama.

By 1989 the restructuring was essentially complete, driven by a corporate vision that sought to make Texaco "one of the most admired companies in the world and the competitive leader in the oil industry." The Eastern Region's

---

*I appreciate the information provided by D. L. Richter, P.E., vice president of Caltex Pacific Indonesia, and G. F. Sandison, assistant division manager, Texaco On-Shore Producing Division, which is used in this case study (Richter, 1993).

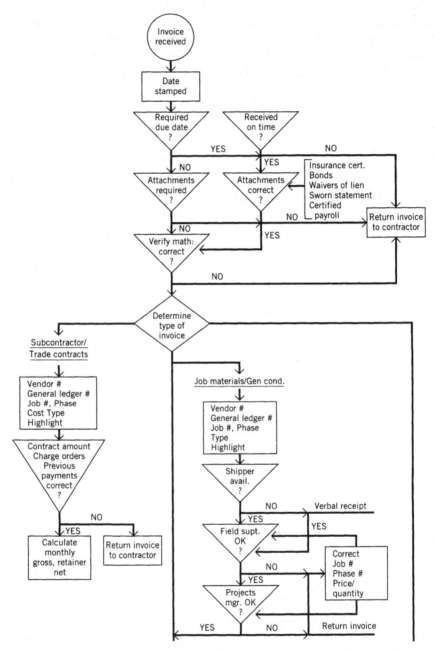

Figure 4.14. Portion of sample invoice processing flowchart, Heartland PCS, Inc. (From Frank, 1993. Reprinted by permission of William O. Frank, Heartland Professional Construction Services.)

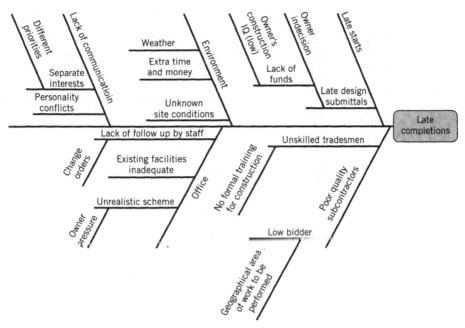

Figure 4.15. Fishbone (cause-and-effect) diagram for project completion time, Heartland PCS, Inc. (From Frank, 1993. Reprinted by permission of William O. Frank, Heartland Professional Construction Services.)

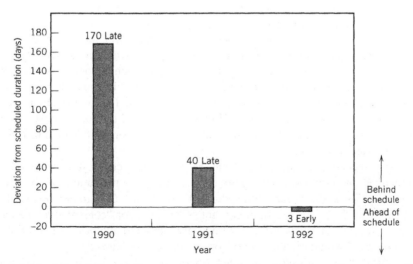

Figure 4.16. Schedule completion improvement, Heartland PCS, Inc. (From Frank, 1993. Reprinted by permission of William O. Frank, Heartland Professional Construction Services.)

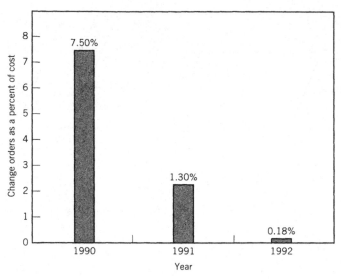

Figure 4.17. Change order improvement, Heartland PCS, Inc. (From Frank, 1993. Reprinted by permission of William O. Frank, Heartland Professional Construction Services.)

objectives were to focus on internal and external customers, improve communications, provide quality products, recognize and reward continuous improvement, and meet environmental and safety concerns. Since there were still such major obstacles as poor employee morale, a preexisting corporate culture and a lack of employee involvement, the region undertook, in late 1989, a quality process to address these problems.

There were several areas of concern. First, there was the essential need for committed management leadership. "The success of this process is dependent on an unwavering commitment at the top. In any organization, the values of the leader will drive the change process" (Richter, 1993). In some cases this meant "changing out" managers who were not supportive and not willing to accept the process. In those cases, a sincere effort was first made to "coach constructively" toward positive change, via performance review, informal discussion, training, and upward appraisal.

The second area of concern was the very real chaos that existed in the organization during the transition to and the implementation of the quality process. Since the organization chose to roll out TQM to all levels in the same year, there was a certain amount of disarray (or "constructive change"!), while the several levels adjusted to new freedom and authority. In addition, the workload of managers and supervisors increased significantly.

Third, it was difficult to break down the traditional functional barriers, since the reward system at Texaco was based on individual or department accomplishments. When the cross-functional team becomes a central element in the quality process, a different reward system is needed, and implementation may be difficult. Finally, there was concern because of difficulty of maintaining morale and a can-do spirit in the midst of the continuing downsizing of the U.S. oil business. The management team tried to maintain open and honest communications and to convince employees that continuous improvement would assure a stable work environment for as many employees as possible.

Implementation of the quality process proceeded in six steps, as follows:

- *Selection of Internal and External Consultants.* External consultants introduced the process and supported it through initial implementation. Internal consultants assisted managers in adopting a new culture and a new set of management skills and practices and improving the productivity and quality performance of the organization. Well-respected and highly motivated employees were selected and assigned these tasks for periods averaging one year.

- *Workforce Interviews.* A formal "vision and values" climate survey was conducted to ascertain attitudes, feelings, and concerns by means of additional informal interviews. The results, which included such negative factors as feeling overwhelmed by the workload, feeling frustrated by the new structure and feeling isolated from management, were used to provide a focus for management's efforts. A primary issue that surfaced from the interviews was communication, with a critical communication "gap" between management and hourly employees. For example, only 27% of hourly workers felt they received adequate information from supervisors about the performance of their division, versus 84% of management employees.

- *Dealing with Isolated Locations.* A special condition at Texaco's On-Shore Producing Division was the geographically dispersed nature of its operations. The approach that was adopted established teams by work groups and specific area wherever possible.

- *Quality Kickoffs.* An introductory presentation for all employees high-lighted the need for a quality process, explained the goals, and the an-nounced the timetable. The "show" was taken "on the road" for all mem-bers of the division, and was attended by division management to empha-size top-level commitment to the process.

- *Formation of Teams and Team Meeting Schedules.* Teams were part of an interlocking network, which meant that they could request assistance if needed. Generally, teams met weekly for an hour or so. Office teams were easy to set up, but field team meetings were often difficult because of variations in field work schedules. A typical team meeting included recognition of members, information sharing, and problem solving focused on team performance measures. Each member left every meeting with a copy of the team's action record.

- *Training.* The program accommodated the training needs of the internal consultants, the team leaders, and the team members. All employees received training in team management, customer focus, problem-solving techniques, meeting management, and performance. After every team meeting, team leaders were coached by the internal consultants.

In the midst of unnerving industry conditions, Texaco's On-Shore Producing Division identified several positive results from the quality process. First is the *establishment of objective, quantifiable team measures* for *all* teams. To be able to develop the performance measurement charts that are the basis of a team's decision-making process, each team had to identify its key products and services and to interview its internal and external customers. This process, especially for "service" parts of the organization, was often a struggle, but usually the results were very appropriate customer-driven measures of performance and a genuine understanding of what the real products are.

A second success was *improved communication* and "breaking down the barriers" both inside and outside the organization. Customers and suppliers (internal and external, alike) were forced to get together and talk. Once this dialogue had begun, people were less reluctant to maintain regular communication. Other benefits included "more visible contact with leaders, more management presentations, cross-functional teams used to solve problems, more frequent performance appraisal reviews with employees and an upward appraisal process employees evaluated supervisors. All these helped employees understand the business better, understand their performance better and understand the importance of communicating across the organization" (Richter, 1993).

Employee involvement was enhanced considerably through the team meetings. The often comfortable "old way," in which employees were reluctant to render opinions or take on risks, gave way to an approach wherein the teams worked out, and often implemented, solutions to their own problems. And as success became apparent, employees increased their involvement. Progress in this area seemed slow at times, but management adopted the realistic attitude that a corporate culture often takes a long while to change. Two measures of improved employee satisfaction are shown in Figure 4.18, which plots trends in employee absenteeism and resignations between 1990 and 1992.

A fourth area of improvement was in *process analysis*. Mapping of all work processes was carried out, with some maps covering four walls of a large room. Processes were reviewed for elimination or improvement of all steps. A continuous assortment of processes with minor or major improvements was the result. The flowcharts in Figure 4.19 illustrate progress in decreasing the time for a plant shutdown (from 653 hours to 64 hours). Such improvements were accomplished with no sacrifice of safety or environmental standards.

Finally, the implementation of Texaco's quality process is having a positive impact on the bottom line: the process has been good for business. Although the commitment to continuous improvement means that all success does not happen at once, already many money-making ideas have been proposed and implemented. A measure of that success is the benchmarking chart in Figure 4.20, which measures performance of eight indicators between 1987 and 1991 against a dozen other world-class petroleum companies.

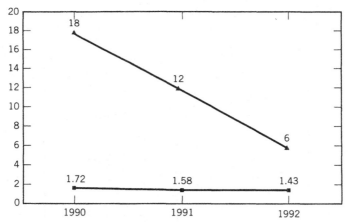

Figure 4.18.   Improvement in human resource statistics at Texaco, On-Shore Producing Division, 1990–1992. Squares, days absent per employee per year, triangles, resignations per year. (From Richter, 1993. Reproduced by permission of D. L. Richter, Texaco, Inc.)

4.6   THE QUALITY ROAD AHEAD

While the future of total quality management applications in engineering organizations is impossible to predict, we can say for certain that there is much opportunity for increased use of TQM. The results of a survey conducted by the American Consulting Engineers Council and reported in late 1992 indicated that the total quality idea was catching on slowly with engineering firms. Twenty-four percent of those surveyed have TQM programs, while another 65% are familiar with the concept. Larger organizations seem to be more active—the average size of firms with such programs was 298, while the average staff size of those without programs was 94. Firms with TQM programs rated themselves higher than others in leadership skills, quality output, and customer satisfaction ("Total Quality Management Continues to Get Lukewarm Reception," 1992).

Another 1992 study, this one conducted by Ernst & Young and the American Quality Foundation, examined quality practices in four major industries in Canada, Germany, Japan, and the United States. The study found that 58% of Japanese organizations regularly translate customer expectations into the design of new products or services, while the corresponding percentages in the other countries were as follows: Germany, 40%; United States, 22%; Canada, 14%. The use of technology in meeting customer expectations was of primary importance to about half the Japanese companies and of less than one-quarter of the U.S. firms. Furthermore, process simplification is regularly used by 50% of the Japanese companies surveyed and by 12% of the U.S. companies (Bowles, 1992; *Quality by Design,* 1992).

The message seems to be twofold: although many organizations around the world, including a large number of engineering organizations, have already profited by the approaches outlined in Chapters 3 and 4, many others have yet to take advantage

New Hope Plant Shutdown

Old Way

Plant shutdown	Cool down & depressurize	Decommission piping valves & vessels	Perform operational & safety maintenance Pressure Safety Valve tests & boiler inspections	Recommission piping valves & vessels	Pump & Mixing-pre-start-up inspection	Start-up
	30 hours	24 hours	563 hours	24 hours	12 hours	

Downtime hours: 653

New Way

Plant shutdown	Cool down & depressurize	Decommission piping valves & vessels	Perform operational & safety maintenance Pressure Safety Valve tests	Recommission piping valves & vessels	PSM-pre-start-up inspection	Start-up
	6 hours	8 hours	34 hours	8 hours	8 hours	

Downtime hours: 64

Figure 4.19. Improvement in shutdown times at Texaco's New Hope plant. (From Richter, 1993. Reproduced by permission of D. L. Richter, Texaco, Inc.)

Performance Indicators		*Tiers*
I. Production costs	1	Highest
II. Depreciation Expenses	2	Above average
III. Projected rates of depreciation, depletion,	3	Average
and amortization	4	Below average
IV. Cash flow per barrel of oil equivalent	5	Lowest
(BOE) produced		
V. Production replacement ratios		
VI. Acquisition, exploration and development costs		
VII. Discounted future net cash flow		
VIII. Upstream returns and reinvestment rates		

Company	I	II	III	IV	V	VI	VII	VIII	Average Score	Overall Position
Amoco	3	2	2	4	4	4	3	4	3.25	9th
Atlantic Richfield	4	3	2	2	4	2	4	2	2.88	6th
British Petroleum	4	4	5	1	5	5	4	5	4.13	13th
Chevron	3	3	2	3	5	3	2	4	3.13	7th
Conoco	3	3	5	3	3	5	2	5	3.63	12th
Exxon	3	4	3	1	5	3	4	2	3.13	7th
Kerr-McGee	5	5	5	2	3	4	1	3	3.50	11th
Marathon	3	4	4	2	3	4	3	3	3.25	9th
Mobil	2	2	3	4	3	2	3	2	2.63	5th
Phillips	2	1	1	5	2	1	5	1	2.25	3rd
Royal Dutch/Shell	3	2	4	4	1	1	NA	2	2.43	4th
TEXACO	1	1	3	3	3	2	1	3	2.13	2nd
Unocal	2	3	1	2	2	3	1	2	2.00	1st

Figure 4.20. Performance benchmarking chart for the petroleum industry, 1987–1991. (From Richter, 1993. Reproduced by permission of D. L. Richter, Texaco, Inc.)

of this new emphasis on quality processes, products, and services. At the beginning of the decade, Miller wrote, "In ten years your consulting firm will be transforming to Total Quality Management or it will be out of business" (Miller, 1990). A dire prediction? What will the year 2000 bring?

4.7 PROBLEMS AND DISCUSSION QUESTIONS

1. Consider a manufacturing operation in which the weight of the product is a critical factor. Suppose we collect eight samples of five sequential observations of weight in grams, as follows:

Observation Number

Sample Number	1	2	3	4	5
1	27.3	26.9	26.8	27.0	27.3
2	26.7	26.8	27.3	27.0	27.4
3	27.5	26.9	27.3	27.3	27.1
4	27.0	27.1	27.2	26.9	27.1
5	27.6	27.3	27.4	27.2	27.4
6	26.8	27.1	27.1	27.0	27.0
7	27.0	27.6	27.3	27.3	26.8
8	26.9	27.5	27.4	27.0	27.1

Develop control charts for $\bar{x}$ and $\bar{R}$ by finding and plotting averages and upper and lower control limits for each.

Now suppose the following observations are made subsequent to preparation of the charts. Which samples fall outside the control limits for $\bar{x}$ and $\bar{R}$?

Observation Number

Sample Number	1	2	3	4	5
27	26.7	27.4	27.4	27.7	27.6
28	27.1	27.5	27.3	27.2	27.0
29	26.7	27.1	27.0	26.8	26.6
30	27.1	27.5	26.9	27.4	27.2

2. Plot a histogram for the 40 observations of product weight from the first eight samples in question 1.

3. Suppose you conduct a study to ascertain why your fellow electrical engineering students do not participate in the IEEE student chapter. The primary reasons given are as follows:

Reason	Number Giving This as Primary Reason
Inconvenient meeting time	21
Programs not interesting	8
Programs not relevant	36
Too expensive	3
Poor leadership	7
I don't have enough time	8

Plot these results in a Pareto chart. Comment on these results.

4. Plot a run chart for the means of the first eight samples in Question 1. Do you observe any trend?

5. Suppose the cost of furnishing and placing asphalt pavement as a function of the number of square meters placed is as shown in the following table, for your company's last 14 such projects:

Project	Square Meters	Total Cost
7	114,000	$456,000
8	150,000	$540,000
9	126,000	$516,600
10	241,500	$845,250
11	76,000	$387,600
12	160,000	$761,600
13	29,500	$153.400
14	87,500	$433,000
15	205,000	$850,000
16	165,000	$745,000
17	237,500	$827,500
18	36,000	$149,300
19	46,500	$257,000
20	196,000	$758,500

Develop a scatter diagram of these data. If you have sufficient statistical background, find the best linear regression line and the regression coefficient for the data. Tell whether there is strong correlation. What factors are neglected in this analysis?

6. Draw a cause-and-effect diagram that shows the various factors that may lead to lack of success for engineering students.

7. Suppose you have just completed two years as the project manager for a very successful building construction project. You have the luxury of spending a day trying to classify the reasons for your projects sucess. Make any assumptions you wish, and define "success" in any way you wish. Draw a cause-and-effect diagram that indicates the reasons for the project's success.

8. A member of the board of directors of your worldwide engineering design firm has proposed that all the company's operations be consolidated and located in Stockholm. At a board retreat, you have been selected to lead a discussion of the advantages and disadvantages of this proposal. How might you use force field analysis to aid the board in its discussion? Sketch some of the positive and negative forces.

9. Distinguish between brainstorming and the nominal group technique and give an example of an appropriate use of each in engineering management.

10. How might one of the departments in your College of Engineering use benchmarking to improve its service to students and the professional community? As the basis for your answer, use the four-step process suggested by Hunt.

11. Conduct a modest survey among engineering organizations in your area to determine interest and level of participation in total quality management. Each student will contact one or two such organizations. Try to ascertain what "quality" means to these organizations and how they are managing their operations to achieve quality. Write a brief report on your findings. In class, compile the results to show the proportion of engineering organizations that are actively involved, modestly involved, and not at all involved in quality management.

4.8 REFERENCES

Bowles, J. 1992. "Is American Management Really Committed to Quality?" *Management Review*, Vol. 81, No. 4, April, 42–46.

Bowles, J., and J. Hammond. 1991. *Beyond Quality: New Standards of Total Performance That Can Change the Future of Corporate America*. New York: Berkley Books.

Brocka, B., and M. S. Brocka. 1992. *Quality Management: Implementing the Best Ideas of the Masters*. Homewood, IL: Business One Irwin.

Cecil, R., and R. Ferraro. 1992. "IEs Fill Facilitator Role in Benchmarking Operations to Improve Performance." *Industrial Engineering*, Vol. 24, No. 4, April, 30–33.

Deming, W. E. 1986. *Out of the Crisis*. Cambridge, MA: Massachusetts Institute of Technology Center for Advanced Engineering Study.

Eastman News. 1992. The Eastman Kodak Company, Vol. 47, No. 12, June 11.

Feigenbaum, A. V. 1991. *Total Quality Control*, 3rd ed., rev. New York: McGraw-Hill.

Frank, W. O. "TQM and Continuous Improvement After Two Years of Implementation." 1993 Quality Management Conference, Design and Construction Quality Institute and National Institute for Engineering Management and Systems, San Diego, CA, February 24–26.

"Heartland Cuts Change Orders 77% with TQM Concepts." 1992. *Quality in Construction*, Construction Quality Management, Winter, 3–4.

Hines, W. W., and D. G. Montgomery. 1990. *Probability and Statistics in Engineering and Management Science*, 3rd ed. New York: Wiley.

Hunt, V. D. 1992. *Quality in America: How to Implement a Competitive Quality Program.* Homewood, IL: Business Irwin One.

Ishikawa, K. 1992. *Guide to Quality Control.* White Plains, NY: Kraus International.

Juran, J. M., and F. M. Gyrna. 1988. *Juran's Quality Control Handbook,* 4th ed. New York: McGraw-Hill.

Miller, R. M. 1990. "Quality: Are We Up to It?" *Journal, Florida Engineering Society*, February, 14–20. Reprinted in *Engineering Management Journal*, Vol. 2, No. 3, September, 45–50.

Miller, R. M. 1992. Personal communication, July 7.

Quality by Design. 1992. William M. Hayden, Jr. Consultants, Inc., Vol. 5, No. 4, October, 5.

Quality Management Portfolio. 1990. The National Institute for Engineering Management and Systems, Vol. 1, Issue 1, Spring.

Richter, D. L. 1993. Personal communication, January 18.

Ross, R. J. 1991. "The Development of a Total Quality Management System for a Semiconductor Failure Analysis Operation." Unpublished National Technological University Capstone Project Report, University of Alaska Fairbanks, MB 780-G/ESM 684, Fall.

"Total Quality Management Continues to Get Lukewarm Reception." 1992. *Engineering Times,* National Society of Professional Engineers, Vol. 14, No. 11, November 13.

Tribus, M. 1988. "Managing to Survive in a Competitive World." *Quality First,* National Society of Professional Engineers, Publication No. 1459, August.

VTT, the Technical Research Center of Finland, 1992. "Ongoing and Lately Concluded Research Projects in the Section of Construction Management." Tampere: Building Production Laboratory.

Walton, M. 1990. *Deming Management at Work.* New York: Putnam.

Yazici, H. 1990. "Implementation of SPC Techniques in the PVC Pipe Industry. *Engineering Management Journal,* Vol. 2, No. 3, September, 59 64

5

The Human Element in Engineering Management

In the first four chapters, we have dealt with organizational and quality aspects of the management of engineering activities. We now turn to some of the "people issues" that are key to making organizations "work right" and are thus basic to the overall success of the technical enterprise. First we deal with *motivating employees* to work toward the organization's objectives and to fulfill their personal goals as well. Then, we consider the topic of *leadership*—what it is, how it differs from management, and some characteristics of successful leaders.

Since successful management often requires the manager to delegate, we present a brief section on *delegation*. A discussion of the *development of engineering personnel,* to fulfill both their personal goals and those of the organization, is followed by a presentation on *recognition systems*. Finally, we discuss briefly *personnel administration functions,* such as hiring, training, evaluating, and disciplining, that are typical responsibilities of the engineering manager.

Each topic could be the subject of an entire chapter, and whole books have been written on each! The purpose here is to summarize key concepts and to stress their importance in the management of technical activities.

5.1 MOTIVATION

Why do some individuals perform better than others? What causes one design engineer to finish a project in one week, while her equally capable colleague takes three weeks to finish a project of similar magnitude? Why does one survey crew chief lead his crew to lay out ten stations in a day while another accomplishes half as much? Although every situation is unique and many factors are involved, the answers, in large measure, relate to employee motivation. As engineering managers, we can help our organizations achieve success by understanding how to motivate our employees to be high performers.

Does motivation make any difference in employee performance? From his extensive work with many technical organizations, including consulting engineering, design, and construction, Mel Hensey (1992) is convinced that it does indeed make a large difference. While it is difficult to measure productivity, particularly for knowledge workers, managers can assess relative output. As they have done it, particularly in groups, they have noted the following: *High morale and high performance groups or individuals with equal skills resources will outperform those that are just marginally acceptable by a factor of almost 3 to 1.* Motivation makes a *big* difference in group output, from marginal to high performing.

Connolly (1983) suggests that high-performing employees, when asked why they were working so hard, tend to give explanations for their motivation that are related to the past (e.g., "Father always insisted that I not waste time."), to the present ("This programming problem is both fun and challenging."), or to the future ("I know that if I work hard I'll have a good chance for that promotion next year."). Such motivational factors will be a central consideration in the discussion that follows.

Motivating people involves stimulating them to action. Amos and Sarchet (1981) define motivation as "a force or drive causing some action, behavior or result." Motivation results from needs that the "action, behavior or result" will help to satisfy. Amos and Sarchet (1981) presented a list of examples of needs common to many engineers.

- *Desire to Achieve:* likes to see a goal that can be achieved.
- *Self-Expression and Creativity:*   is constantly creating, designing, and constructing.
- *Challenge:* is crossing new frontiers with opportunities for challenge.
- *Diversity of Problems:* is involved with work that has little routine; new problems arise.
- *Pride in Accomplishment:* takes satisfaction in finished product—a system, a building, a drawing.
- *Independence:* makes decisions based on personal knowledge and experience.
- *Practice of Technical Knowledge and Skills:* finds pleasure in using rare talents and skills.
- *Recognition:* desires to have accomplishments recognized.
- *Professional Status:* desires recognition of stature in the profession.

The challenge for the engineering manager is to provide a means to allow the engineer to satisfy those needs. Thus, high motivators for engineers in the workplace include opportunities to attack problems directly, recognition of accomplishments, association with competent coworkers, the chance to use technical knowledge and skills, independence, and opportunities to work on challenging and different projects.

Another writer has suggested much the same thing, as follows (Raudsepp, 1977):

First, engineers want to feel their work is interesting: they want a feeling of personal professional growth.... Secondly, an engineer wants a salary at least equal that of other engineers on his level.... Third, engineers want status that is more than symbolic; they do not value grandiose titles and regard them as placating devices to deflect their attention from more concrete symbols of status and higher salaries. Fourth, they want effective communication with other components of the organization; they want to feel that top management knows what they are doing; they want to be kept informed about company policies, plans and operations outside their work areas. Fifth, engineers want effective managers who can become involved in the engineers' projects.... Sixth, engineers want modern, up-to-date facilities and ample working space.

The particular job motivational factors that tend to have a special influence on a particular work group may vary from one organization to another. Cleland and Kocaoglu (1981) developed a list of 29 such factors (Figure 5.1). The items have been used by various engineering organizations to determine what motivates people to do their best work. Participants complete a questionnaire based on the list, results are summarized and announced, and a lively discussion follows, as the participants learn what motivates them and their peers, as well. Cleland and Kocaoglu report that the following factors (given alphabetically) have emerged most frequently:

Chance for promotion
Chance to turn out quality work
Feeling my job is important
Getting along with others on the job
Good pay
Large amount of freedom on the job
Opportunity for self-development and improvement
Opportunity to do interesting work
Personal satisfaction
Recognition of peers
Respect for me as a person

This writer has used the same questionnaire over several years in a class of senior engineering students, asking students to identify the factors they believe to be especially important to (1) themselves, (2) their peers, and (3) those whom they currently supervise or will supervise in the near future. The results, in order of importance, are summarized as follows:

Factors Most Important to Themselves

Opportunity to do interesting work
Personal satisfaction
Good pay
Respect for me as a person
Chance for promotion

Figure 5.1. Job motivational factors. (From Cleland, D. I., and D. F. Kocaoglu, *Engineering Management*, New York. Copyright © 1981. McGraw-Hill, Inc.)

Please indicate the five items from the list which you believe are the most important in motivating you to do your best work.

_____	1 Steady employment
_____	2 Respect for me as a person
_____	3 Adequate rest periods or coffee breaks
_____	4 Good pay
_____	5 Good physical working conditions
_____	6 Chance to turn out quality work
_____	7 Getting along well with others on the job
_____	8 Having a local house organ, employee paper, bulletin
_____	9 Chance for promotion
_____	10 Opportunity to do interesting work
_____	11 Pensions and other security benefits
_____	12 Having employee services such as office, recreational, and social activities
_____	13 Not having to work too hard
_____	14 Knowing what is going on in the organization
_____	15 Feeling my job is important
_____	16 Having an employee council
_____	17 Having a written description of the duties in my job
_____	18 Being told by my boss when I do a good job
_____	19 Getting a performance rating, so I know how I stand
_____	20 Attending staff meetings
_____	21 Agreement with agency's objectives
_____	22 Large amount of freedom on the job
_____	23 Opportunity for self-development and improvement
_____	24 Chance to work not under direct or close supervision
_____	25 Having an efficient supervisor
_____	26 Fair vacation arrangements
_____	27 Unique contributions
_____	28 Recognition of peers
_____	29 Personal satisfaction

Factors Most Important to Their Peers

Good pay

Chance for promotion

Steady employment

Opportunity to do interesting work

Personal satisfaction

Factors most important to those supervised

Good pay

Steady employment

Respect for me as a person

A word of caution is in order. Unless the manager intends to utilize the results in a positive way to improve motivational aspects of the workplace, such a questionnaire should not be used. If employees expect their input to be taken seriously and, after the responses have been tabulated, these expectations are not fulfilled, the situation will probably be worse than before the exercise. If management is serious about learning what motivates employees and just as serious about acting on the results, however, the exercise can be very satisfactory.

We now turn to some theories that relate to employee motivation and satisfaction of needs. Three are discussed; several others are simply mentioned and referenced.

Abraham Maslow (1943, 1954) suggested a "hierarchy of human needs" to explain how persons are motivated to action. The psychologist grouped human needs into five broad categories: one group dominates an individual at any point in time, and a need at a higher level in the hierarchy achieves dominance only when the previously active one has been satisfied. These categories, from lowest to highest, are as follows:

1. *Physiological Needs:* food, water, sex, shelter, clothing. The survival needs required to maintain life.
2. *Safety Needs:* safe neighborhood, steady job, life insurance, "moat" or "stockade." The security needs that help a person hold onto what has been attained in level 1.
3. *Social Needs:* friendship, affection, belonging, love, acceptance, social activity. The need to relate and belong to others as individuals and in groups.
4. *Esteem Needs:* status, recognition, prestige, respect; holding and using power. The need to be looked up to.
5. *Self-Actualization Needs:* self-fulfillment, self-realization, sense of accomplishment. The need to achieve one's own unique self in a real, "actual" way, irrespective of what others may think.

As we consider how this need hierarchy theory may apply to the engineering profession, we think of people at different stages of their careers. The engineering student or recent graduate may be mostly concerned about physiological and safety needs, with some concern about social needs. The vice president for engineering is probably most concerned about social, esteem, and self-actualization needs, because the lesser needs have been met.

This leads us to make two observations about Maslow's theory. First, despite its imperfections, it reminds us that a satisfied need is no longer a motivator of action. If your social needs are satisfied, you probably will not be interested in participating on the company softball team! Second, it is not true that only one need is operant at a time, even though one need may be dominant. The engineer seeking a master's degree may be motivated by all five needs to some extent!

The work of Frederick Herzberg is interesting in part because it was based on interviews with a large group of professionals, including engineers. His approach was to ask employees to describe work situations that aroused their emotions: What led them to feel especially satisfied or especially dissatisfied about their jobs. After tabulating the responses (Herzberg, 1968), he proposed a "two-factor" theory of work motivation.

Satisfiers or "Motivators"

Achievement
Recognition
The work itself
Responsibility
Advancement
Growth

Dissatisfiers or "Hygiene Factors"

Company policy and administration
Supervision
Relationships with supervisors, peers, and subordinates
Working conditions
Salary
Personal life
Status
Security

The dissatisfiers, or "hygiene factors," may decrease motivation, and therefore productivity, if they are not met, but they will not improve motivation if they are met. That is, increased satisfaction with these factors, according to Herzberg, has little positive effect on performance, but *decreased* satisfaction has a negative effect. On the other hand, the satisfiers, or "motivators," can have a positive effect on performance if they are present, because, unlike the hygiene factors, they can yield a real sense of satisfaction.

Herzberg's satisfiers tend to be associated with the *content* of the job; they seem to be allied with the self-actualization needs in the Maslow hierarchy. The dissatisfiers tend to be associated with the job's *context* and seem to relate to Maslow's security and social needs.

The expectations managers have for their employees' performance can have a major impact on the motivation of those employees. In 1960 Douglas McGregor put forth the notion of "Theory X and Theory Y" in his book *The Human Side of Enterprise*. While these ideas were really not new theories, they did help to categorize the attitudes that people in general, and managers in particular, hold about human nature. (McGregor, 1960). These two "theories" are summarized below (as quoted in Cleland and Kocaoglu, 1981):

Theory X

1. The average human being has an inherent dislike of work and will avoid it if possible.

2. Because of this human characteristic of dislike of work, most people must be coerced, controlled, directed and threatened with punishment to get them to put forth adequate effort toward the achievement of organizational objectives.

3. The average human being prefers to be directed, wishes to avoid responsibility, has relatively little ambition and wants security above all.

Theory Y

1. The expenditure of physical and mental effort in work is as natural as play or rest. The average human being does not inherently dislike work. Depending upon controllable conditions, work may be a source of satisfaction (and will be voluntarily performed) or a source of punishment (and will be avoided if possible).

2. External control and the threat of punishment are not the only means for bringing about effort toward organizational objectives. People will exercise self-direction and self-control in the service of objectives to which they are committed.

3. Commitment to objectives is a function of the rewards associated with their achievement. The most significant of such rewards, the satisfaction of ego and self-actualization needs, can be direct products of effort directed toward organizational objectives.

4. The average human being learns, under proper conditions, not only to accept but to seek responsibility. Avoidance of responsibility, lack of ambition, and emphasis on security are generally consequences of experiences, not inherent human characteristics.

5. The capacity to exercise a relatively high degree of imagination, ingenuity, and creativity in the solution of organizational problems is widely, not narrowly, distributed in the population.

Although few people probably hold strictly to one theory or the other, each of us certainly tends toward one of them. McGregor helps us understand that employees' motivations depend on the attitudes they perceive in their leaders. Managers who display the "people are no good" approach can expect employees to respond accordingly. Managers who take the opposite approach are likely to find employees will respond with enthusiasm and creativity. Amos and Sarchet (1981) suggest that employees who work for Theory X managers will tend to become "hygiene seekers" (in the sense used by Herzberg), while those who work for Theory Y managers will tend to become "motivation seekers." Some other studies and theories of employee motivation of interest to the engineering manger include the following:

• The so-called Hawthorne studies of effects of variation of lighting levels, ventilation, and other physical factors on employee performance at Chicago's Western Electric Company between 1927 and 1931 (Roethlisberger and Dickson, 1939).

- Gellerman's studies of motivation in management, including the two major management styles of (1) coercion and threat of dismissal and (2) compensation and reward (Gellerman, 1963).
- Skinner's operant conditioning theory, which suggests that people respond to their environment and will perform if they receive a reward, rather than because of an internal drive or attitude (Skinner, 1953).
- McClelland's work of developing a methodology for determining whether a person was primarily motivated by a "need for achievement," a "need for affiliation," or a "need for power" (McClelland, 1961).
- Vroom's preference–expectation theory, which suggests that workers have preferences for certain rewards related to their performance and that these preferences are tempered by their expectations that the rewards will be forthcoming (Vroom, 1964).

Let us now turn to the specific motivator of compensation, in the form of salary, fringe benefits, bonuses, and/or profit sharing, and ask whether compensation is a proper and effective motivator. Hensey (1992) reports that compensation research studies in industrial and technical organizations show that such compensation *does* provide strong motivation for people to perform, *if and only if* the employees see that their pay is clearly and closely dependent on job performance. "Elements that are needed to improve the administration of salaries or bonuses," says Hensey "include clear-cut performance goals, workable performance measures, honest performance appraisals and then paying for performance and contribution only." Hensey suggests that compensation systems often do not follow this model, with managers "fiddling with their pay-for-performance system to keep everyone happy." He also reminds us that Herzberg classified compensation as a dissatisfier or hygiene factor because of poor administrative practices, not because pay is unimportant.

Amos and Sarchet (1981), in concluding their presentation on motivational techniques for engineers, describe eight "creative actions that motivate people." Although no single set of actions will work for everyone, the concepts, as summarized below, seem especially helpful for those involved in motivating technical professionals.

1. People must know where and how they fit into the accomplishment of goals or objectives.
2. The engineer manager must always promote and assist subordinates' advancement in the organization.
3. It is essential for the engineer manager to get out of the office and on the production line or project, to learn what is going on.
4. The engineering manager should consider keeping in contact with employees' families and letting them know when a project is especially well done or when efforts are required during evenings or weekends.

5. Delegation of activities to be performed by others makes work more meaningful.

6. Participation in decisions and actions affecting their work or workplace plays a large part in gaining commitments from employees.

7. Communication is the key to developing better relationships with people.

8. Job enrichment is a technique used to build into the job a higher sense of challenge and achievement.

Some of the ideas just expressed are covered in greater detail in Sections 5.4 and 5.6.

While it often is assumed that professional people whose task is to be creative will not be motivated to do their most productive work unless supervision and control are minimal, some research by William Souder (1974) indicates no such correlation among workers in research and development laboratories. The significance of this finding is discussed in Section 7.3.2 in our chapter on project management.

In some settings, perhaps the issue of motivation is overdone. A representative of NASA, when asked "How do you motivate astronauts?" is supposed to have replied, "We don't motivate them, but we are *very* careful about whom we select!" (quoted in Meredith and Mantel, 1989). Therefore, the matter of "motivation" is sometimes entirely (or nearly so) a matter of the nature of the job and the kind of people who are invited to be members of the organization. Thus as a rationalization for the motivation of astronauts to seek a "higher sense of challenge and achievement," the phrase "job enrichment" may be an understatement.

To conclude this section on motivation, we present in Figure 5.2 the general model of work motivation proposed by Connolly (1983). Although it is "merely a sketch of a highly complex process," the model seems helpful in tying together some of the ideas contained in the theories discussed and referenced herein.

Connolly's logic is summarized as follows: the focus of the model is on the individual's conscious or unconscious choice about how hard she or he will work in the next time period—the level-of-effort choice. This choice is a function of the person's beliefs, perceptions, and desires and aspects of the situation. Over time, the level of effort leads to a level of work performance that is moderated both by the individual's ability and by available resources. In turn, the level of work performance leads to extrinsic and intrinsic work-related outcomes. Then, these work-related outcomes lead, directly or indirectly, to personal outcomes, which are the need satisfactions suggested by Maslow and others. Finally, work performance and work-related and personal outcomes feed back on the situation and the individual to influence the level-of-effort choice in the next time period.

5.2   LEADERSHIP

The current status of the study of leadership is well represented by two paragraphs taken from the opening of Connolly's chapter on the subject:

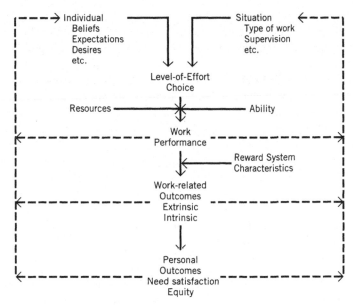

Figure 5.2. Summary model of individual work motivation. (From Connolly, *Scientists, Engineers and Organizations,* Monterey, CA: Brooks/Cole Engineering Division 1983, p. 83.)

Of all the subjects covered by organizational behavior, leadership is perhaps the most frustrating. We see examples of what we believe to be good or bad leadership around us every day. Sports teams play well or badly, and it seems obvious that it has something to do with the coach or manager. Project groups come up with good or bad designs, and it seems obvious that it has something to do with the project leader. Orchestras play well or badly, apparently as a result of the conductor. Whole corporations raise their profits, or head towards bankruptcy, as chief executives come and go. Leadership, in short, seems to affect every area of life, and in a rather clear-cut way.

Yet, when we sit down to digest what is known about this interesting phenomenon, we come up with a complex and unsatisfactory picture. There is a mountain of research—literally thousands of studies—looking at leadership in all sorts of contexts, from leaders of nations to leaders of small discussion groups, from sports teams to juries. Despite this enormous effort, we still lack satisfactory answers to most of the key questions: What exactly is effective leadership? Can we select, or train, people who will be effective as leaders? How much of a difference does leadership actually make? Can we learn to become more effective as leaders? (Connolly, 1983)

What is leadership, and how is leadership distinguished from management? It is impossible to cover this elusive topic thoroughly, and our overview in this section samples definitions that are, by turns, concrete, abstract, number-based, and somewhat mysterious. We begin with David Campbell (1980), who defines leadership as "any *action* that focuses *resources* to create *new opportunities*."

This definition suggests that leadership is active, not passive; that the leader uses resources of many kinds, ranging from time to money to facilities, to make things happen; and that new opportunities can range from new jobs to happiness to higher profits to long life.

Katz and Kahn (1978) define leadership as "the influential increment over and above mechanical compliance with the routine directives of the organization." From this interesting definition we learn that the effective leader will be the one who causes subordinates to do that something extra (the "increment") beyond the routine requirements of their positions.

The future engineering manager would do well to develop a personal sense of the distinction between leadership and management. Stephen Covey's analysis may be helpful:

> Management is a bottom line focus: How can I best accomplish certain things? Leadership deals with the top line: What are the things I want to accomplish? In the words of both Peter Drucker and Warren Bennis, "Management is doing things right; leadership is doing the right things." Management is efficiency in climbing the ladder of success; leadership determines whether the ladder is leaning against the right wall. (Covey 1989)

In short, while management is concerned with the myriad of details—the techniques involved in economic analysis, personnel relations, legal principles, project scheduling, and other matters required for the enterprise to survive and prosper—leadership involves those policy setting, planning and evaluating, forecasting, inspiring, motivational responsibilities that set the enterprise on its short-run and long-run courses.

Covey suggests that we envision a group of machete-wielding workers slashing their way through the jungle. It is they who are the producers, because they are actually cutting out the undergrowth. Behind them are those who develop the schedules, prepare the meals, carry out muscle-building programs, account for costs, arrange housing, and implement safety programs; they are the managers. But then there is another member of the group, the one who climbs the tallest tree, surveys the surrounding territory, and shouts, "Wrong jungle!" That person is the leader! We need both managers and leaders, but there are distinctions in roles and functions.

Warren Bennis, who has written much about leadership, says in *The Unconscious Conspiracy: Why Leaders Can't Lead* that an effective leader is someone "with a kind of entrepreneurial vision, a sense of perspective, and, most of all, the time to spend thinking about the forces that effect the destiny of that person's shop or that institution." Later he talks about knowing how to ask the right questions, being a conceptualist, and being able to look ahead so that right decisions can be made about the organization's future (Bennis, 1976).

The late Admiral Grace M. Hopper used to say, "You lead people and you manage things. We need more leadership and less management" (ESM 450 1990).

So much for an attempt to define leadership and to distinguish it from management. Let us turn to a brief consideration of leadership styles and a review of characteristics that effective leaders seem to exhibit. The Managerial Grid, reproduced as Figure 5.3, has been widely used to identify various combinations of leadership styles (Blake and McCanse, 1991). We have already seen this grid in Chapter 2 in our discussion of organizational effectiveness; now we apply it to individual leadership effectiveness. The originators of this approach suggest that styles can be classified by the degree to which one tends to emphasize concern for people and concern for production. There are five basic styles. Under style 1,1, there is little concern for either people or production; those managers become primarily message carriers from superiors to subordinates. At the other extreme, style 9,9 represents the highest possible dedication to both people and production, so that the "team" performs effectively to achieve the organization's purposes.

The 1,9 style has much concern for the people side of the operation, promoting friendliness and "getting along," with little emphasis on accomplishing a defined goal. The opposite is true of the 9,1 style, where the emphasis is on an efficient operation that "produces," with little or no concern for the people who make that happen. In the middle is the 5,5 style: the leader has a balanced perspective, but such balance may be tantamount to mediocre, adequate, or "just satisfactory" performance, far short of the organization's capabilities.

Clearly the grid approach is a descriptive one, attempting to describe the way managers tend to conduct themselves. The implication is that style 9,9 is preferred,

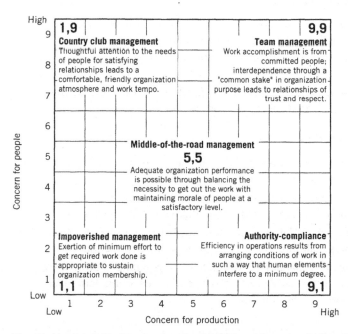

Figure 5.3. The Managerial Grid. [The Leadership Grid® Figure from *Leadership Dilemmas—Grid Solutions,* by Robert R. Blake and Anne Adams McCanse (formerly the Management Grid Figure by Robert R. Blake and Jane S. Mouton) Houston, TX: Gulf Publishing Company, p. 29. Copyright © 1991, by Scientific Methods, Inc. Reproduced by permission of the owners.]

but there is some evidence that other styles may be preferable in certain cases. For example, the most effective choice for the leader of an engineering research laboratory might be a 1,1 style, leaving highly motivated knowledge workers alone to pursue their intense intellectual interests without much attention to their personal lives or meddling to induce "production" of this or that project.

What roles do leaders play? Professor Henry Mintzberg conducted an in-depth study of the chief executive officers of five major corporations. From this study he developed ten major roles, which he further grouped into three broad categories, as follows (Mintzberg, 1973, as quoted in Campbell, 1980):

Interpersonal Roles

- *Leader:* motivating, directing, assigning tasks, assessing performance, coaching, following up, "charisma."
- *Liaison:* dealing with outsiders, communicating with peers, making contacts, scouting for new resources, playing politics ("There are three ways to get to the top of an oak tree: (1) climb it; (2) sit on an acorn; (3) make friends with a very big bird!").
- *Figurehead:* appearing at formal and informal functions, welcoming dignitaries, announcing promotions and new contracts; the ceremonial/social role.

Informational Roles

- *Monitor:* reading, listening, studying reports and journals, walking through your facility, attending workshops, staying up to date.
- *Disseminator:* speaking, writing, informing, sharing, distributing.
- *Spokesperson:* selling, informing, taking a stand, making contacts, representing your people and your organization to the world (different from the ceremonial, or figurehead, role above).

Decision Roles

- *Entrepreneur:* seeking opportunities, taking chances, trying new things, assuming risks.
- *Resource Allocator:* budgeting; deciding who gets what; assigning priorities; giving out dollars, personnel, equipment, time, and space.
- *Disturbance Handler:* reacting to crises, putting out fires, soothing ruffled feathers; calming and solving through decisive action.
- *Negotiator:* bargaining with others, negotiating contracts, swapping resources, buying time, straightening out errors; handling terminations, cancellations, extra work, and failures.

Many writers have suggested their choices for the most desirable qualities, traits, or characteristics of successful leaders. According to David Campbell (1980), leaders are bright, mentally agile, alert people, who seek out responsibility, are good at whatever they are "leadering in," are energetic, are good communicators, and have a sense of humor. That is an ambitious and worthy list for all of us! Rodman Drake (1987) lists the following traits as important for corporate leadership: focuses on goals, maintains simple values, stays in touch, properly manages change, builds a solid management team, has the ability to delegate, and faces up to failure.

Expressing many of the foregoing factors in a 10-point list, Dr. Whitt N. Schultz (1988) gives his set of essential qualities.

Ten Essential Qualities of Successful Leaders

1. They are accurate observers. They observe and absorb. They look at everything as if it's the first and last time they'll ever see it.
2. They are excellent listeners. Good listening skills are essential to the learning process.
3. They take copious notes. They possess alert minds that allow them to remember details they may use at a later date.
4. They welcome new ideas. They're open and responsive to the ideas and suggestions of others.
5. They regard time as a precious commodity. They always spend their time wisely and skillfully.
6. They set regular goals. And, once they've determined their goals, they strive hard to achieve them.
7. They always try hard to understand others. They reserve judgment until after they've understood the other person's point of view.
8. They always anticipate achieving their goals. Then, after achieving them, they develop new ones.
9. They know how to ask clear, courteous, and incisive questions. People who are skilled at asking questions excel at learning things from others.
10. They know how to organize their approach to challenges. They also possess the ability to focus their minds on important and relevant tasks.

We conclude this section with some thoughts of L. P. Williams, quoted in a consideration of "leadership and the professional" by Morgan W. McCall, Jr. (1983):

> Leadership, then, is the ability to tie men and women to one's person and cause. But how is that done? Leadership is obviously a combination of many things... When pressures increase, we turn to those we trust to act in the general interest... Some people simply seem to have an aura around them that inspires trust and confidence. We have faith in them and in their ability to deal with matters in a way that will be satisfactory to us. They inspire confidence and we grant them our loyalty. These are the qualities that unite followers and that leaders, somehow, in some mysterious way, are able to call forth. Quite simply, we know leadership when we experience it.

5.3 DELEGATION

Closely related to leadership and communication is the topic of delegation. It involves, literally, giving people things to do. Mackenzie (1972) makes the point that if this is so and if managing is getting things done through people, then *the manager who does not delegate is not managing!* There are at least four benefits from proper delegation: it (1) expands results from what a manager can do to what he or she can control, (2) releases the manager's time for more important work, (3) develops subordinates' skill, competence, and initiative, and (4) provides for decision making at the lowest possible level.

Despite these advantages, delegation seems to be "the least well-practiced of any engineering management concept" (Amos and Sarchet, 1981). For reasons to be identified later in this section, even though proper delegation can greatly assist in the allocation of one's time, managers often give it lip service but fail to put it into practice.

Three interrelated phases are involved in the delegation process. First, the *duties are assigned* and the *desired results are identified.* Second, *authority to carry out the assignment* is provided. And third, *responsibility to accomplish the duties and assignments* is also transferred.

Let us clarify *authority, responsibility,* and *accountability.* Silverman (1988) gives the following simple definitions:

Authority

The power to make final decisions that others are required to follow.

Responsibility

The obligation, which results from a person's formal role in an organization, to perform assigned tasks effectively.

Accountability

The fact of being answerable for the satisfactory completion of tasks.

An important aspect of effective delegation is the balance between authority and responsibility. When the assignment is made, the authority must be consistent with the responsibility. If the project manager delegates to a junior engineer the responsibility (obligation) for conducting an analysis and deciding a highway routing, the authority (power) to decide how the study will be conducted and to make the decision must be delegated as well. While authority and responsibility can be delegated, accountability, as defined above, cannot (Silverman, 1988). In the case just mentioned, the ultimate "answerability" still rests with the project manager, who is accountable for the project despite having assigned responsibility to the junior engineer.

The following guidelines for effective delegation are based on the work of Amos and Sarchet (1981), Mackenzie (1972), and Roadstrum (1978):

1. Assignments must be defined clearly in terms of the expected outcomes.
2. Adequate guidance, instruction, and training must be provided to the delegatee.
3. Authority must be commensurate with the assigned responsibility and sufficient to make possible the completion of assignments.
4. Abilities and qualifications of subordinates must be considered when delegating, as well as individuals' potential personal and professional growth.
5. Activities must be organized effectively, with proper means for coordinating the results arising from the several subordinates.
6. Supervisor–subordinate relationships must be clear and direct, to ensure that all affected parties understand who is responsible for the various elements.
7. Counseling and advice must be available to subordinates after work has been delegated.
8. Without meddling in subordinates' activities, there must nonetheless be a control system that alerts the supervisor/delegator when deviations in the plan arise.
9. A means for feedback, recognition, and reward must be in place and operating.
10. The delegator must display a large measure of trust in the subordinates; such trust is inspired by competent subordinates and is evidenced by a willingness to "back off" on the part of the delegator.

Note that each of the guidelines contains the word "must"! Perhaps that word is not too strong if we truly desire effective delegation.

Given these guidelines, why does delegation fail so often? Mackenzie (1972) developed a comprehensive list of 27 "barriers to delegation," of which 14 were related most closely to the delegator, 7 to the delegatee, and 6 to the situation. Four barriers seem to be especially prevalent in engineering management. The first is a preference for operating. We engineers seem to prefer to keep our fingers in the pot long after the details should have been delegated. The chemical manager has a hard time staying out of his lab; the nuclear project manager enjoys her technical calculations.

The other three barriers that seem especially prevalent among engineering managers are the demand to "know all the details," the "I can do it better myself" fallacy, and perfectionism. We are, by nature, education, and experience, interested in details and in making the outcome "perfect." As managers, we must learn to modulate those preferences if our delegation is to be effective.

The questionnaire that follows, also attributed to Mackenzie (1972), may help the reader to assess personal skill at delegating. Between two and four "yeses" indicate a need to improve, five or more "yeses" suggest a serious problem.

1. Do you take work home regularly?
2. Do you work longer hours than your subordinates?

3. Do you spend time doing for others what they could be doing for themselves?

4. When you return from an absence from the office, do you find the "in" basket too full?

5. Are you still handling activities and problems you had before your last promotion?

6. Are you often interrupted with queries or requests on ongoing projects or assignments?

7. Do you spend time on routine details that others could handle?

8. Do you like to keep a finger in every pie?

9. Do you rush to meet deadlines?

10. Are you unable to keep on top of priorities?

5.4 PERSONNEL DEVELOPMENT

The engineering manager has an especially important responsibility for nurturing the development and maintenance of competence among the employees of the organization. While it might be suggested that individual employees have sole responsibility for personal development activity, experience suggests that in the higher quality organizations, managers take an active part in the development and maintenance of competence of individuals. Indeed, the most successful outcome will likely result when each member of the employee–employer team has an active part in this responsibility.

We suggest that the employer has a responsibility for encouraging development in two directions. First, employees should develop special competence in one or two areas. Second, employees must keep up to date on the engineering profession as a whole and be aware of major developments. Thus, we urge development of depth and breadth.

With growing technical competence comes the need to apply that competence effectively. A perspective is needed that allows one to see the details within the context of the bigger picture. An understanding of the human dimension allows the engineer to work effectively with associates inside and outside the organization and to assume increasingly responsible managerial positions.

Among the development activities to be considered as joint projects of employer and employee are the following:

- *On-the-Job Training.* Some employers provide structured training programs, such as the Engineer-in-Training (EIT) program of the Alaska Department of Transportation and Public Facilities, in which new EITs work in several divisions of the department for approximately six months each over a period of three years.

- *Formalized Graduate Education.* The possibilities are numerous, in both technical and management areas. Many universities offer a master's degree in engineering management as well as various technical master's degrees, on an after-hours, part-time basis, or on a full-time basis. The National Technological University offers master's degree programs in engineering via satellite television to more than 400 industrial sites across the United States, where employers provide funding and support personnel to assist their employees in earning these graduate degrees.

- *Professional Registration.* Many employers assist and encourage their young engineers by providing funding and released time to prepare for the PE examination and by paying their annual registration fees. Professional registration is covered further in Chapter 14.

- *Noncredit Short Courses and Workshops.* These opportunities are often available in house at the workplace. Also, engineers can participate in outside courses and, increasingly, in satellite-based televised courses sponsored by professional societies and other organizations.

- *Professional Society Activity.* Activities range from the monthly luncheon, featuring a technical or professional program and contacts with other professionals, to week-long meetings and specialty conferences. Often employers cover some or all of the cost.

- *Reading.* A wide variety of books, journals, and magazines cover almost every useful technical or professional topic. Often employers provide libraries, in part to assist the engineer in maintaining competence.

5.5   INCENTIVE, RECOGNITION, AND REWARD SYSTEMS

Think about this bold statement made by Mel Hensey (1992): "*Achievement* leads to motivation, not the other way around." That is, a successfully completed project motivates the project team to do well on the next project, a set of calculations that help a design to accomplish its purpose motivates the analyst to perform well again, a successful lecture encourages the lecturer to make another one, and so forth. Furthermore, Hensley says that recognition for achievement is nearly as important as the achievement itself in motivating the individual and the team toward high future performance.

The key to proper recognition is some sort of feedback system from the ultimate users of our work products. Although such comments and suggestions may be fed through supervisors, they need to originate with consumers, client groups, or others who receive the products or services. "And they need to be specific, timely and focused on performance goals and results" (Hensey, 1992). Among methods used to obtain such feedback are face-to-face meetings and short questionnaires at project end to elicit candid and specific evaluations.

How, then, might the engineering manager deliver recognition to a group of employees? Hensey, once again: "Our practice would suggest that simple verbal recognition and acknowledgment in organizations is the greatest untapped, underused, effective, yet inexpensive method of rewarding and motivating people. It is underused by a factor of 10 or more in most organizations."

The same author suggests the following guidelines:

1. Reinforce randomly, not too often, and soon after performance.
2. Use minimal negative criticism, only when necessary, to correct especially bad performance.
3. Tangible recognition in the form of awards, brass plaques, and the like is appropriate for extraordinary performance and is most effective when presented by a respected leader, soon after performance and with clear sincerity. (Hensey, 1992)

The value of nonmonetary incentives was demonstrated by a program developed for a large polyethylene plant designed and constructed by Fluor Daniel Inc. for Chevron's Cedar Bayou facility in Baytown, Texas (Fischer and Nunn, 1992). The project's mission statement included the following: "We will foster an environment that promotes and rewards teamwork by sharing responsibility and recognizing accomplishments. Our successful project will emphasize personal growth as individuals and employees." A committee was established to develop and coordinate the recognition program, which consisted of the following elements: (1) for each special group accomplishment, a "well-choreographed presentation" in which management participated, (2) memos and framed certificates for each member of the group, (3) special mementos and plaques, as appropriate, presented at a lunch or at refreshment time, and (4) displays in a prominent, central location of posters citing accomplishments and pictures of the honored group.

During the project 52 awards were given to a total of 674 persons, primarily Chevron/Fluor Daniel personnel but also including outside inspectors and vendor teams. Total cost was about $18,000; the $27 per person can hardly be considered a "monetary" incentive!

The Chevron/Fluor Daniel nonmonetary incentive program demonstrated several key ingredients:

- Recognize specific accomplishments or behaviors, such as challenging milestones met, innovative responses to problems, and steps taken to secure the support of others.
- Recognition by itself is rewarding, with long-lasting effects.
- Keep the program timely, consistent, fun, and focused on personal accomplishments.
- Recognition is meaningful only if it is so perceived by the recipients; feedback is essential.
- Do not neglect to recognize such out-of-the-mainstream departments as office service, accounting, and vendor information control.
- Develop a broad base of support, including wide management interest and participation and a grassroots committee with considerable freedom in its operation.
- Focus on accomplishments of the group and of the task force.
- Publicize the recognition widely.

This author once distributed silly "diplomas" to some of his employees who had volunteered for an organizational paint-up/fix-up day; the large number of these simple pieces of paper that appeared on office doors was a good indication that the recognition was appreciated. One company has developed a "Golden Banana" award for special recognition. This award originated when the company president, moved on the spur of the moment to give a deserving employee something— anything—tangible, looked in his desk drawer, and discovered a leftover banana, which he presented in an impromptu and laugh-filled "ceremony."

Aubrey C. Daniels, in "Parties—With a Purpose," (1992) describes some innovative recognition opportunities:

- A skit performed by a Honeywell division team to illustrate how the members had set up a successful new production line (and, incidentally, to recognize them for their good work).
- An Eastman Chemical Company presentation of engraved pocket knives to employees who had kept the company "on the cutting edge" in performance improvement.
- A potluck pool party for employees of an AG Communication Systems group and their families, after the workers had met an important milestone in their quality road map.
- A midafternoon ice cream cone presented by the president of Preston Trucking Company, at a shop beside the road, to a driver whose driving had just been noticed as outstanding.
- A breakfast for production operators in Eastman Kodak's paper-sensitizing division, served by the manger and his staff, in tuxedos, to show appreciation to employees who had broken a quality record.

As noted, some organizations grant rewards to teams rather than to individuals. While such incentive systems bring peer pressure on low performers to do better and on new team members to learn quickly, they may create frustration among the team's top performers, who are not singled-out for distinction as individuals. Some consulting and design firms have dual incentive systems in which recognition is accorded for both individual and team performance.

5.6   THE PERSONNEL ADMINISTRATION FUNCTION

We now turn from topics traditionally considered to be within the area of "personnel administration" to several related topics that are typically of interest to the engineering manager. Although the personnel departments of many organizations employ specialists who handle routine and nonroutine matters in these areas, the active manager/supervisor needs to be aware of them and should know how to seek guidance if needed. We shall note the areas of recruitment; screening, selection, and placement; performance evaluation and appraisal; promotion and transfer; layoff and termination; compensation and benefits; discipline; collective bargaining; and grievances and arbitration. We then conclude with a brief discussion of current and future trends in personnel administration. The interested reader is referred to a number of excellent source materials (Beach, 1985; Carrell and Kuzmits, 1986; Miner and Miner, 1985; Pigors and Myers, 1981).

Recruitment involves attracting applicants who are available and qualified to fill positions in the organization. Applicants may be drawn from those already employed in the organization as well as from the population outside. Methods for generating interest include posting jobs for internal applicants, direct unsolicited applications, employee referrals, campus recruiting, use of private employment agencies, and advertising through a variety of media. Responsibility for recruitment usually is shared by the manager and the personnel department.

After the recruitment period, the process of screening, selecting, and placing begins. Screening/selection sorts out those judged unqualified to meet job and organizational requirements, normally through a series of application reviews, tests, interviews, and background investigations. Neglecting the details of the screening process, such as failure to conduct thorough reference checks, often leads to unhappy and unproductive matches between employee qualifications and expectations and employer needs and expectations. Organizations must abide by affirmative action guidelines and legislation related to equal pay, civil rights, age discrimination, vocational rehabilitation, and military veterans. Placement involves matching what the organization thinks the new employee can do with what the job demands, and what it offers. Orientation is an important part of the placement process.

Performance appraisal is the systematic evaluation of employees with respect to job performance and to their potential for future development and performance. Despite suggestions by Deming (1986) and other writers on total quality management that such systems are inappropriate because they focus "on the end product, at the end of the system, not on leadership to help people," appraisals are nonetheless an important aspect of personnel administration today. One writer has even said that "companies need not sacrifice performance appraisal on the altar of total quality but should work at building a stronger performance appraisal, one that is equally effective for individuals or work teams" (Guinn, 1992).

The supervisor has a major role to play in this process, usually with assistance from the personnel office. Many methods are used, including rating scales, checklists, comparison with agreed-on objectives, interpersonal comparisons, and essays. In a well-designed personnel system, the career development program and the performance appraisal process are closely linked, with each appraisal session the setting for comparing progress with objectives and setting new development objectives for the next period. In a project setting, appraisal of performance can be especially cumbersome and often is poorly done because normally the employee is responsible to both the project manager and the functional, or discipline, manager. Some sort of joint appraisal is required.

A *promotion* is an advancement of an employee to a job with greater responsibilities, greater skill, more status, and a higher rate of pay. *Transfer* refers to the movement of an employee from one job to another on the same occupational level and at about the same level of pay. Issues of full disclosure of opportunities, nondiscrimination, and seniority versus competence are important in promotion decisions. Transfers may be made to staff areas of increasing activity, to replace workers who leave and to utilize workers' talents more efficiently. It is important that an employee be informed as fully as possible about the reasons for the promotion or transfer.

Layoffs and terminations are difficult for employee and manager alike. While both terms mean that someone is out of a job, "layoff" implies that there is some hope the employee will be rehired if the economy picks up. Reasons for terminations range from lack of work to incompetence to outright fraud or other criminal activity. In any case, standards of openness, timely notice, and fairness must be adhered to. With layoffs, issues often involve seniority, affirmative action, separation pay, and employer-assisted placement outside the organization.

The matter of *employee compensation and benefits* is another complicated topic. Compensation issues especially require personnel, financial, and legal expertise. Included in this category are the need to attract and retain good employees, the evaluation of jobs for proper wage or salary, the pay increase process, and such controversial topics as wage and salary compression (decreasing the differentials between higher and lower pay grades) and comparable worth. Benefits include pensions, as well as paid time off for vacations, holidays, sick time, and other leaves; insurance; profit sharing; and such employee services as education expenses, food services, and social activities.

Discipline is essential to the workplace, yet it is difficult to know how to approach cases of employees who have stepped beyond the bounds of established practice. Let it suffice here to indicate that the solution can be either the negative, "big stick" approach or a more positive application of "constructive discipline," in the hope of developing a willing adherence to the company's rules and procedures. The manager must be aware of rules and related penalties and must be sure they are communicated thoroughly. He or she may become involved in grievance procedures, appeals, charges of inconsistency, and the like.

Collective bargaining is another area that is a specialty unto itself. The manager may be involved in applying work rules, pay and benefit programs, and other matters involving the workplace that have been developed during bargaining between the employer and representatives of the bargaining unit. The manager may also be involved in the contract bargaining process. Furthermore, disputes will probably arise, and strikes may occur. The topic is mentioned here as another of the myriad areas that may involve the engineering manager in either the private or public sector.

Even in the best-managed organization, employee discontent, gripes, and complaints will arise. A *grievance* can be thought of as any dissatisfaction or feeling of injustice in connection with one's employment that is brought to the attention of management. Grievances are surely not confined to organized labor. Though the process for handling grievances may vary depending on whether there is a union, a clear procedure known in advance by all parties is essential in either case and must be consciously followed. *Arbitration,* utilizing the services of an outside person familiar with the process, who hears both sides and renders a decision, is a common means of resolving grievance cases.

What of current and future trends in the management of personnel? In the area of personnel welfare and convenience, there is a trend toward more flexible work schedules, the provision of child care services, and an increased emphasis on employee fitness, safety and health, and related environmental issues in the workplace. The workplace will continue to see increasing proportions of females and racial/ethnic minority workers, and affirmative action and equal employment opportunity will continue as important social issues. Other aspects of governmental regulation of business, ranging from payroll reporting to environmental standards, will impact the personnel function. It is likely that the workplace in general will see a further decline in the importance of employee unions, although service industries, including engineering, may show an opposite trend. Areas that are likely to affect engineering management include continued rapidly advancing technology (accompanied by the need for more technically competent workers) and the rising importance of the total quality management trend, with its emphasis on continuous improvement.

5.7 DISCUSSION QUESTIONS AND CASES

1. Consider a job you hold now or held in the past. To what extent were your needs met by this job? What was the relationship between having your needs met and your motivation to help the organization succeed?

2. Consider three kinds of supervisory positions: conductor of a professional symphony orchestra, lead carpenter on an apartment building project, and design crew chief for a portion of a space shuttle project. How might (a) Maslow's hierarchy of needs theory and (b) McGregor's Theory X/Theory Y ideas be applied differently by each of these supervisors?

3. Write a paragraph that explains your concept of the connection between leadership and motivation.

4. As you assume increasing management responsibility, you are likely to become responsible for technical activities you do not completely understand. Discuss the importance to the engineering manager of technical competence versus people, planning, and organizational competence.

5. Machiavelli asked whether it is better for a leader to be loved or feared. Give your opinion.

6. Of the several leadership *roles* listed in Section 5.2, which seem to be especially important to the first-level engineering manager? How might your list change as the engineer moves up in the organizational hierarchy?

7. Of the several *traits* suggested in Section 5.2 for successful leaders, which seem to be especially important to the first-level engineering manager? How might your list change as the engineer moves up in the organizational hierarchy?

8. It has been suggested that the increasing importance of self-directed teams, as discussed in Section 2.2.6, has decreased the importance of effective delegation. Do you agree? Why?

9. How might a reward program you design for your company's architects and engineers differ from that for its stock clerks and telephone operators? Would group versus individual rewards be given different emphasis for the two groups?

The Case of the Discouraged Employee

Susan is about ready to quit her job. She was hired by a mechanical engineering design firm a year and half ago, fresh out of State University with a new B.S. and an EIT. During her interview, she mentioned her interests and achievements in creative writing. Upon being hired, she was assigned to a group that prepares proposals, status documents, and closeout reports. After 18 months of this activity, she wonders if she made the right choice. Other new engineers who were hired at the same time have had a variety of projects, and several have received promotions. Being stuck in the editorial office, Susan rarely gets involved with projects, and she thinks she might have seen the company president twice.

Last summer's company picnic seemed to be dominated by the "old guard." The volleyball game, located adjacent to the beer keg, got out of hand, and Susan went home early.

Her one-year performance evaluation, submitted 6 months late, consisted of a short fill-in form completed by her supervisor. At a 12-minute conference, Susan learned that the supervisor was pleased with her work and that a 3% raise would be forthcoming in another year if her performance continued to be satisfactory.

Being a new engineer and a reserved, rather shy person, Susan hesitates to complain. But she feels less than fully motivated and thinks a change of employment might be best for her. Unfortunately, there has been a downturn in the economy, and jobs are scarce. She would be competing with this year's graduates, who have the same design experience as she does, which is *none*. And starting salaries may be below what she makes now.

From Susan's perspective, and from that of the design firm, what are the problems, and what are the solutions? In addition to your responses to these questions, list some additional information you would want to have before answering this question fully.

The Case of the Division in Crisis

The Players

William: Division chief for the Pollution Control Design Division; four department heads report to him.

Horacio: Company vice president for engineering, William's supervisor.

The Location

Horacio's office, door closed, 2:30 P.M.

The Conversation

Horacio: This morning I had a rather unsettling meeting with three of your department heads, at their request. They told me things aren't going very well in your division.

William: Really? As I have told you several times this month, things are actually going rather well. Sure, we've missed a few deadlines, and that budget cut was pretty severe, but, overall, I think things are fine. The problems in my division are caused by situations over which I have no control.

Horacio: From what I just heard, however, it sounds like morale is pretty low, communication is strained or nonexistent, and several of the best people are considering quitting. Sam and Ted and Ethel lay the blame squarely on your shoulders.

William: Couldn't they have discussed this matter with me directly, rather than wasting your time with it? I'm sure it's something we can easily fix within the division.

Horacio: They say they have tried to discuss it with you, but you seem to be uninterested and unavailable. They are concerned that you have been bypassing them and going directly to the design engineers for information on project status, technical detail, and client relations.

William: How else can I obtain the straight story? The department heads always seem to filter the information before it gets to me, and then it is essentially useless. Besides, I often need information immediately, and an extra step in the communication chain is too time-consuming. If I'm in charge of this division, I need to know what's going on.

Horacio: I also was told that these three are deeply upset over the way you are undermining their authority. It seems to be commonplace for you to overrule their decisions, without consulting them.

William: But I *am* the chief of my division and an acknowledged expert. Should I just let errors go without correcting them?

Horacio: Of course not, but maybe your communication could be improved. I must say that the technical aspects of the work in your division continue to be first rate. All the engineers in your division have great respect for you as an engineer.

William: Thanks! I take great pride in my own technical expertise. That ASME Pollution Control Engineer of the Year Award has a prominent place in my office.

Horacio: Do you have any special incentive or reward programs in your division?

William: Only the company's standard merit salary increases. As you know, merit increases haven't been available for the past two years. Company policy, you know.

Horacio: Do you ever get together with your department heads outside of work, in a more social setting?

William: We don't seem very compatible socially. Besides, my wife and I entertain very seldom. Look, I'm not sure this conversation is getting us anywhere. I thought a manager's job was to get the work done correctly, on time, and within budget. If I'm wrong about that, maybe you need a different division chief.

(a) Evaluate William's leadership style and effectiveness.

(b) How effective is Horacio as a leader, based on this conversation?

(c) Suggest an appropriate course of action for Horacio.

5.8   REFERENCES

Amos, J. M., and B. R. Sarchet. 1981. *Management for Engineers.* Englewood Cliffs, NJ: Prentice-Hall.

Beach, D. S. 1985. *Personnel: The Management of People at Work,* 5th ed. New York: Macmillan.

Bennis, W. 1976. *The Unconscious Conspiracy: Why Leaders Can't Lead.* New York: AMACOM Books.

Blake, R. P., and A. A. McCanse. 1991. Leadership Dilemmas—Grid Solutions, Houston, TX: Gulf Publishing Company.

Campbell, D. 1980. *If I'm in Charge Here, Why Is Everybody Laughing?* Niles, IL: Argus Communications.

Carrell, M. R., and F. E. Kuzmits. 1986. *Personnel: Human Resource Management,* 2nd ed. Columbus, OH: Merrill.

Cleland, D. I., and D. Kocaoglu. 1981. *Engineering Management.* New York: McGraw-Hill.

Connolly, T. 1983. *Scientists, Engineers and Organizations.* Monterey, CA: Brooks/Cole Engineering Division of Wadsworth.

Covey, S. 1989. *The Seven Habits of Highly Effective People.* New York: Simon & Schuster.

Daniels, A. C. 1992. "Parties—With a Purpose." *Sky Magazine,* Vol. 21, No. 12, December, 16–20.

Deming, W. E. 1986. *Out of the Crisis.* Cambridge, MA: Massachusetts Institute of Technology Center for Advanced Engineering Study.

Drake, R. L. 1987. "Traits for Leaders." *Creative Management.* December, 3.

ESM 450. 1990. Economic Analysis and Operations Lecture Notes, 1990. University of Alaska, Fairbanks.

Fischer, G. W., and N. P. Nunn. 1992. "Nonmonetary Incentives: It Can Be Done." *Journal of Management in Engineering,* American Society of Civil Engineers, Vol. 8, No. 1, January, 40–52.

Gellerman, S. W. 1963. *Motivation and Productivity.* New York: AMACOM Books.

Guinn, K. A. 1992. "Successfully Integrating Total Quality and Performance Appraisal." *The Human Resources Professional,* Vol. 4, No. 3, Spring, 19–25.

Hensey, M. 1992. *Collective Excellence: Building Effective Teams.* New York: American Society of Civil Engineers.

Herzberg, F. 1968. "One More Time: How Do You Motivate Employees?" *Harvard Business Review,* Vol. 46, 54–62.

Katz, D., and R. L. Kahn. 1978. *The Social Psychology of Organization,* 2nd ed. New York: Wiley.

Mackenzie, R. A. 1972. *The Time Trap: How to Get More Done in Less Time.* New York: McGraw-Hill.

Maslow, A. H. 1943. "A Theory of Human Motivation." *Psychological Review,* Vol. 50, 370–396.

Maslow, A. H. 1954. *Motivation and Personality.* New York: Harper & Row.

McCall, M. W. 1983. "Leadership and the Professional." In Connolly, T. *Scientists, Engineers and Organizations.* Monterey, CA: Brooks/Cole Engineering Division of Wadsworth.

McClelland, D. C. 1961. *The Achieving Society.* Princeton, NJ: D. Van Nostrand.

McClelland, D. C., and D. H. Burnham. 1976. "Power Is the Great Motivator." *Harvard Business Review,* Vol. 54, No. 2, March–April, 100–110.

McGregor, D. 1960. *The Human Side of Enterprise.* New York: McGraw-Hill.

Meredith, J. R., and S. J. Mantel Jr. 1995. *Project Management: A Managerial Approach,* 3rd ed. New York: Wiley.

Miner, J. B., and M. G. Miner. 1985. *Personnel and Industrial Relations: A Managerial Approach,* 4th ed. New York: Macmillan.

Mintzberg, H. 1973. *The Nature of Managerial Work.* New York: Harper & Row.

Pigors, P., and C. A. Myers. 1981. *Personnel Administration,* 9th ed. New York: McGraw-Hill.

Raudsepp, E. 1977. "Motivating the Engineer: The Direct Approach Is Best." *Machine Design,* November 24, pp. 78–80.

Roadstrum, W. H. 1978. *Being Successful as an Engineer.* San Jose, CA: Engineering Press.

Roethlisberger, F. J., and W. G. Dickson. 1939. *Management and the Worker.* Cambridge, MA: Harvard University Press.

Schultz, W. N. 1988. "Successful Leaders: Ten Essential Qualities." *Successful Supervision,* August 22, p. 2.

Silverman, M. 1988. *Project Management: A Short Course for Professionals,* 2nd ed. New York: Wiley.

Skinner, B. F. 1953. *Science and Human Behavior.* New York: Macmillan.

Souder, W. E. 1974. "Autonomy, Gratification and R&D Output: A Small-Sample Field Study," *Management Science.* Vol. 20, No. 8, April.

Vroom, V. H. 1964. *Work and Motivation.* New York: Wiley.

6

Communication in the Engineering Organization

One of the most important aspects of the transition from engineer to engineering manager is the greatly increased importance of communication, both oral and written. Communication is an important part of any employee's activities. However, the responsibility for providing oral and written direction to individuals and groups, for listening, for writing and reading reports and other communications, for conducting meetings, and for engaging in a variety of other forms of communication is a major part of the engineering manager's job. Indeed, it has been estimated that the engineering manager spends about 90% of his or her time communicating in one form or another (Luthans, 1973).

This chapter begins with a brief note on the communication process and then turns to a discussion of communication in organizations. Next, we consider oral communication, identifying three major elements of the process, citing some important skills, and emphasizing the importance of listening skills. We then outline some aspects of written communication from the viewpoint of the technical manager. Next, we offer a plea for improved skills in leading meetings, an activity that usually has potential for much improvement. We conclude with some guidelines for communication during negotiations and some general thoughts on improving the communication process.

6.1 THE COMMUNICATION PROCESS

Those who study communication theory suggest that at its most basic level, communication consists of three elements—a transmitter, a transmission medium or channel, and a receiver. Figure 6.1, taken from Connolly (1983), captures these basic ideas and adds the encoding and decoding processes and the concept of noise, which is always present to some extent. Engineers will understand that at least in its simplest form, communication is a serial, rather than a parallel, process. Thus, a disruption at any point in the path will cause a disruption in the overall success of the communication.

We know that the probability of success of a multiple-step serial process is the product of the probabilities of success of the individual steps. Consider three

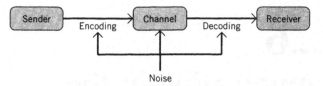

Figure 6.1. The three basic elements of a communication link. (From T. Connolly, *Scientists, Engineers and Organizations,* Monterey, CA: Brooks/Cole Engineering Division, 1983, p. 90.)

probabilities: (1) that the sender sends the correct message, (2) that the transmission channel works correctly, and (3) that the receiver receives and interprets the message correctly, and assume that in each case the value is 0.8. We can very easily find that the probability that the communication will be successful is

$$P_{(success)} = 0.8^3 = 0.512$$

Note that this concept applies equally to oral and written communication and to communication either between individuals or within groups.

6.2 COMMUNICATION IN ORGANIZATIONS

Let us consider some of the communication links that are common in techni-cal organizations. Although to say communication is a multidimensional process is almost trite, Cleland and Kocaoglu point out that many engineering managers overemphasize the importance of communication with subordinates to the exclu-sion of communication in other directions in the organization. The manager must communicate in at least three directions—with subordinates, as noted above, with supervisors, and laterally with peers (Cleland and Kocaoglu, 1981).

Connolly provides some insight into the content of communications in these three directions, as follows:

1. *Downward communication* is likely to involve instruction, both on the subordinate's specific job and on the practices and procedures to be followed doing it. Second, bosses give their subordinates feedback on how they are doing. Finally, though this aspect is often neglected, they provide the subordinate with a rationale for the job, why it is important, how it fits into the larger picture of the organization's activities and what these larger purposes are.

2. *Upward communication* is largely the mirror image of these content categories. Subordinates report on what they are doing and what problems they are encountering, how existing policies and procedures are working, and their ideas as to what needs to be done and how it should be done. They may also pass on similar information about the activities of their peers, though this is inhibited by the informal taboo on "ratting to the boss."

3. *Lateral communication*—communication between two people at the same organizational level—includes both information concerned with coordinating their interlocked work activities and information communication for social and emotional support. People at the same organizational level may share many interests other than work coordination: they are often of similar ages and backgrounds, or are at similar stages of their careers. They may also share a common interest in diluting the boss's power over them by comparing notes, sharing secrets, and pooling their information (Connolly, 1983).

Project managers in technical organizations often are required to communicate in an important fourth direction—outside the organization. This task will primarily involve the client/owner, although interfacing with the public at large through news media, public testimony, addresses before citizen groups, and the like may be part of the responsibility. Figure 6.2, taken from Stuckenbruck (1981), shows the communication links important to the engineering project manager.

The impact of various group sizes and organizational patterns should also be considered. In a structure that entitles all members to communicate with each other, the number of communication links is a quadratic function of the number of members, with

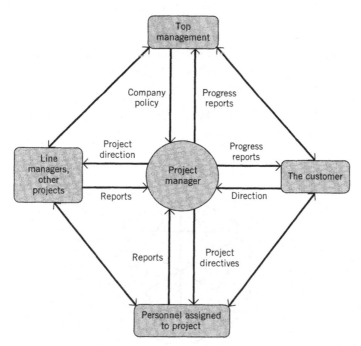

Figure 6.2. The project manager's communication links. (From Lynn C. Stuckenbruck, *The Implementation of Project Management,* © 1981 by Addison-Wesley Publishing Company, Inc. Reprinted by permission.)

$$l = \frac{(n-1)(n)}{2}$$

where l = number of links

n = number of members

The advantage, of course, as indicated in Figure 6.3, is unrestricted access to every other member; the disadvantage is potential chaos! We return to the concept of groups (nodes) in linked scheduling networks in Chapter 7.

Thomsett (1980) describes a training exercise in which five-person groups compete to assemble a puzzle in the shortest time. Members may communicate only with hand signals and only with the other member or members with whom they are formally linked; furthermore, they may see only the pieces held by the predetermined persons and may pass the pieces only along those channels. Normally three different group structures are utilized, as shown in Figure 6.4: leader and subordinates, hierarchical, and adaptive or free network. The reader can imagine the advantages and limitations that tend to be associated with each form, including the chaos characterizing the adaptive structure in the absence of a strong leader.

Rossini (1983) has contributed an excellent study of the impact of organizational structure and other factors on communication and effectiveness in interdisciplinary research and development teams. The paper appears as Reading 4 in Connolly's *Scientists, Engineers and Organizations* (1983).

There is an old parlor game called "gossip" (in the United States; the name is "whispers" in England), in which a message is whispered in the ear of one person, who then whispers the same message (or an approximation thereof!) to the next person, and so forth, until the message is announced aloud by the final receiver and compared to the original message. The results are often hilarious. The message "The bulldozer lost its blade and is broken" may come out as "The bull dozing in the glade seems to be smoking."

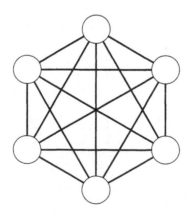

Figure 6.3.   An adaptive communication network: fifteen links connect six group members.

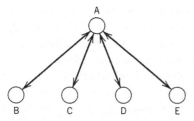

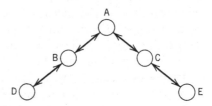

1. Chief programmer/technical leader structure: B. C, D, and E can communicate only to A.

2. Hierarchical structure: D can communicate only to B. B can communicate to D or A, and so on.

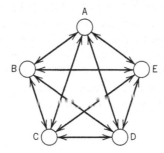

3. Adaptive structure: Everyone can communicate with one another.

Figure 6.4. Three organizational structures. (From Rob Thomsett, *People and Project Management*, © 1980, p. 27. Reprinted by permission of Prentice-Hall, Englewood Cliffs, New Jersey.)

The game illustrates the difficulties inherent in the communication process, but an interesting characteristic is that the results seem to "make sense" more often than might be expected. If the message loses 50% of its meaning with each repetition (not unlikely in some parlors!), our routine probability calculation suggests that less than 1% of the message would remain after seven repetitions. In fact, the results have much more meaning than might be expected. Connolly says, "In serial repetition, it is not just repeated garbling, but repeated sense-making, that shapes the final version" (Connolly, 1983). The caution is that in large organizations especially, serial repetition can lead to large communication distortions.

The importance of communication in technical organizations, and the unfortunate results of breakdowns in communication, are illustrated by a study of structural engineering problems experienced by a large nationwide industrial firm (Hensey, 1987). Of the 40 problems identified at different locations throughout the United States (none of which, fortunately, was catastrophic), 25% were due at least in part to communication coordination failures. In one case walls were not designed for blast pressures as required for process safety and would have failed at the design blast pressure. The technical cause, which resulted in reconstruction of the walls with reinforced concrete, was "intended function of walls not understood by designer." However, the basic cause identified by the study was *"lack of communication* between owner's process engineer and structural designer." Other such basic causes reported by the study included "failure to communicate a chronic, known problem," "lack of cooperation in getting field measurements for structural members," and "lack of site visit by designer or communication with plant management."

While every organization has some form of formal organization, as discussed in detail in Chapter 2, every organization likewise has some form of *informal communication system*. We usually refer to such a system as "the grapevine." The communication channels may be different for different messages, with the boss's secretary starting the process by telling an assistant project manager a particularly juicy tidbit in one case and the lead structural designer passing a rumor to the head custodian in another. Furthermore, the settings for and methods of communication will take on different forms (during lunch or coffee break, while walking to the parking lot, before the start of a formal meeting, by telephone at home in the evening). In any case, such transmission of "information" (true or false) is a real part of organizational life, and the engineering manager must be both aware of this system and prepared to work with it. The idea is to combat false rumors. Trying to eliminate the grapevine is futile—it won't go away!

6.3   ORAL COMMUNICATION

The Book of Lists (Wallace, 1980) says that the greatest fear of Americans is the fear of speaking in public. Forty-one percent reported this as their greatest fear, putting it well ahead of heights, insects, taxes, financial problems, and driving in Los Angeles. (Death was reported as fear number 8.) A study by Motivational Systems found the following primary reasons for fears about public speaking: fear of embarrassing mistakes (81% of respondents), fear of damaging one's career (77%), fear of "freezing up" (63%), fear of being dull or boring (58%), and fear of appearing nervous (52%) (Golder, date unknown).

Nevertheless, engineers are often required to make a variety of public speaking appearances, and fortunately it is possible to learn to overcome that "butterfly stomach, cotton mouth" sensation. We should note that "public speaking" ranges from presentations to small groups while seated at a conference table to oral briefings with flip charts and other visual aids to formal addresses to large gatherings.

We might divide what happens during oral communication into three elements—the verbal, the vocal, and the visual:

- *Verbal:* data, information, words, content; what you say.
- *Vocal:* voice, expression, resonance, tempo, volume; the way what you say is heard.
- *Visual:* eye placement, posture, gestures, facial expressions, visual aids, blackboard, how you look, what you wear; what is seen by the listener.

Some research conducted at UCLA confirmed the importance of nonverbal elements in successful oral communication. When asked to what extent the impact and believability of the typical spoken message depended on each of these three elements, respondents tended to vote overwhelmingly in favor of the vocal and the visual. The relative importance of each was reported as follows (Longfellow 1970):

Verbal	7%
Vocal	38%
Visual	55%

These proportions reflect the typical listener's weighting of these factors in getting a message across. Whether presenting a paper at a professional meeting, reviewing a design with a client, or explaining the company's new pension plan to the project team, the engineer will do well to remember that the message itself is but a small part of the communication effort.

College students take courses in oral communications, and there are many training organizations willing to assist in polishing speaking skills. Rather than reprising in depth those many pointers, we consider the following suggestions, taken from a variety of sources, including an article by Antoni Louw (1992).

Prior to the Speech

1. Do not hesitate to prepare notes on index cards, but try not to read from a prepared manuscript.
2. Work hard on the opening and closing, perhaps memorizing the first and last half-minute or so.
3. Practice in front of a mirror. Use a piece of furniture as a lectern, and practice gestures and other body movements. If possible, practice in the exact location of your scheduled speech, and also practice by sitting where the listener will sit.
4. Use different words in each rehearsal to express the same thoughts, so that (a) the actual presentation can seem more "unrehearsed" and (b) you concentrate more on ideas than on words.
5. Tape your presentation, and critique tempo, inflection, and the like by playing back the audio or video recording.
6. Try it out on family members, roommates, or close friends, working hard on eye contact.
7. Remind yourself that you have a right to be there and are an expert in your subject.
8. Consider joining a club that is devoted to public speaking, such as Toastmistresses or Toastmasters.

During the Speech

1. Breathe to calm down. Inhale deeply, exhale through your mouth once or twice, and then start your speech.

2. Maintain eye contact with your listeners, between 3 and 6 seconds at a time in a random pattern. Consider eye contact with silence, or repetition, after a point you want to stress.

3. Don't hesitate to use gestures, provided they are natural. They will help to increase blood flow and will make you appear more dynamic.

4. Be careful with your posture, striking a good balance between too casual and too formal.

5. Pay attention to the volume, tempo, and resonance of your voice, and provide some variety to your voice and to your tempo.

6. Make sure that your dress and overall appearance are appropriate for the occasion.

7. Watch you choice of language and keep nonwords ("um," "uh," "and-a," "like" as an interjection) to a minimum.

8. Counteract stage fright by "embracing" the audience: place your attention on the back of the room and then convince yourself that the entire space is yours.

9. Cultivate ways to involve your listeners. A quick audience response "survey" might be an effective technique, for example.

10. Inject humor where appropriate; note that this does not necessarily mean telling jokes.

11. Be your natural self; don't try to be perfect.

The distinctions between one-way and two-way communication are important to the engineering manager. A simple experiment that can be a classroom exercise demonstrates the important differences (Haynes and Massie, 1969). One member of the group is given a paper bearing a pattern of four to six rectangles. The communicator describes the pattern without showing it to the group, and each other member of the group draws what has been described. The speaker faces away from the group. Members are not allowed to speak at all, to ask questions or for any other purpose. The situation is repeated with another communicator and an equally complex set of rectangles. This time the speaker faces the group, and members are encouraged to ask questions until they are satisfied that they have succeeded in reproducing the pattern. Two measures of effectiveness are applied to this one-way and two-way communication exercise—the accuracy of the listeners' diagrams and the time required for the communication.

As might be expected, some rather definite findings have emerged under experimental conditions, as follows:

1. One-way communication is considerably faster.

2. Two-way communication is more accurate.

3. The receivers of two-way communications are more sure of themselves and make more correct judgments of right and wrong.

4. The sender feels more psychologically under attack in the two-way system.

5. The two-way method is relatively noisy and disorderly.

One should not conclude that two-way communication is always to be preferred; if you have a large amount of information to transmit quickly, it may be most effective to tell you audience, "Sit down, shut up, and listen!" (or words to that effect). Each mode has its place, and the engineering manager must decide which is more appropriate in any particular situation.

Professional engineering societies sometimes offer guidance on communication challenges. An example is a small brochure produced by the National Society of Professional Engineers entitled "We Have Ways of Making You Talk!" The brochure contains simple, commonsense guidance for the engineer making an appearance on television, including how to prepare (both your mind and your person!) and how to conduct yourself while on the "hot seat" (National Society of Professional Engineers, no date).

The use of effective *visual aids* can enhance any oral presentation. While 35 mm slides, flip charts, and overhead projection transparencies are still the most widely used materials in the majority of technical presentations, videotapes and computer graphics, either on large monitors or projected on screens, are becoming increasingly popular. I am indebted to media expert Dr. Edmund S. Cridge for the materials in Table 6.1, which provide guidelines for visual materials that can assist the engineering manager in getting the oral message across effectively (Cridge, 1993).

We cannot leave this section on oral communication without a brief considera-tion of the importance of *listening*. Like most skills, good listening can be studied and learned. Americans tend to speak at about 150 words per minute; we are ca-pable of listening to about 1000 words per minute. Is that 85% idle time spent in reviewing and assimilating what has been said, developing an argumentative rebut-tal or brilliant reply, or thinking about an unrelated topic?

Davis (1967) provides some guidance with his Ten Commandments for Good Listening, as follows:

1. Stop talking!
2. Put the talker at ease.
3. Show the talker you want to listen.
4. Remove distractions.
5. Empathize with the talker.
6. Be patient.
7. Hold your temper.
8. Go easy on arguments and criticism.
9. Ask questions.
10. Stop talking!

Another author, who focuses on effective listening during formal one-way communication (i.e., when there is no chance for interaction), suggests:

TABLE 6.1 Guidelines for Visual Aids

In General	The Presentation Sequence	The Physical Environment
• Allow adequate time for preparation. • Allow adequate time for rehearsal. • Use visuals and other technologies only if they will improve the delivery and clarify the message.	• Display the title and your name in the first frame. • Provide an overview of the presentation in the second frame. • Present seven or fewer points in one presentation. • Allow at least one minute for each visual. • Display the overview again as the last visual, to support your oral summary. • If the same slide is to be used more than once, make a duplicate for each use. • When using a videotape, preset the tape to the beginning of the sequence, not to the beginning of the tape.	• Arrange seating so that no one is more than eight times the width of the screen away from it. • Avoid facing the audience toward windows or other light sources. • Know the location of light switches and other controls. • Keep some light on in the room. • Provide adequate time for equipment setup.

Listen positively and attentively, allowing the speaker to make his or her point. Analyze the speaker's attitude and frame of mind. Is the person an optimist or pessimist? Generally reliable or unpredictable? Try to reach beyond the speaker's words to his or her meaning. When in doubt, rephrase the speaker's words with a "Do I understand that ... ?" Take notes of essential points unless that inhibits communication. Finally, consider the speaker's nonverbal language as well (Babcock, 1991).

6.4 WRITTEN COMMUNICATION

Just as with oral communication, engineers have learned and practiced writing as a major part of their education, and it is not our purpose to duplicate that effort here. We wish simply to emphasize that the technical manager, more than the technically oriented engineer, is required to write, and those who write well tend to be better managers.

The principles of clarity, brevity, and simplicity offer a good place to start. Of course, "always be clear" requires clear thinking, followed by clear expression. Also, it may be truly impossible to word a message so clearly that everyone will understand it. E. B. White appealed for brevity and simplicity in a classic statement,

TABLE 6.1   continued

Legibility	Layout	Color
• Use bold, simple lettering such as Gothic or Roman sans serif.	• For margins, leave the equivalent of two characters on each side and 1 1/2 line heights at top and bottom. Too much margin is better than too little.	• Keep color schemes simple.
• Use a mixture of upper- and lowercase letters as in regular writing.		• Use pastel background with dark characters; avoid dark background with white lines or characters.
• Leave 1 1/2 times letter height between lines and 1 1/2 times letter width between words.	• Provide the following aspect ratios (frame shape in height-to-width):	• Use bold characters sparingly to underline or highlight a word or phrase.
• Use bullets, letters, or numbers at the beginning of each major point.	Overhead transparencies 4 units × 5 units	• Be consistent when using colors to distinguish main features.
• Make lowercase letter height, as displayed, a minimum of 1/2 inch high; for any viewing distance above 10 feet, increase the height by 1/2 inch per 10 feet.	2 in. × 2 in. slides 2 units × 3 units	
	Television 3 units × 4 units	
	• Use five lines or fewer per screen.	
	• Use four words or fewer per line.	
	• Present only one concept, point, idea, or issue on a single screen.	
	• Include no more than 3 curves or 20 words per screen.	

SOURCE: Cridge (1993).

"Omit needless words!... This requires not that the writer makes all sentences short ... but that every word tell" (White, 1959).

The legal profession and the banking and insurance industries are infamous for confusing "legalese." Clients were considerably less confused when the National Bank of Washington, D.C., changed one of its loan forms. The old form read as follows:

> Upon default hereunder, whether by acceleration or otherwise, the holder shall have the right, immediately and without notice of any kind, to set-off any and all sums hereunder against any account of, or any other property of, the undersigned, held by the holder, regardless of whether such accounts or property be individually or jointly owned.

This is the revised wording:

> If for any reason I fail to make any payment on time, I shall be in default. The bank can then demand immediate payment of the entire remaining unpaid balance of this loan, without giving anyone further notice. (Pigors and Myers, 1981)

Technical writing manuals and texts urge writers to consider such matters as economy and emphasis, avoiding wordiness and so-called careless pronouns and careless modifiers. Some examples from Conway are useful:

Emphasis

Original: There are eight pieces of equipment available at the job site.

Revision: Eight pieces of equipment are available at the job site.

Careless Pronouns

Original: Consolidated made the sale and delivered the software package; it was a significant one.

Revision: Consolidated made a significant sale and delivered the software package.

Careless Modifiers

Original: Operating the equipment at optimum speed, the goal was easily reached in the first quarter.

Revision: Operating the equipment at optimum speed, company employees easily reached the goal in the first quarter. (Conway, 1987)

The late Malcolm Baldrige had a penchant for simplicity in writing and a severe dislike for redundancy, imprecision, and overused words. He instructed his employees to "reduce gobbledygook and doublespeak." When he was secretary of commerce, the department's word processors were programmed to respond with *Do not use this word!* when an employee typed clichés or jargon like *viable, parameter,* or *prioritize* (Harty and Keenan, 1987).

Robert Half, in an essay entitled "Memomania," quoted Blaise Pascal, who once wrote "I have made this letter longer than usual, because I lack the time to make it short." Half then reports that 22% of average executive's time is spent writing or reading memos, that executives believed 39% of all memos to be unnecessary or wasteful, and that 76% thought most memos were too long. "If you must write a memo," says Half (1987), "keep it short and to the point.... The basic rule for memo writing is when in doubt, don't!" Often, a telephone call or face-to-face communication will take less time and will afford the chance for two-way communication.

The 1990s have seen an explosive growth in electronic means of communication. Engineers are both the providers and the users of such media. Both facsimile transmission (fax) and computer-based electronic mail (e-mail) utilize telephone systems that formerly transmitted only voice communication. "Instant" transmission and delivery of information worldwide happens millions of times each day. Local area networks (LANs) tie offices together for transmission of the written word, drawings, and other graphical information. The development of a design drawing using computer-aided drafting and the subsequent electronic forwarding of that drawing to a client or a construction site halfway around the world are commonplace.

The principles of clarity, brevity, and simplicity surely ought to apply to construction specifications. Goldbloom (1992) reminds us of the fundamental legal principle that when there is disagreement over specification language, the courts will most likely interpret the language against the party who authored it. He identifies four traps into which writers of specifications fall:

- *Sentences of the Wrong Length.* Fewer words generally mean less risk of misunderstanding, but incomplete sentences should be avoided.
- *Improper Use of Terms.* Note well the distinctions between "will" and "shall," "inspection" and "supervision," "consists of" and "includes," "calendar day" and "working day," and such technical terms as "grout" and "mortar."
- *Ambiguous Wording.* Examples of words or expressions to be avoided include "reasonable," "workmanlike," "to the extent necessary," "reasonable time," "first-class workmanship," "to the satisfaction of the engineer," and "in an approved manner."
- *Poor Word Choices.* A contract said, "Place the buried pipe on unmade ground . . . ." "Undisturbed ground" would have been clearer. The same contract required the backfill to be compacted "to a density at least equal to that of the contiguous fill," whereas "adjacent fill" probably would have been easier to understand.

The challenge for the engineering manager in the future will be to develop integrated communication systems that utilize the best of the new technology along with more traditional methods in ways that do not neglect human needs and sensitivities, serving to enhance, without excessive depersonalization, the effective operation of human organizations.

6.5 CHOOSING THE APPROPRIATE METHOD

When is each of the several oral and written communication methods appropriate? Table 6.2, adapted by Babcock (1991) from the work of Timm (1986), provides some guidance for seven communication methods. It is interesting that the informal conversation seems to have many features to recommend it, at least when the subject matter is simple and no record is needed.

TABLE 6.2 Characteristics of Common Communication Methods

Communication Method	Speed	Feedback	Record Kept?	Formality	Complexity	Cost
Informal conversation	Fast	High	No	Informal	Simple	Low
Telephone conversation	Fast	Medium	No	Informal	Simple	Low–medium
Formal oral presentation	Medium	High	Varies	Formal	Medium	Medium
Informal note	Medium	Low	Maybe	Informal	Simple	Low
Memo	Medium	Low	Yes	Informal	Low	Low–medium
Letter	Slow	Low	Yes	Formal	Medium	Medium
Formal report	Very slow	Low	Yes	Very formal	Complex	High

SOURCE: Daniel L. Babcock, *Managing Engineering and Technology,* © 1991, p. 335. Reprinted by permission of Prentice-Hall, Englewood Cliffs, New Jersey. Adapted from Paul R. Timm, *Managerial Communications: A Finger on the Pulse,* 2nd Edition. Copyright Prentice-Hall, Englewood Cliffs, New Jersey, 1986, p. 59.

Hensey (1987) emphasizes that each form of communication has a proper place in the process of coordinating technical projects. He suggests the following guidelines for selection:

1. Face-to-face communication in meetings or consultations should be used for addressing issues, problems, or complex matters; gathering ideas interactively; and initiating important actions or decisions.
2. Telephone conversations or conferences should be used for soliciting information; providing sensitive information; and as an urgent substitute for face-to-face communication.
3. Written communication (e.g., memos, letters, reports) should be used for requesting or transmitting factual information; providing updates or routine changes; and confirming verbal communications or decisions. The only effective way to be sure of receiving the complete communication is to summarize your understanding of it and see if there are any corrections.

6.6 MEETINGS

Another topic that involves (or should involve!) communication is the topic of meetings. Many, if not most, people think that most meetings are a waste of time. In their book on time management, Douglas and Douglas (1980) call meetings "Time Waster No. 6." They suggest that we approach the management of time spent in meetings from two standpoints—those we call and lead, and those we attend. In both cases, we must be concerned about activities that take place before, during, and after the meeting. Entire books have been written on meeting management, including Carnes's *Effective Meetings for Busy People: Let's Decide It and Go Home* (1980). Carnes writes with a delightful light touch that includes such "canons" and other tidbits as "Never use a bullwhip when a smile will do better," and "To get the message, listen between the lines of the unwritten handwriting on the wall."

We conduct meetings for a variety of purposes, including telling, selling, solving, and educating. Silverman (1987), in writing about project management meetings, distinguishes between information-disseminating meetings, problem-solving meetings (including the last problem-solving meeting that closes out the project), and customer review meetings. Each type has a specific purpose and should follow a distinct format.

Effective and efficient communication can take place in meetings if leader and participants alike are prepared, if all understand and are serious about the meeting's purpose, and if some rather simple guidelines are followed. The following points, adapted from Cleland and Kocaoglu (1981), may be helpful.

1. Hold a meeting for a definite purpose and only when it is likely to achieve results.

2. Be thoroughly prepared before the meeting, giving the necessary attention to agenda and other materials.

3. Establish firm time limits and adhere to those limits.

4. Start with a statement of the objective of the meeting, and reinforce the purpose of the meeting as required. Make sure the agenda is understood and agreed upon.

5. Arrange for a lively discussion, with balanced participation. Limit discussion to the meeting's purpose. Encourage and control disagreements.

6. Summarize progress or lack thereof at times during the meeting; evaluate whether the discussion could be more effective.

7. Bring the meeting to a definite conclusion by summarizing points made or actions taken and arranging for followup.

8. Distribute summary, action-item minutes very soon after the meeting ends.

In his lively little book, *Up the Organization,* Robert Townsend (1970) says this about the weekly staff meeting:

The Weekly Staff Meeting
Purpose: information, not problem solving.
For: all division and department heads
Takes place same time same place, like TV news. Starts on the dot no matter who's missing. Goes around the room: reports on problems, developments (a crossed wire is handled by Joe saying to Pete: "I'll see you after the meeting on that."). A number of people should and will say "Pass."
Ends on the dot (or sooner).
No attendance taken.
No notice of meeting sent in advance.
No stigma for non-attendance.
One-page minutes dictated, typed and circulated the same day (The chief ought to write this. In the worst companies, the chief's assistant-to does it, and undoes all the trust created by the meeting.)
Every six months have a secret ballot—"Do we need a weekly staff meeting?"

6.7 NEGOTIATION

One specialized aspect of communication that has great importance to some engineering managers is negotiation. Among the possible objects of negotiating are consulting contracts, collective bargaining agreements, and project completions. In fact, typical project managers negotiate almost everything they do, from start-up to the performance phase to project closeout.

Researchers have discovered that when people negotiate, they tend to act in a very predictable manner. Thus when developing a negotiating strategy, it is important to have a set of guidelines for effective negotiation. Leo P. Reilly, in an address to an assembly of professional engineers, suggested the following

("PEI Governors at Oak Brook," 1992; "The Basic Elements for Effective Negotiating," 1992):

1. Be patient.
2. Be positive.
3. Know your bottom line.
4. Know your opening offer.
5. Gather information.
6. Float trial balloons.
7. Know your status.
8. Limit your authority.
9. Offer a nibble (small concession that will result in closure).
10. Never reward intimidation tactics.

Tirella and Bates (1993a) suggest that negotiating is a "game" that is unique because given sufficient patience and persistence, both sides can win. These authors list several of the "nastier trick plays and traps" employed by opponents, including anger, mixing real issues and straw issues, the deadlock, and the flat-out lie, and offer advice on dealing with each. Their book, *Win–Win Negotiating: A Professional's Handbook* (Tirella and Bates, 1993b), provides practical guidelines for the engineering manager/negotiator.

6.8   SOME FINAL THOUGHTS ON IMPROVING THE COMMUNICATION PROCESS

We conclude this chapter with some engineering terminology applied to communication. This material is based on Connolly's suggestions for overcoming communication breakdowns; it seems to apply to some degree to any type of communication. The author begins by asking how we seem to be able normally to communicate effectively if it is so vulnerable a process. He concludes that we are able to utilize a number of error-reducing strategies, such as the following.

- *Improving the Signal-to-Noise Ratio.* Move to a quieter location; clean the blackboard.
- *Channel Switching.* Use a telephone call if a letter seems ineffective; use a drawing if a telephone call isn't working.
- *Serial Redundancy.* Repeat the message several times in slightly different terminology, to compensate for channel noise and coding noise.
- *Parallel Redundancy.* Use several channels at once, such as facial expressions, gestures, a visual aid, and the spoken word.
- *Feedback.* Nod and smile; pause for questions; request a written response. (Connolly, 1983)

6.9 DISCUSSION QUESTIONS AND CASE

1. Suppose the probability of a successful oral communication attempt is found to be 0.60 (use the simplified model in Figure 6.1). Suppose further that we want to develop a written communication scheme that has the same probability of conveying the same message successfully. If the probability that the message is written correctly (step 1) is 0.75 and the probability that the transmission medium works properly (step 2—delivery of the letter, e.g.) is 0.80, what must be the probability that the receiver reads and interprets the message correctly?

2. Consider a group of seven employees. If they all have formal communication links with one another (as would likely be the case in a self-directed team), how many such communication links are there? Now suppose you decide, instead, to develop a hierarchical structure with three levels. Make a sketch showing how these seven employees might be linked. How many formal communication links are there in this revised structure? Cite some advantages and disadvantages of this structure compared to the self-directed team, from the standpoint of communication.

3. Study the three elements of oral communication—verbal, vocal, and visual—and the results of the UCLA study in Section 6.3. Consider yourself in the role of a project manager who must make an oral presentation of the team's proposed schematic design to representatives of your client chemical manufacturing firm. What specific things will you do to assure that each of these three elements is successful during your presentation? Do you think the percentages from the UCLA study are accurate for this type of oral presentation?

4. Describe two types of communication situations for which one-way communication might be appropriate for the engineering manager, and two types for which two-way communication might be appropriate.

5. Rewrite the following section of a performance bond (West Ridge Natural Sciences Facility, 1992) so that it achieves the goals of clarity, brevity and simplicity:

   > Now therefore, if the Principal shall well, truly and faithfully perform its duties, all the undertakings, covenants, terms, conditions and agreements of said contract during the original term thereof, and any extensions thereof which may be granted by the Owner, with or without notice to the Surety, and if he shall satisfy all claims and demands incurred under such contract, and shall fully indemnify and save harmless the Owner from all costs and damages which it may suffer by reason of failure to do so, and shall reimburse and repay the Owner all outlay and expense which the Owner may incur in making good any default, then this obligation shall be void; otherwise to remain in full force and effect.

6. Rewrite the following section of a technical specification (West Ridge Natural Sciences Facility, 1992) to impart the characteristics of clarity, brevity, and simplicity:

   > "[Exterior Insulation and Finish Systems] Provide systems complying with the following performance requirements: ...

C. Weathertightness: Resistant to water penetration from exterior into system and assemblies behind it or through them into interior of building that results in deterioration of thermal-insulating effectiveness or other degradation of system and assemblies behind system including substrates, supporting wall construction and interior finish.

7. In light of your new understanding of effective written communication principles, explain at least two ways in which a report you wrote during the past year could have been made more useful to the reader with limited time.

8. At the end of Section 6.4, we mention the need for developing "integrated communication systems." For a relatively large construction project involving, say, 100 field workers, or another engineering enterprise of similar magnitude, describe the elements of what you consider to be an appropriate integrated communication system and explain how the elements are integrated.

9. For an organization with which you are familiar, describe the operation of the informal communication system, or grapevine. Why is such a system important to the engineering manager?

10. Write a brief description of a meeting you attended recently that was effective; tell why it was effective. Write a similar statement about a meeting you considered to have been ineffective.

11. Suppose you are presenting a briefing of the status of a school design project for the local school board. Your oral presentation is supplemented with charts, drawings, specifications, design calculations, and models. After the first several minutes of your presentation, it is clear that your points are not being understood; many board members do not have sufficient technical background to grasp some key concepts. Since you believe it is important to convey this information successfully, you ask for a brief recess to permit you to reconsider your presentation technique. Tell how you might employ each of Connolly's error reducing strategies in Section 6.8 to improve your chances of successful communication.

The Case of the Meeting Nobody Attended

Michelle Ericsson obtained a BS in manufacturing systems engineering from State University 15 years ago. After graduation, she joined the company, which soon recognized her special interest in and aptitude for automatic controlled production machinery. She spent time at several of the company's locations, during which she earned an MS in engineering management by part-time evening study. She moved to her present location 2 years ago and after a year and a half was promoted to facility manager.

At the urging of the board of directors, the facility management decided to install a series of new production machines that would be shared by several of its self-directed teams. Thus, planning for this major undertaking was begun.

Michelle called a meeting of the engineering staff and the leaders of the teams. After presenting her own ideas for the design concept and several details, she invited suggestions. A spirited discussion ensued, during which several alternative concepts were proposed. One of these seemed especially viable, and the group spent the next 45 minutes brainstorming for some tentative design details. At the end of the meeting, it was agreed that Michelle and her staff would "flesh out" a detailed design and have it available for the next meeting, at which time the discussion would focus on matters related to the installation of the machinery — sequence, schedule, personnel, impacts on other operations, and the like.

Two weeks later, the second meeting was held. It soon became apparent that the design details were based on Michelle's original plans, with essentially no mention of the proposal that had surfaced at the earlier meeting. Since the main purpose of the second meeting was to consider the installation phase, Michelle presented her suggested schedule, personnel assignments, coordination scheme with other activities, and other details. She then invited comments and suggestions for alternative approaches. Although these proposals elicited less response than the ideas at the earlier meeting, several valuable ideas were offered.

All facility personnel were invited to attend the third meeting, which was to be a discussion of the entire process — design, installation and operation. Michelle began by describing the process that had led to the details she was about to announce. She explained the design concept and some of its details and then told how the installation would be accomplished. Team leaders recognized that the installation work would be proceeding as Michelle had originally proposed. With respect to operation of the machinery, she said, in effect, "Here is how we shall operate the new machinery...." When a Team D member raised his hand, Michelle closed the meeting, saying that the gathering was too large to permit questions to be answered effectively and suggesting that questions be presented to her in writing.

During the weeks that followed, Michelle often communicated with team leaders and others by e-mail, but she was rarely seen on the production floor or in the engineers' offices, and she never participated in coffee breaks or other informal gatherings. The facility had always had a rather active informal communication system; it became even more so during this period. Rumors circulated about the new machinery, about management's in-house activities, and even about the executives' personal lives.

When the grapevine delivered a completely bogus message, Michelle decided to send everyone an official memorandum. It began:

> The rumors that have been circulated around this facility during the past month must cease. Only official written documents will be used in the management of this organization. The purposes of this memorandum are (1) to clarify procedures that I have developed for assuring the proper operation of the new production machinery and (2) to set forth the methods by which we shall handle employee grievances in the future....

Instead of folding up, the informal communication network became increasingly active. Absenteeism increased, and productivity declined. As matters continued to deteriorate, Michelle decided to call meetings of the teams, to help her understand their concerns. The first such meeting, of the Assembly B procurement–manufacturing–packaging team, was called for 9:00 on Monday morning. When Michelle walked into the conference room at 9:02 A.M., an embarrassed team leader was the only one present. He explained that the other team members had said they had more important things to do.

(a) What is the problem? What "concerns" will Michelle be confronted with if she is able to have some team meetings?

(b) Identify as many examples of poor communication as you can. For each, suggest a better way.

(c) Can you find some examples of effective communication?

(d) What should happen next?

6.10   REFERENCES

Babcock, D.L. 1991. *Managing Engineering and Technology*. Englewood Cliffs, NJ: Prentice-Hall.

"The Basic Elements for Effective Negotiating". 1992. *Industry Engineer,* National Society of Professional Engineers, Vol. 10, No. 1, October/November, 4.

Carnes, W.T. 1980. *Effective Meetings for Busy People: Let's Decide It and Go Home.* New York: McGraw-Hill.

Cleland, D.I., and D. Kocaoglu. 1981. *Engineering Management.* New York: McGraw-Hill.

Connolly, T. 1983. *Scientists, Engineers and Organizations.* Monterey, CA: Brooks/Cole Engineering Division of Wadsworth.

Conway, W.D. 1987. *Essentials of Technical Writing.* New York: Macmillan.

Cridge, E.S. 1993. "Guidelines for Using Visuals in a Presentation." Unpublished paper, University of Alaska, Fairbanks, June 25.

Davis, K. 1967. *Human Relations at Work,* 3rd ed. New York: McGraw-Hill.

Douglas, M.E., and D.N. Douglas. 1980. *Manage Your Time, Manage Your Work, Manage Yourself.* New York: AMACOM Books..

Goldbloom, J. 1992. "Improving Specifications." *Civil Engineering,* Vol. 62, No. 9, September, 68–70.

Golder, Herman, Company. "Executives' Digest." No date.

Half, R. 1987. "Memomania." *American Way,* November 1, pp. 21–24.

Harty, K.J., and J. Keenan. 1987. *Writing for Business and Industry.* New York: Macmillan.

Haynes, W.W., and J.L. Massie. 1969. *Management: Analysis, Concepts and Cases.* Englewood Cliffs, NJ: Prentice-Hall.

Hensey, M. 1987. "Communication Lessons from Structural Problems." *Journal of Management in Engineering,* American Society of Civil Engineers, Vol. 3, No. 1, January, 20–27.

Longfellow, L. A. 1970. "Body Talk: The Game of Feeling and Expression." *Psychology Today,* October.

Louw, A. A. 1992. "Break Your Barriers and Be a Better Presenter." *Training and Development,* Vol. 46, No. 2, February, 17–22.

Luthans, F. 1973. *Organizational Behavior.* New York: McGraw-Hill.

National Society of Professional Engineers. no date, "We Have Ways of Making You Talk!" Publication No. 2118.

"PEI Governors at Oak Brook Meeting Learn the Basic Elements for Effective Negotiating." 1992. *Industry Engineer,* National Society of Professional Engineers, Vol. 9, No. 6, August/September, 1–2.

Pigors, P., and C. A. Myers. 1981. *Personnel Administration,* 9th ed. New York: McGraw-Hill.

Rossini, F. A. 1983. "Working on Interdisciplinary Teams," Quoted in Connolly, T., *Scientists, Engineers and Organizations.* Monterey, CA: Brooks/Cole Engineering Division of Wadsworth.

Silverman, M. 1987. *The Art of Managing Technical Projects.* Englewood Cliffs, NJ: Prentice-Hall.

Stuckenbruck, L. C. 1981. *The Implementation of Project Management.* Reading, MA: Addison-Wesley.

Thomsett, R. 1980. *People and Project Management.* New York: Yourdon Press.

Timm, P. R. 1986. *Managerial Communications: A Finger on the Pulse,* 2nd ed. Englewood Cliffs, NJ: Prentice-Hall.

Tirella, O. C., and G. D. Bates. 1993a. "Trick Plays and Traps." *Civil Engineering,* Vol. 63, No. 6, June, 70–71.

Tirella, O. C., and G. D. Bates. 1993b. *Win–Win Negotiating: A Professional's Handbook.* New York: ASCE Press.

Townsend, R. 1970. *Up the Organization.* Greenwich, CT: Fawcett.

Wallace, I., et al. 1980. *The Book of Lists.* New York: Morrow Press.

West Ridge Natural Sciences Facility. 1992. University of Alaska Fairbanks, Project Manual, March 16.

White, E. B. 1959. *The Elements of Style by William Strunk, Jr., with Revisions, an Introduction and a New Chapter on Writing.* New York: Macmillian Company.

7

Management of Engineering Projects

More than many other disciplines, engineering work is often accomplished by short-term, task force–type organizations. They are formed to accomplish a specified objective—design a building, build a rocket, develop a new computer—and then they are disbanded.

Some of the ideas discussed in this chapter have already been considered in our treatments of such topics as organizational structure, motivation of personnel, and leadership. Now we emphasize how they are related to the activities of these temporary project organizations. We begin by defining projects and the distinctions between project management and functional management. We present the notion of the "triple constraint" of project management success and describe the typical life cycle of a project.

Several types of project management organization are presented and evaluated. Then we discuss the roles, responsibilities, and required skills of that key person, the project manager. Next, we present special methods for planning and controlling the project schedule, including both bar charting and network approaches. We end with a review of the basic elements of planning and controlling the cost aspects of projects.

7.1 THE NATURE OF PROJECTS AND PROJECT MANAGEMENT

Just what *is* a project? And how does project management differ from the management of activities carried out in more traditional functional organizations? Here we suggest answers to these questions, give a simple threefold framework for viewing project management, and describe the useful concept of project life cycle.

7.1.1 Some Definitions

Meredith and Mantel (1995) suggest that a project is "a specific, finite task to be accomplished ... [usually] a one-time activity with a well-defined set of desired ends." Other writers characterize projects in much the same way. For example, Babcock (1991) says:

A project represents a collection of tasks aimed toward a single set of objectives, culminating in a definable end point and having a finite life span. A project is a one-of-a-kind activity, aimed at producing some product or outcome that has never existed before. (There have been earlier aircraft or oil refineries or office buildings, but none of them were exactly like the one being created by *this* project.)

The significant characteristics of projects, then, can be listed as follows:

1. They are *temporary* undertakings, with definite start and end dates.
2. They consume the organization's *scarce resources* and thus compete with other parts of the organization for those resources.
3. They have *specific objectives*.
4. They produce *unique outcomes* by means of a combination of activities that also is unique.
5. They usually have *budgets of their own*.
6. As a rule, *one person is assigned overall responsibility* for project success.

Usually we distinguish projects from *programs* by saying that a program is a collection of similar or related projects, as in the Space Platform Program or a company's capital building program.

Given these characteristics of projects, what is *project management?* Instead of simply saying that it is "the management of projects," let us record Kerzner's (1989) definition: "the planning, organizing, directing and controlling of company resources for a relatively short-term objective that has established complete specific goals and objectives." Included in another valid definition is the goal of meeting established objectives "in time, dollars and technical results" (Spinner, 1981). Silverman (1988) says that project management "is typified by the use of specialized control techniques," while Stuckenbruck (1981) proposes that project management be defined by the appointment of a single person with the responsibility for the success or failure of one of these temporary undertakings.

Already we begin to see distinctions between project management and the more traditional and well-known functional management. Viewed from the standpoint of those who do the managing:

> *Functional managers* are responsible for operating the company and optimizing the use of resources on an overall basis. They are the managers whose jobs go on "forever." Typically, they are the chief engineers, the controllers, the sales managers and the production managers, who supervise the daily and continuing operations of any company.... *Project managers*, on the other hand, have a specific goal, a specific time, and a specific budget. When the project is finished, their job is over. They are also concerned with optimizing resources but only as applied to their projects. Sometimes there are clashes between these two managerial species because their definitions of "optimization" may differ. (Silverman, 1987)

Indeed, as we shall see, an inherent tendency of project organizations is the *conflict* that often characterizes their relationships. An old but classic paper on project management helps in understanding the distinctions between functional and project management, as follows:

> The essence of project management is that it cuts across, and in a sense conflicts with, the normal organizational structure. Throughout the project, personnel at various levels in many functions of the business contribute to it. Because a project usually requires decisions and actions from a number of functional areas at once, the main flow of information and the main interdependencies in the project are not vertical but lateral. Up-and-down information flow is relatively light in a well-run project; indeed, any attempt to consistently send needed information from one functional area up to a common authority and down to another area through conventional channels is apt to cripple the project and wreck the time schedule. (Stewart, 1965)

When is project management needed? Stewart's classic paper also helps to answer this "when is a project a project?" question. Briefly, this author suggests four criteria for judging whether an organization should implement project management.

1. *Scope:* "definable in terms of a single, specified end result and ... where substantially more people, more dollars, more organizational units and more time will be involved than on any other infrequent undertaking in the organization's experience...."
2. *Unfamiliarity:* "a unique, or infrequent, effort by the existing management group ... [and] ... a greater than usual sense of uncertainty about the realism of initial cost estimates, time commitments, or both."
3. *Complexity:* "the degree of interdependence among tasks. If a given task depends on the completion of other assignments in other functional areas, and if it will, in turn, affect the cost of timing of subsequent tasks...."
4. *Stake:* "The company's stake in the outcome of the undertaking. Would failure to complete the job on schedule or within budget entail serious penalties for the company?" (Stewart, 1965)

7.1.2 The "Essence" of Successful Project Management

Sometimes a basic classification helps one understand a complicated process. So it is with project management. While it may be an oversimplification, it is helpful to think of the prime objectives of project management as technical achievement, elapsed time, and cost. Rosenau (1984) calls these three the "triple constraint" and suggests that the essence of successful project management consists of satisfying the triple constraint, or meeting these three goals simultaneously.

Figure 7.1 shows both Rosenau's three-dimensional graph, where the three objectives combine to form a cumulative target for a particular project, and Babcock's three-legged stool (Babcock, 1991). Meredith and Mantel (1989) comment that Rosenau's graph "implies that there is some 'function' (not shown on the graph) that relates them, one to another." Such functions are different for

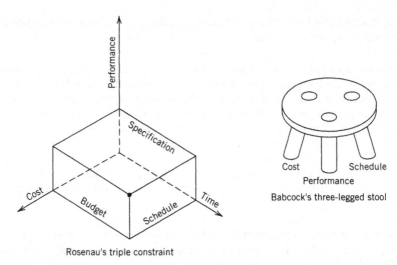

Figure 7.1. Two representations of the three aspects of successful project management: Rosenau's triple constraint and Babcock's three-legged stool. (From M. D. Rosenau, Jr., *Successful Project Management,* © 1981. Publisher Van Nostrand Reinhold, New York. All rights reserved. Daniel L. Babcock *Managing Engineering and Technology,* © 1991, p. 280. Reprinted by permission of Prentice-Hall, Englewood Cliffs, New Jersey.)

different projects, and they are different at different stages of any one project. The primary challenge to the project manager is to manage these relationships, or *tradeoffs.* For example, within the conceptual three-dimensional solid in Figure 7.1, the project manager may need to increase the project cost in the interest of saving time. Or, an improvement in some aspect of technical performance may justify a time extension and/or a budget increase.

Silverman (1987) refers to our three primary objectives as "golden limits," because they define the project's boundaries: "Any solutions that don't affect these limits could be acceptable, but exceeding those boundaries requires corrective action at a higher (i.e., 'golden') organizational level." Finally, we have modified Kerzner's time–cost–performance triangle (Kerzner, 1989) to show that the three primary objectives are set within the constraints of environmental, legal, and safety factors and good customer relations. Figure 7.2, based on the cover of Kerzner's book, illustrates this idea.

7.1.3   The Project Life Cycle

One of the important and interesting characteristics of projects is their tendency to go through similar stages between initiation and termination. We call the total of these several phases the project's "life cycle." Although the number and titles of the phases will vary for projects of different types, the following general listing describes many projects and the kinds of activity that take place during each phase:

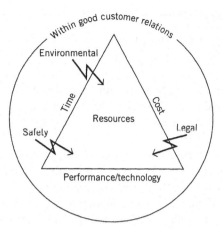

Figure 7.2. The project objectives and constraints. (Based on Harold Kerzner, *Project Management,* 3rd edition, p. 5. Copyright © 1989. Publisher Van Nostrand Reinhold, New York. All rights reserved.)

Conceptual Phase

Establish goals.
Identify in general how goals will be accomplished.
Estimate how long it will take and how much it will cost ($\pm 20\%$ for each).
Select project organization.

Definition, Detailed Engineering, and Design Phase

Staff project organization.
Determine specific processes required to accomplish goals.
Update time and cost estimates ($\pm 10\%$).

Production or Construction Phase

Build facility or produce prototype.
Continue design if appropriate.
Update time and cost estimates ($\pm 5\%$).
Develop documentation and training.

Operation, Turn-on, and/or Start-up

Make product or turn over to customer.
Conduct training.
Integrate project into existing system.
Carry out project review and analyze feedback.

Divestment and Tear-Down

Transfer technology and documentation.

Terminate projects organization.

Other designations of project phases might be formation, buildup, production, phaseout, and final audit (for a manufacturing project) and conceptual, planning, definition and design, implementation, and conversion (for a computer programming project) (Kerzner, 1989).

Figure 7.3a shows a typical distribution of project effort as a function of time, with the most intensive effort occurring during the production/construction and operation phases. "Level of effort" may be measured in time or funds expended or physical progress. Figure 7.3b is a cumulative completion curve, which can be thought of as the integral of the curve in part (a). Often such a curve shows percent complete or total funds expended to date as a function of time. The "S curve" shape of the graph in part (b) is typical of project progress—that is, a relatively small rate of progress early in project, a more rapid rate in the middle phases, and a slowing down toward the end.

7.2   PROJECT ORGANIZATIONS

In Chapter 2, we presented an introduction to the various organizational structures available to the engineering manager. Whereas that discussion was a general overview, we now focus on how *projects* in particular might be conducted within different structures.

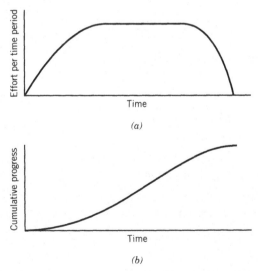

Figure 7.3.   Project completion as a function of time: *(a)* level of effort versus time and *(b)* progress to date versus time.

The initial fundamental decision has to do with whether we establish any kind of project management organization. That is, we can manage projects from within the functional organization, or we can establish a special structure for managing projects. After a brief discussion of this choice, we present the several organizational types, along with their advantages and disadvantages. Since the matrix organization in some form is often used, we focus on its effectiveness and the inevitable role conflicts in such organizations. To complete this section, and as an introduction to the section on the roles and skills required of the project manager, we cite some ideas on the proper place of this key person in the organization.

7.2.1 An Important Decision: Project Organization or Not

If a special project organization is not established, there are a variety of possible ways by which projects can be managed from within the *functional organization*. In general, project management responsibility falls on a functional manager who, by definition, has other management tasks as well. The focus of the project's organization, communication, evaluation, and so forth tends to be on the functional grouping rather than on the project.

In contrast, the *project organization* approach focuses on the project. An important key is the appointment of a project manager who can bring this point of view to the utilization of the company's resources to achieve the goal of completing the project on time, within budget, and in accordance with the technical specifications. The project manager coordinates resources that are "borrowed" temporarily from the various functional lines.

Advantages of the project management approach include the following:

* More responsive to the client
* Early identification and correction of problems
* Timely decisions about tradeoffs between conflicting projects goals
* Focus on optimizing project goals rather than suboptimizing individual tasks
* Better control
* Shorter development times
* Higher quality and reliability
* Greater flexibility
* Good horizontal communications
* Challenge and satisfaction for project personnel

On the other hand, there are disadvantages, as well:

* Greater organizational complexity
* Potential for violation of organizational policy
* Weakening of functional organizations

- Disruption of main line product support
- Increased likelihood of intracompany conflicts
- Higher overhead costs due to additional management personnel
- Greater stress for project personnel at project completion (lists taken in part from Meredith and Mantel, 1995)

When we discuss various organizational structures in more detail, we will point out the advantages and disadvantages of each. For now, we can take both warning and comfort from these words of Meredith and Mantel (1995):

> Project management is difficult even when everything goes well. When things go badly, project managers have been known to turn gray and take a hard drink. The trouble is that project organization is the only feasible way to accomplish certain goals. It is literally not possible to design and build a major weapon system, for example, in a *timely and economically acceptable manner,* except by project organization. Tough as it may be, it is all we have—and it works.

7.2.2   The Basic Three: Functional, Projectized, and Matrix

7.2.2.1   *The Functional Organization.* Projects that are relatively small, of short duration, not complex, and noncritical timewise are often conducted successfully from within the functional organization. When the task involves a single department or a few closely allied departments, a coordinator from an affected department can be designated to lead the project. Or, a task force located outside any one department may be appointed. The difficulty with most such arrangements is their rather unofficial status, their lack of focus on the project, and the difficulty of communication across functional lines.

Several variants of project management within functional organizations have been used. The functional organizational chart in Figure 7.4 highlights the engineering function. We show four possible options for coordinating projects within such a structure, based on some work by Cable and Adams (1982). One possibility relies on the *project leader,* a designated staff position within the functional organization. In our example, this person is a member of the structural engineering department; he or she would have no chain-of-command authority under such an arrangement.

Under the *task force* approach, each functional group involved in the project provides members for such a coordinating body. Members of the task force in our example are drawn from power and HVAC engineering, with the task force reporting to the vice presidents for manufacturing, engineering, and finance and administration. Here, the communication channels bypass the chief engineers.

Creation of a temporary *liaison office* is an attempt to facilitate horizontal communications during the course of project. This new functional group is positioned on the same level as the existing functional groups. While it often makes interdepartmental communication more efficient, the problem of lack of authority remains.

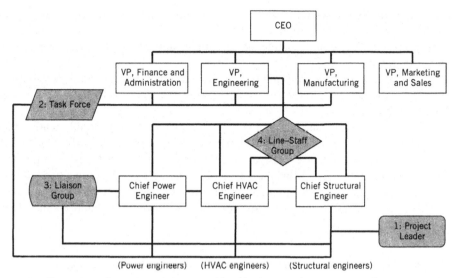

Figure 7.4. Project management within the functional organization: four options.

Finally, we show a *line–staff group* above the level of the functional departments involved in the project effort. In our chart (Figure 7.4), this group coordinates the project activities of the structural and HVAC departments and reports to the engineering vice president. Such an arrangement provides a desired project focus, but the authority of a line position is still lacking.

7.2.2.2 The Projectized Organization.
The pure project organization (called "projectized" to distinguish from the general term "project organization") is at the other end of the organizational spectrum from the functional organization. In this type of arrangement, the entire focus is on the project. It is a self-contained unit with its own technical staff and its own administration, "tied to the parent firm by the tenuous strands of periodic progress reports and oversight" (Meredith and Mantel, 1995). For large, complex, costly projects requiring advanced technology and involving many different organizations, this form is very appealing. It gives the project manager full authority over the project and has a project team fully dedicated to the project's goals and activities. Communication is clear and rapid, and decisions can be reached quickly.

On the other hand, this complete focus on project goals may cause specialists to lose touch with their technical disciplines. Alternatively, it may result in the project's striking out on its own, far from the company's policies and goals. Another potential downside to total focus on project goals is a tendency to duplicate resources and support services unnecessarily: Should the project have its own data processing center? mail room? travel agency? vehicle pool? (In the purest form of projectization, these would be required.) Finally, there often is severe anxiety at project termination, since the personnel have no ties to a functional "home" to which they can return.

Figure 7.5 shows an example of an organizational chart for a projectized organization.

Babcock has conveniently summarized the major advantages and disadvantages of the functional and projectized organizational structures, which we reproduce here as Table 7.1.

Although we do not show a sample organizational chart, we should mention the *pure product* organization as an option appropriate in some settings. Similar to the projected form, it focuses exclusively on the product itself. It tends to enjoy the same benefits and suffer from the same disadvantages as the projectized form. Automobile manufacturing and consumer product organizations (e.g., Pontiac Division of General Motors; Crest Division of Procter & Gamble) are often structured in this way.

7.2.2.3 *The Matrix Organization.* In an attempt to take advantage of the best features of both the functional and projectized organization, an organization called the *matrix* was developed. We can say that it falls somewhere in the middle of the spectrum between these other two extremes. In graphic form, the organization chart resembles a two-dimensional mathematical matrix, with one dimension representing the functional part of the organization and the other the project part.

Probably most organizations that do project work use some form of matrix organization. Figure 7.6 charts a simple matrix organization at a research institute in Finland. Each project manager (left-hand side of the chart) reports to upper management and has an authority that cuts across the various functional areas. The intersections of the project lines and the functional lines represent one or more individuals from the functional organization who are assigned to the project. The potential for conflict when these individuals are expected to report to two bosses is apparent immediately. Figure 7.6 includes the names of those assigned to the various positions to indicate that one person may have more than one role in such an organization.

To clarify the reporting relationships, Cleland and King (1983) propose that the project manager be made responsible for decisions about what, when, and why the task is to be accomplished; this person also decides the allocation of resources and evaluates overall project success. The functional manager is responsible for the who, where, and how questions, and for evaluating how well the functional input has been incorporated into the project.

The matrix type of organization can take on many flavors. Some writers (Babcock 1991; Larson and Gobeli, 1987; Meredith and Mantel, 1995) have classified them as the functional matrix, the balanced matrix, and the project matrix. Others (Kerzner, 1989; Stuckenbruck, 1981) identify various shades of weak and strong matrices. In both these classifications, the distinctions are based on the extent to which the project, versus the functional organization, is emphasized, and the relative degree of authority and responsibility vested in the project and functional managers.

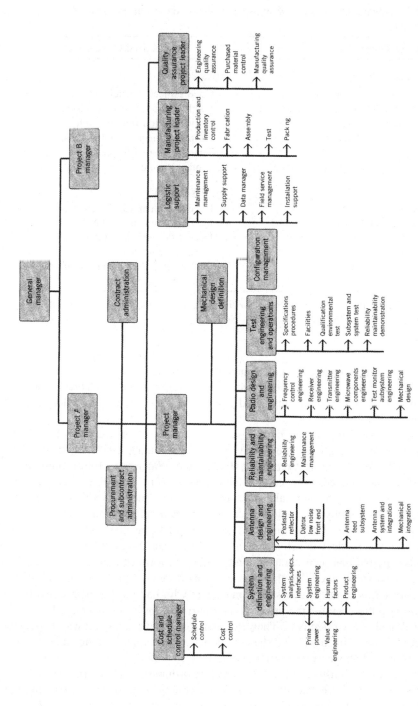

Figure 7.5. Example of a projectized organizational structure (Adapted from Daniel L. Babcock, *Managing Engineering and Technology*, Copyright © 1991, page 304. Reprinted by permission of Prentice-Hall, Englewood Cliffs, New Jersey. Adapted from Russell D. Archibald, *Managing High-Technology Programs and Projects*, © John Wiley & Sons, Inc. New York, 1976, pp. 104–105. Reprinted by permission of John Wiley & Sons, Inc.)

179

TABLE 7.1   Advantages and Disadvantages of Projectized and Functional Organizations

Functional Organization	Projectized Organization
ADVANTAGES	ADVANTAGES
Efficient use of technical personnel	Good project schedule and cost control
Career continuity and growth for technical personnel	Single point for customer contact
Good technology transfer between projects	Rapid reaction time possible
Good stability, security, and morale	Simpler project communication
	Training ground for general management
DISADVANTAGES	DISADVANTAGES
Weak customer interface	Uncertain technical direction
Weak project authority	Inefficient use of specialists
Poor horizontal communication	Insecurity regarding future job assignments
Discipline (technology) rather than program oriented	Poor crossfeed of technical information between projects
Slower work flow	

SOURCE: Daniel L. Babcock, *Managing Engineering and Technology,* © 1991, p. 304. Reprinted by permission of Prentice-Hall, Englewood Cliffs, New Jersey.

Table 7.2 lists advantages and disadvantages of the matrix organization in general, while Table 7.3 gives comparative advantages and disadvantages of the three types of matrix structure just cited. Both tables are from an excellent paper by Larson and Gobeli (1987).

The matrix organization is both loved and hated. Peter Drucker described this organizational design as "fiendishly difficult" (Cleland, 1982). In their very popular book, *In Search of Excellence,* Peters and Waterman (1982) claim that the trend toward increasingly complicated management structures "reaches its ultimate expression in the formal matrix organization structure [which] regularly degenerates into anarchy and rapidly becomes bureaucratic and non-creative."

Larson and Gobeli (1987) conducted a survey of more than 500 experienced managers in the product development area. More than three-quarters of the respondents reported that their company had used some form of matrix organization, and 89% of those believed that this type of structure would probably or definitely continue. In terms of effectiveness, the managers who had used a matrix structure had a strong preference for the project-type matrix, rating it, on average, between "effective" and "very effective." They rated the balanced matrix midway between "ineffective" and "effective," and the functional matrix slightly better than "ineffective" (but well above "very ineffective").

We have spent considerable time on the two-dimensional matrix. However, some organizations have had good experience adding a third, or even a fourth, dimension. A third dimension might be geographical location— that is, a company conducts projects in different places, drawing from the functional resources at each

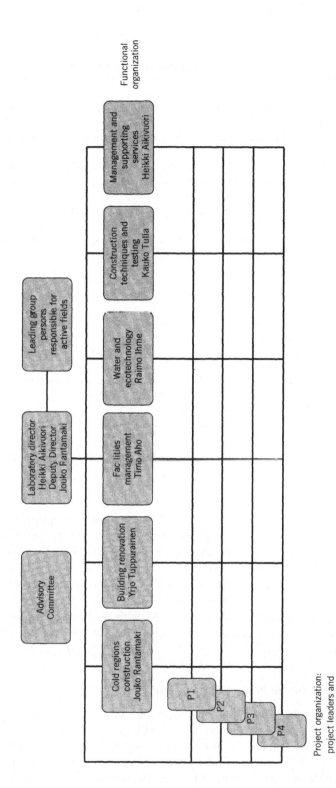

Figure 7.6. Example of a simple matrix organization: Building Laboratory of the Technical Research Center of Finland, Oulu, Finland.

181

TABLE 7.2   Advantages and Disadvantages of a Matrix Organization

ADVANTAGES

+ Efficient use of resources—Individual specialist as well as equipment can be shared across projects.
+ Project integration—There is a clear and workable mechanism for coordinating work across functional lines.
+ Improved information flow—Communication is enhanced both laterally and vertically.
+ Flexibility—Frequent contact between members from different departments expedites decision making and adaptive responses.
+ Discipline retention—Functional experts and specialists are kept together even though projects come and go.
+ Improved motivation and commitment—Involvement of members in decision making enhances commitment and motivation.

DISADVANTAGES

− Power struggles—Conflict occurs, since boundaries of authority and responsibility deliberately overlap.
− Heightened conflict—Competition over scarce resources occurs, especially when personnel are being shared across projects.
− Slow reaction time—Heavy emphasis on consultation and shared decision making retards timely decision making.
− Difficulty in monitoring and controlling—Multidiscipline involvement heightens information demands and makes it difficult to evaluate responsibility.
− Excessive overhead—Double management by creating project managers.
− Experienced stress—Dual reporting relations contributes to ambiguity and role conflict.

SOURCE: Larson and Gobeli (1987) in J. R. Meredith and S. J. Mantel, Jr., *Project Management: A Managerial Approach.* Copyright © 1989. Reprinted by permission of John Wiley & Sons, Inc.

location. A fourth dimension might be project types. As noted in Section 2.2.5, the international engineering and planning firm of CH2M-Hill reports considerable success with a matrix structure that involves the dimensions of discipline, region, and project (Poirot, 1991). Since it is difficult to show three dimensions on a single chart, we include Figure 7.7, which diagrams the overall company organization with its districts/regions and its disciplines, and Figure 7.8 which depicts the interactions of disciplines to accomplish projects. As with most organizations, the structure of CH2M-Hill's matrix changes over time; this structure was current in 1992 (Poirot, 1993).

7.2.3   Role Conflict in the Project Organization

Silverman (1987) describes *role conflict* as occurring "when one person receives differing directions about how to behave in a specific job or position simultaneously from two or more persons." Only in the purest functional or projectized organization is there no chance of role conflict among members of the project team! We need first to recognize that most of the project organizations we create

TABLE 7.3 Comparative Advantages and Disadvantages of Three Types of Matrix Structure

	Matrix Type		
	Functional	Balanced	Project
ADVANTAGES			
+ Resource efficiency	High	High	High
+ Project integration	Weak	Moderate	Strong
+ Discipline retention	High	Moderate	Low
+ Flexibility	Moderate	High	Moderate
+ Improved information flow	Moderate	High	Moderate
+ Improved motivation and commitment	Uncertain	Uncertain	Uncertain
DISADVANTAGES			
− Power struggles	Moderate	High	Moderate
− Heightened conflict	Low	Moderate	Moderate
− Reaction time	Moderate	Slow	Fast
− Difficulty in monitoring and controlling	Moderate	High	Low
− Excessive overhead	Moderate	High	High
− Experienced stress	Moderate	High	Moderate

SOURCE: Larson and Gobeli (1987) in J. R. Meredith and S. J. Mantel Jr., *Project Management: A Managerial Approach.* Copyright © 1989. Reprinted by Permission of John Wiley & Sons, Inc.

have the potential for such conflict and second to understand some ways to avoid or minimize it. We have already noted Kerzner's guidance for the kinds of direction and advice both functional and project management should provide. But it may be necessary to formalize such mechanisms.

Silverman (1987) advocates the preparation of a *project manual* for all but the simplest projects. He suggests that such a manual include a section on the "structural interfaces" between various elements of the project, outlining clearly the various reporting relationships. Without such definition, he believes that upper management will tend to resolve role conflicts in favor of functional management.

Even with such a section in our project manual, however, role conflicts will arise. Silverman outlines a hierarchy of steps the individual might take, beginning with "*I* am the 'expert' in the work that I do, and I am working on a current priority list. Therefore, when I receive conflicting inputs, *I* will start the interpretation process with what *I* think is the most important thing to do." The list contains 10 other items, including negotiating to win key issues and yield on others, and ends with a major decision point:

If all else fails:

(a) You can consider escalation (going up another level to the boss-in-common).

(b) You can threaten escalation.

(c) You can escalate.

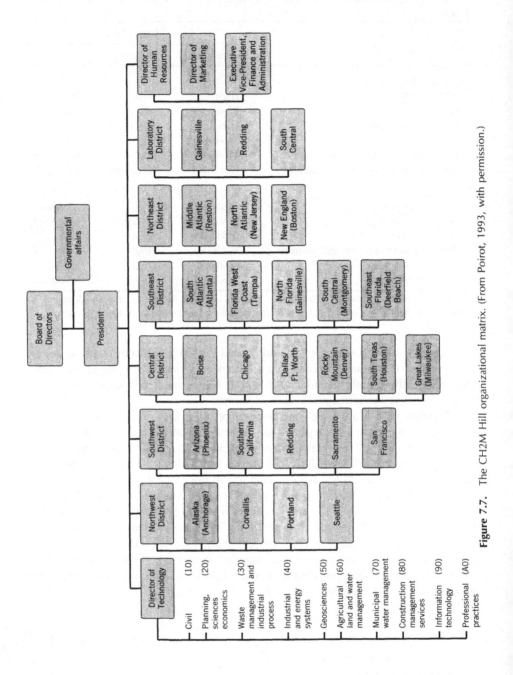

Figure 7.7. The CH2M Hill organizational matrix. (From Poirot, 1993, with permission.)

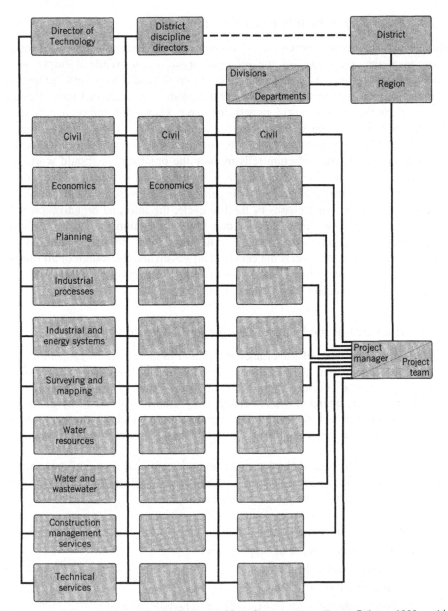

Figure 7.8. The CH2M Hill project team/discipline matrix. (From Poirot, 1993, with permission.)

Before traveling this road, however, consider your timing. How much testing and negotiating should be done before calling for senior support? Does the top leadership want to be involved? When will they support and encourage your approach? Does escalation represent failure?

In short, role conflict is a complex but very real condition in most project organizations. Forewarned is forearmed.

7.2.4 Don't Forget the Informal Structure!

Before we leave our discussion of organizational structures, let us mention once again the importance of the *informal organization*. While it might seem that temporary groupings established to do projects would have less "grapevine communication" than more permanent organizations, these informal structures are very real in the project world nonetheless. We know that informal structures develop within the formal organization on the basis of friendships, common interests and backgrounds, immediate needs, and the like. It has been suggested that whereas the formal structure tells us how the organization should work, the informal structure, if it were written down, would describe how it really works.

In a project setting, wherein the organization assumes different sizes and characteristics at different phases of the life cycle, the informal organization will, of course, change over time. But we may find, for example, that the assistant project manager's second assistant clerk, who was one of the first to be hired and has lived in the town forever, is the center of all gossip and the source of all rumors that turn out to be true. When considering the effective management of our project, we cannot deny or neglect such communication channels.

7.2.5 The Place of the Project Manager in the Organization

There is one more topic of importance in relation to the project organization, and it leads nicely into our next section on the job and background of the project manager. When we consider how the particular position of project manager fits into the project organization, we are concerned primarily with authority granted to the project manager and the level in the organization to which the project manager reports. Indeed, the reporting level is often a good indication of the degree of authority.

Goodman (1976) compiled a list of 14 authority areas as a way of determining this degree of authority. The question is, "Does the project manager have the *final* authority to make crucial decisions in these areas?"

1. Initiate work in support areas.
2. Assign priority of work in support areas.
3. Relax performance requirements.
4. Authorize total overtime budget.
5. Authorize subcontractor to exceed cost, schedule, or scope.
6. Contract change in schedule, cost, or scope.
7. Make or buy.
8. Hire additional people.
9. Exceed personnel ceilings when a crash effort is needed.
10. Cancel subcontracts and bring work in-house.
11. Select subcontractors.
12. Authorize exceeding company funds allocated to the project.

13. Determine content of original proposal.
14. Decide initial price of proposal.

All this relates to our earlier discussion of the types of matrix organization—functional matrix, balanced matrix, and project matrix, or weak versus strong matrix. The greater the authority granted to the project manager, the more the organization is of the project matrix or strong matrix type. As advocates for project success, we tend to favor the strong matrix with a high degree of project manager authority. And the project managers in the study by Larson and Gobeli (1987) felt that such an arrangement is important to project success. But only upper management can make this decision, although individual project managers can help their own case by insisting on a project manual that defines clearly the authority the person will have.

To whom should the project manager report? Silverman (1987) makes a strong case for reporting to a level in the organization at least one step above that from which his or her personnel are drawn. He says that any other choice results in building in a trap that will cause delays and possibly failure, since conflicts involving assignment of personnel and resources will tend to be resolved against the project manager. Figure 7.9 suggests that the proper reporting level is the common supervisor of all functions involved in the project (Stuckenbruck, 1981).

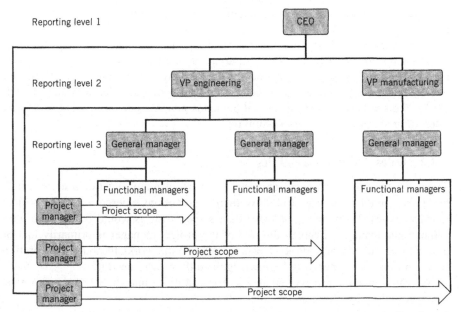

Figure 7.9. Project management reporting levels. (From Lynn C. Stuckenbruck, *The Implementation of Project Management.* Copyright © 1981 by Addison-Wesley Publishing Company, Inc. Reprinted by permission.)

7.3 THE PROJECT MANAGER

7.3.1 A Different Kind of Position

Silverman (1987) has given a four-point description of his first job as a project manager, which had followed a period of service as a functional manager:

1. I had control over neither the technical objectives nor the selection of the project team members. The specifications had already been agreed upon, and top management had assigned the team members to the project. Those team members still reported to their original bosses.
2. I had no control over the length of time for the project. The contract defining that had been signed just before I arrived.
3. I had no control over the costs. The accounting department allocated funds and measured expenditures.
4. But I was responsible for it all! When project goals were not achieved or budgets were exceeded, I was supposed to explain and correct the problems! Those were interesting days.

The project manager must plan, motivate, control, and do all those other things managers are supposed to do, in a temporary environment, where communication channels are complex, authority lines often are confusing, and "success" is measured in terms of all three components of the triple constraint. Thus the job is different and difficult, and the project manager is truly the person in the middle. Fortunately, Silverman reports improvements over the years in his lot as a project manager!

Among the special demands placed on the project manager are acquisition of adequate resources, acquisition and motivation of personnel, dealing with crises and other obstacles that are very much a part of any unique, one-time effort, and managing tradeoffs among cost, schedule, and performance (Meredith and Mantel, 1995). We shall see how these special demands give the person in this position an interesting set of roles and responsibilities.

7.3.2 Roles and Responsibilities

Typically, the project manager oversees the work of many functional areas. Since he or she cannot be an expert in all of them, two notions emerge. First, the project manager is usually more a *generalist* than a specialist, as would be the case with a functional manager. Second, the role of the project manager is primarily that of a *facilitator* rather than a direct supervisor. He or she works to obtain cooperation between those with specialized knowledge and those who need it.

We can divide the project manager's responsibilities into three broad categories, as suggested by Meredith and Mantel (1995):

1. *Responsibility to the Parent Firm:* including wise use of resources, competent management of the project, and timely and accurate communications.

The importance of regular reporting "up the line" cannot be overemphasized; there must be no surprises to the project manager's supervisor.

2. *Responsibility to the Project:* despite the conflicting demands of the several parties involved, the project manager (PM) must "sort out understanding from misunderstanding, soothe ruffled feathers, balanced petty rivalries and cater to the demands of the client. One must, of course, remember that none of these strenuous activities relieves the PM of the responsibility of keeping the project on time, within budget and up to specifications" (Meredith and Mantel, 1995). Note well that this responsibility includes the customer, whether the project is for an external client or for a part of the organization. It cannot be overemphasized that defining and satisfying the customer is the essence of total quality management.

3. *Responsibility to Members of the Project Team:* by providing definition of the individual's role and expectations in the project, clear direction, a supportive attitude, interest in personal development and well-being, and active concern for where the team members will go upon project completion. There seems to be no evidence that highly creative project team members such as engineers and scientists are different from others in their preference for managers who exhibit interest in them as individuals. The notion that the research engineer should be "neglected and left alone to create and be productive" is not validated in practice.

Having said all that, it is still important that the project manager's responsibilities be clearly delineated before the PM is selected. Every time a project manager is assigned to a project, there must be a complete understanding of the amount of responsibility the person will have in such areas as profit, decision making, customer interfacing, interfacing between internal groups, conflict resolution, "bending" organizational policy, and defining the meaning of "on time, on budget, and within specification" for the particular project.

7.3.3   What Kind of Person Can Fulfill That Role?

Now comes the task of trying to identify the attributes we need to look for in this "superhero." Meredith and Mantel (1995) provide a list of the qualities that have been sought in project managers—strong technical background; hard-nosed manager; mature individual; currently available; on good terms with senior executives; can keep the project team happy; experienced in several different departments; can walk on (or part) the waters. Then they say:

> These reasons for choosing the PM are not so much wrong as they are "not right."...Above all, the best PM is the one who can *get the job done!* (emphasis added)...Of all the characteristics desirable in a PM, this *drive to complete the task* is the most important.

The project manager needs both technical and administrative *credibility*. The first requirement is rather tricky, for we have already suggested that a generalist, who cannot be expected to be expert in all the functional areas involved, is necessary for this important slot. Even so, a level of technical competence that is respected by the customer, the team, and upper management is essential. Administrative credibility includes many talents we have already mentioned— planning and control of schedule and budget; personnel, equipment, and material planning; communication among all the project elements; effective management of tradeoffs.

The project manager needs political, interpersonal, and technical *sensitivity*. Such sensitivity will aid in recognizing, in their early stages, such organizational phenomena as conflict between individuals and between organizations, technical failures that have been "covered up," and the shifting sands of upper management and client priorities.

The project manager needs *leadership,* as defined and explained in Chapter 5. In addition, "the project manager must know how to get others to share commitment to the project" (Meredith and Mantel, 1995)

Kerzner (1989), in a modestly successful attempt at comic relief, gives the following guidance for selecting project managers:

> The ideal project manager would probably have doctorates in engineering, business and psychology, and experience with ten different companies in a variety of project office positions, and would be about twenty-five years old.

Finally, Table 7.4 lists project management skills developed from a survey conducted by Barry Z. Posner (1987). The respondents were nearly 300 project managers from technology-oriented organizations. Among the questions was an open-ended one: "What personal characteristics, traits or skills make for 'above average' project managers?" Responses were assigned to six different clusters; Table 7.4 shows both the skills represented in each cluster and the percentage of project managers whose response was included in the cluster. While nearly half the respondents cited technological skills, the greatest percentage (84%) mentioned "being a good communicator" as an essential project manager skill. Enough said!

7.4   PROJECT PLANNING, SCHEDULING, AND CONTROL

We now consider the management of the time aspects of projects—sequencing of project activities, establishing their start and end dates, and monitoring project schedule progress. Because project time and cost dimensions are so closely related, we start with a simple but powerful method for breaking the project into subelements, which can form the basis for planning and control of both the schedule and the budget. We then discuss a traditional method for scheduling based on bar charts, followed by a more extensive presentation of network scheduling methods. The section ends with a review of the use of computer

TABLE 7.4   Project Management Skills

1. Communication skills (84)	4. Leadership skills (68)
Listening	Sets example
Persuading	Energetic
2. Organizational skills (75)	Vision (big picture)
Planning	Delegates
Goal setting	Positive
Analyzing	5. Coping skills (59)
3. Team-building skills (72)	Flexibility
Empathy	Creativity
Motivation	Patience
Esprit de corps	Persistence
	6. Technological skills (46)
	Experience
	Project knowledge

NOTE: Numbers in parentheses represent the percentage of project managers whose response was included in this cluster.

SOURCE: J. R. Meredith and S. J. Mantel, Jr., *Project Management: A Managerial Approach,* © 1989. Reprinted by permission of John Wiley & Sons, Inc.

software for assisting the engineering manager in planning and controlling project schedules.

7.4.1   Work Breakdown Structure

The work breakdown structure, or WBS, is a "picture" of the project subdivided into hierarchical units of work. It can be thought of as "an organizational chart with tasks substituted for people" (Meredith and Mantel, 1995). Figure 7.10 shows a small portion of such a structure for a hypothetical house-building project. Note that the top level is the project itself; each successive level shows further details about the various project elements.

The value of such an approach is manifold. At the lowest level in the hierarchy, we can (1) plan and schedule the work and provide data to the overall scheduling effort, (2) estimate costs, (3) authorize work and assign responsibilities, and (4) track schedule, cost, and quality performance (the triple constraint, again). Furthermore, we can summarize any of the above at whatever level we wish. Thus, for a house-building project we might need information on the total budget for the ground floor, or the total number of carpenter-weeks expected to be needed for all elements of ground floor framing. Note well that this structure provides a framework for both planning and monitoring the project.

The exact organization of the WBS, as well as the number of levels and thus the amount of detail, will depend on the type of organization and the type of project. Kerzner (1989) suggests the following general design for the work breakdown structure hierarchy:

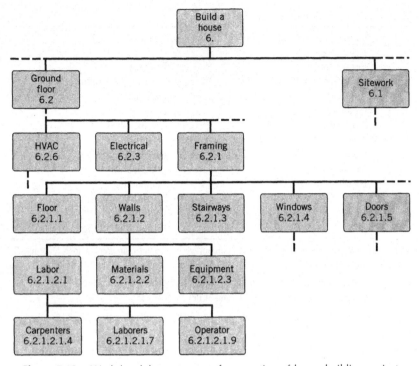

Figure 7.10. Work breakdown structure for a portion of house-building project.

LEVEL	DESCRIPTION
1	Total program
2	Project
3	Task
4	Subtask
5	Work package
6	Level of effort

SKANSKA, Sweden's largest construction firm, uses a work breakdown structure that includes cost account codes with as many as 17 digits. (Hogård, 1992; SKANSKA, n.d.) In general terms, the structure is as follows:

$$DDDD.PPP.CCC.RRR.XXXX$$

where

$$DDDD = department$$

$$PPP = project$$

$$CCC = cost\ center$$

$$RRR = resource$$

$$XXXX = worksite\ details$$

If we apply to a typical project, the "project" and "details" fields, totaling seven digits, we might have the following:

- P = Harbor Development Project
- PA = All Buildings in Project
- PA1 = Building One of the Project
- PA1.3 = Third Floor of Building One
- PA1.34 = Piping System for Third Floor of Building One
- PA1.342 = All Floor Drains for This Piping System
- PA1.3421 = Floor Drain One for This Piping System

For a complete design–build project, the coding system is first applied at the initial design stage. As the project details are developed, codes are subdivided into the sort of structure just exemplified. All design, procurement, and construction documents utilize this common coding system.

While an entire project might not be broken down this level of detail, such a structure allows the firm to plan and control the schedule, cost, amount of labor, responsible party, document requirements, purchasing procedure, and/or other relevant aspects of each project element.

7.4.2 Bar Charting Methods

Bar charts are a simple yet effective method for displaying the elements of a project schedule. Sometimes called Gantt charts, after Henry I. Gantt, an early nineteenth century management theorist, they plot planned progress on a horizontal time scale for each of a series of project tasks. They can also be used to show actual progress for those same tasks and thus provide a means of comparison.

Figure 7.11, a bar chart for a simple eight-task project, includes both the original "baseline" schedule and an indication of the progress of each task, if any, as of June 10. Note the work breakdown structure designator for each task, the horizontal time scale, and the start and finish dates shown both graphically (as the beginning and end, respectively, of each bar) and in writing.

Not only can such charts be used as scheduling tools on their own or in conjunction with a work breakdown structure, but also they can display the results of a network analysis of the type to be discussed in Section 7.4.3. In fact, the bar chart in Figure 7.11 is a computer-generated report based on such a network schedule analysis. Furthermore, they can be used to display summaries at any desired level of the work breakdown structure. For example, in the SKANSKA WBS system, we could generate a bar chart that shows the status of each building as a single bar.

The bar chart approach to project scheduling offers the following advantages:

1. Simple to construct and easy to comprehend
2. Convenient organization by work breakdown structure elements

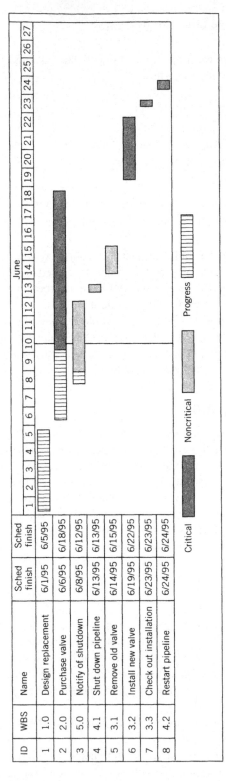

ID	WBS	Name	Sched finish	Sched finish	June
					1 2 3 4 5 6 7 8 9 10 11 12 13 14 15 16 17 18 19 20 21 22 23 24 25 26 27
1	1.0	Design replacement	6/1/95	6/5/95	
2	2.0	Purchase valve	6/6/95	6/18/95	
3	5.0	Notify of shutdown	6/8/95	6/12/95	
4	4.1	Shut down pipeline	6/13/95	6/13/95	
5	3.1	Remove old valve	6/14/95	6/15/95	
6	3.2	Install new valve	6/19/95	6/22/95	
7	3.3	Check out installation	6/23/95	6/23/95	
8	4.2	Restart pipeline	6/24/95	6/24/95	

Critical   Noncritical   Progress

Figure 7.11. Bar chart for an eight-activity pipeline valve replacement project.

3. With computer capability, easy to update and show project progress
4. Relatively inexpensive to use
5. Summaries at any desired WBS level

There are disadvantages, as well, however:

1. No indication of the interrelationships amount activities. For example, is there a technological reason that "Remove old valve" must follow "Shutdown" in Figure 7.11? If the completion of "Purchase and deliver" is delayed, will that delay the start of "Install new valve"? The answer to both these questions is "yes," but this cannot be answered with the bar chart as shown. (Some bar charting programs include lines or arrows between related activities, but they tend to be difficult to interpret.)
2. In developing the bar chart, the analyst tends to fit the tasks into the available time frame rather than analyzing closely how much time will actually be required.
3. The chart shows expected time requirements but not level of effort requirements. If a task requires 5 person-weeks of effort spread over 8 weeks, we learn only about the 8 weeks from the bar chart.
4. Bar charts are impractical for complex projects with many tasks.
5. Absent a computer-driven plot generator, updating of the schedule is time-consuming.

7.4.3 Network Scheduling Methods

7.4.3.1 The Network Approach. In about 1960, a new method for planning and controlling project schedules began to be used. Rather than displaying the project elements as bars on a bar chart, the method utilizes a network to show the interrelationships, timewise, among project activities. Development of this approach was motivated by needs in two independent settings, the behind-schedule and out-of-control status of the U.S. Navy's Fleet Ballistic Missile Program (the "Polaris" Program) and the E. I. du Pont de Nemours and Company's desire to improve the turn-around times for its factory renovation projects.

Two separate and independent groups of developers produced two very similar methods for schedule analysis and control. In the case of the Polaris Program, the result was a method based on multiple estimates of each activity duration, which permitted prediction of the probability of meeting specified completion times for each project phase. This method came to be known by the acronym PERT, or project evaluation and review technique. The other method, at Du Pont, was based on a single time estimate for each activity; this arrangement reflected concern for the cost impacts of schedules of various lengths. The technique was called the Critical Path Method (CPM), or Critical Path Scheduling (CPS) because it determines the critical sequence of project activities whose total length is equal to the project duration (the "critical path") (Malcolm et al., 1959; Kelley and Walker, 1959).

Both these methods begin by sectioning a project into a series of identifiable activities. The work breakdown structure is thus a convenient basis for starting the analysis. Then, one draws a network that relates the activities in the order in which they are expected to be performed. Following the construction of the network, an estimate (in the case of CPM) or its three estimates (for PERT) is made for each activity duration. Calculations are performed to determine the earliest that each activity can start and finish, based on the activities that must precede it; a second series of calculations determines the latest that each activity must start and finish to meet the required completion schedule.

In a typical analysis, some activities will have considerable flexibility in their scheduling. That is, there will be differences between their early times (when they would be finished under best-case conditions) and their late times (when the MUST be completed). On the other hand, it usually happens that some activities have identical early and late dates; we call those activities "critical," and we say that the sequence of those activities is the "critical path." If the time of accomplishing any activity on that path is later than its early time, and no compensating steps are taken for later activities, the overall project can be expected to overrun its original completion schedule.

The CPM approach allows the generation and analysis of various alternative schedules quite easily, especially if a computer program is used. Once the basic analysis has been accomplished, the engineering manager can use it to generate cash flow estimates over time, report and control costs, analyze resource needs, and monitor schedule progress.

Sections 7.4.3.2 and 7.4.3.3 present two basic approaches to network scheduling. In the first case, the activities are represented by nodes and connected together by lines or arrows, while in the second method the activities are shown as arrows and the points in time corresponding to the start and finish of activities are represented by circles.

The sample projects use one estimate for the duration of each activity, since most engineering projects seem to use this method rather than the three-time-estimate (PERT) approach. Details on this latter method are found in such references as Anderson, Sweeney, and Williams (1991) and Rosenau (1984). In fact, over the years the terminology has become somewhat less specific. Now it seems that CPM, PERT, and PERT/COST are often used interchangeably to refer to any network scheduling method.

We conclude our discussion on network scheduling with a brief section on precedence networking, which is a variant of the activity-on-node approach, a mention of the use of this method for resource and cost planning and control, and a discussion of computer software for analyzing network schedules.

Both "activities" examples utilize a hypothetical sounding rocket project, whose WBS is shown in Figure 7.12. The rocket will be used to investigate conditions in the upper atmosphere. We assume that the project is "complete" right after launch; it does not include data gathering and analysis. The project consists of the development of scope and specifications; the procurement of the motor and its fuel; the design, procurement, and assembly of both the payload and the tracking system; and the final assembly, testing, and launch of the rocket.

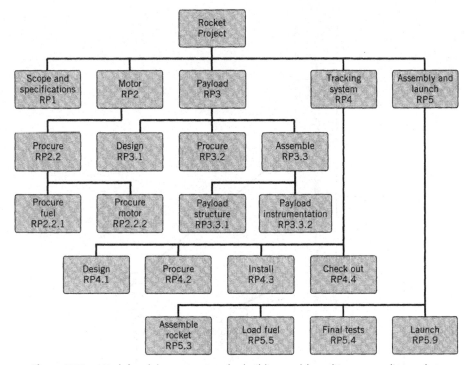

Figure 7.12. Work breakdown structure for building and launching a sounding rocket.

Our schedule will utilize for activities the most detailed elements at the end of each branch in the work breakdown structure. We have, of course, made many simplifications for purposes of this illustration. In real life, we would typically provide much more detail by adding one or more levels to the WBS. Upon completion of the analysis based on this large number of activities, summary schedule information could be generated at any desired level of detail.

7.4.3.2 Activity-on-Node Method. Figure 7.13 contains an activity-on-node schedule network for the 15 activities of our rocket project. Note that each activity corresponds to a WBS element located farthest down its respective branch. There is no requirement that all branches have equal amounts of detail, and in fact the several branches vary in level of detail.

In constructing the network, we use the following guidelines:

1. Each activity is represented by a node. We used a rectangular form; a square, a circle, an ellipse, or other shape would have been acceptable.
2. The activities are connected by lines, with the flow in time going from left to right. Arrows are sometimes used to clarify the order of activities.
3. For any given activity, a line or arrow connected to the node's left side indicates that the activity at the other end of the line must be completed

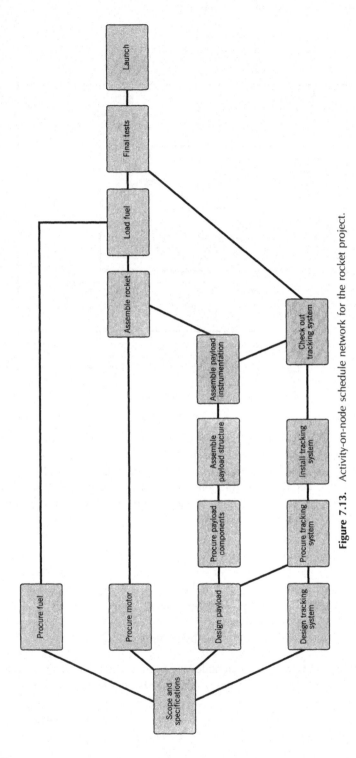

Figure 7.13. Activity-on-node schedule network for the rocket project.

before the activity in question can begin. For example, "Final tests" cannot begin until both "Load fuel" and "Check out tracking system" have been completed. If no line is connected to the left side of a node, that activity (e.g., "Scope and specifications") can begin at the beginning of the project.

4. For any given activity, a line or arrow attached to the node's right side indicates that the activity at the other end of the line cannot begin until the activity in question has been completed. For example, "Design payload" must be completed before either "Procure payload components" or "Procure tracking system" can begin. If no line is connected to an activity's right side, it is a final activity in the project—in this case, "Launch."

5. We build the network without considering the time requirements for the activities. The placement of the nodes is not related to a time scale but only to the technological sequencing required to accomplish the project.

Another way to envision the process of network development is to ask, for each activity, the three questions:

• What activity or activities must immediately precede this activity?
• What activity or activities must immediately follow this activity?
• What activity or activities can be done concurrently with this activity?

It is important to understand that the network in Figure 7.13 represents the approach chosen by one project manager or project team to carry out this project. Many of the assumed sequences could be challenged. In the end, the result is the best thinking of those whose will carry out the project, based on information available at the time the schedule is developed. It can be changed if new information or a new approach emerges. In any case, many engineering managers have found that the project network is an excellent communication tool when used by the project team to plan and control its project. It literally "builds the project on paper," using the suggestions of all concerned.

Once the network has been drawn, it is time to add the time dimension. We begin that phase by estimating the time that will be required to complete each activity. Considerations include the quantity of work the activity represents, the number of persons, machines, and so on expected to be assigned to the activity, the methods to be used, the anticipated weather and other environmental conditions, and any other relevant factors. This is not an easy task, and judgement and experience play a big part in the analysis. As mentioned earlier, this example utilizes only one time estimate for each activity, rather than the optimistic, pessimistic, and most likely estimates required by the probabilistic approach.

In Figure 7.14, we have added time estimates, in working days, for each activity of our rocket project, together with the results of calculations to be explained shortly. The legend indicates that each node contains the activity's identification number, WBS code, description, duration, early start time, early finish time, late start time, and late finish time. The calculations are quite straightforward and consist of two parts, a "forward pass" and a "backward pass."

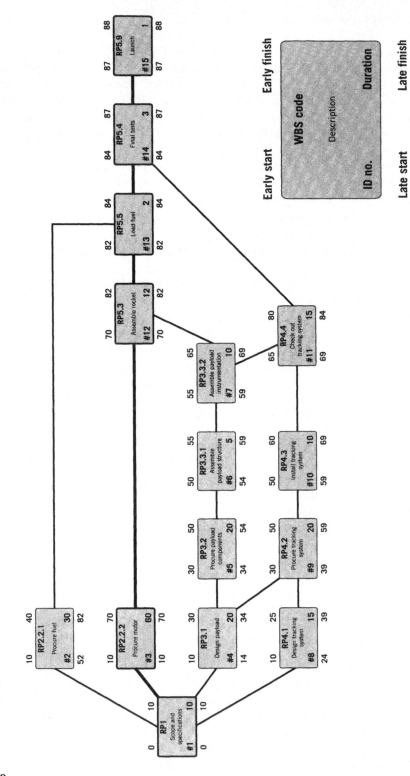

Figure 7.14. Activity-on-node schedule network for the rocket project, with duration and calculations added.

The forward pass begins by assuming that the project begins at time zero. Thus, the early start of the first activity is 0. Its early finish is equal to its early start plus its duration. So, the "Scope and specifications" activity has an early finish of $0 + 10 = 10$. The early start for any other activity is the latest early finish of all its immediate successors. For the four activities that follow the first activity directly, the early start time is 10. The method proceeds in like manner from the beginning to the end. Note that the "Assemble rocket" activity has two immediate predecessors: "Procure motor" and "Assemble payload instrumentation." Since both must be completed before this activity can begin, we use 70 (the larger early finish time) for the early start time of this activity. This sort of comparison must be made to determine early start values for four other activities in this project. To summarize,

early start = largest value of early finishes for all immediate predecessors

early finish = early start + duration

Upon finishing the forward pass calculations, we determine that the early finish time for "Launch" is 88. Thus, it appears that if the activities proceed as planned, the project can be completed in 88 working days.

There is sometimes confusion over the number representing "point in time" and the number representing "day." For example, "point in time one" is the end of the first day, or the beginning of the second day, since our convention is to say that the project begins at time zero. Thus, when we say the project can be completed "at time eighty-eight," we mean at the end of the 88th day.

Now it is time to perform our "backward pass" calculations to determine the latest that each activity must start and finish. These "must" times are relative to some project completion date: that is, our late start and finish times are the times that must be met to ensure that an assumed project completion requirement is satisfied. What should this time be? In some cases, there will be a contractual requirement that specifies when the project is to be completed. In other cases, we are doing an initial analysis using some arbitrary completion time. Often, at least for a preliminary "first cut," planners assume that the early finish for the last activity is the project completion time. In that case, our late start and finish times for each activity will be calculated relative to that finish date. We shall proceed under that assumption, using 88 days as the finish time.

If the project must be completed in 88 days and "Launch" takes one day, then this activity must begin at time 87, the late start time. Thus, the preceding activity, "Final tests," must be completed by time 87, and because it lasts 3 days, it must begin no later than time 84. We proceed backward in this manner until we have reached the first activity.

If the start of two or more activities depends on the completion of a single activity, as is the case with #4, "Design payload," and #1, "Scope and specifications," a comparison and choice must be made. For example, the two immediate successors of "Design payload," "Procure payload components" and "Procure tracking system," must be started no later than 34 and 39, respectively.

We choose 34, the smaller of these times, as the late finish for "Design payload." That is, "Design payload" MUST be completed no later than 34 if "Procure payload components" is to begin by this time and thus not delay the project completion beyond time 88.

Similar comparisons are made for the "Assemble payload instrumentation" and "Scope and specifications" activities. In this latter case, the late finish is 10, and the late start is 0, which should be expected, since the project must be launched at time zero to ensure completion at the end of the 88th working day. To summarize:

late finish = smallest value of late starts of all immediate successors

late start = late finish − duration

Finally, we determine the value of "slack" (sometimes called "float") for each activity. Found as the difference between the activity's late and early finish times, it represents the activity's degree of "criticality," or the degree of flexibility the engineering manager has in its schedule. Actually, it can be found in any of three ways:

late finish − early finish

late start − early start

late finish − early start − duration

The slack for #10, "Install tracking system," is thus 69 − 60, or 59 − 50, or 69 − 50 − 10 = 9. The slack for #3, "Procure motor," is 70 − 70 = 0.

When we use the project's early finish date as its late finish date for the backward pass calculations, there will be a sequence of activities from beginning to end of the network that will have zero slack. These activities are called "critical," and they form the "critical path." In Figure 7.14, we have highlighted this sequence, which consists of "Scope and specifications," "Procure motor," "Assemble rocket," "Load fuel," "Final tests," and "Launch." The total length of this path is 88 days, our project duration. The activities on the critical path are critical in the sense that they must occur at the times they are scheduled. If one of them slips to a start time and/or finish time later than those indicated on the network, and no compensating action is taken later in the project, our project will overrun its 88–day target completion time. Furthermore, if it is desired to shorten the project duration, we must squeeze some time out of an activity or activities on the critical path.

Other activities, of course, have some positive values of slack, calculated as just indicated. Slack is best defined as the amount of time an activity can slip beyond its early times without causing the total project duration to be extended. It is easy to calculate slack values; they could even be added to the network in Figure 7.14 if desired.

Although we define slack for individual activities, it is important to understand that its values are not additive along a path. For example, note that each of the activities #9, #10, and #11, related to procuring, installing, and checking out the tracking system, has a slack of 9 days. This does NOT mean that there is a total of 27 days of slack available in this path! Using the notion that slack is the amount of time an activity can slip from its EARLY times, we see that if activity #9 slips 9 days, or "uses up all its slack," then activities #10 and #11 will be forced to proceed at their late times; the slippage of #9 has caused the other two to become critical!

Sometimes it is helpful to prepare a tabulation of information about activity times and slack values. As one might expect, computers are good at that sort or thing, as we shall mention later. We leave the preparation of a table of results to the next section, where we discuss the activity-on-arrow method.

7.4.3.3 *Activity-on-Arrow Method.* Another way to show the activities in a project schedule network is to represent them as arrows, with their directions indicating the flow of work and the intersections symbolizing starts and finishes. Actually, this method was developed first, with both the Polaris Program and the Du Pont effort utilizing similar activity-on-arrow networks. Later, the activity-on-node method seems to have become the more popular.

Figure 7.15 depicts an activity-on-arrow network for our rocket project. It is important to understand that the networks in Figures 7.13, 7.14, and 7.15 show essentially the same information about activity sequences. For an activity-on-arrow network, the following guidelines are used.

1. Each activity is represented by an arrow, with the flow of work such that the tail of the arrow is at the activity's beginning and the head at the end.
2. The points in time corresponding to the start of some activities and/or the finish of some activities are called "events" and are represented by nodes — circles, ellipses, or the like.
3. At any event, all activities that enter the node must be completed before any of the activities that leave the node can be started.
4. The arrow lengths have no relation to the time requirements for their respective activities. Like the activity-on-node method, we draw the network first, before any consideration of the time dimensions of the project.
5. The events are numbered as a means of identifying the activities by their unique begin event–end event combinations. For example, we can refer to "Procure fuel" as "activity 3–19."

The dashed arrow between the end of "design payload" and the beginning of "tracking system" (activity 5–13) is called a "dummy" or a "constraint." It is used to assure the proper sequencing of activities. Reference to Figure 7.13 indicates that "Procure tracking system" cannot begin until both "Design payload" and "Design tracking system" have been completed, whereas the start of "Procure payload components" depends only on the completion of "Design payload." With the activity-on-arrow method, a dummy must be provided to assure proper sequencing in such a case. Otherwise, that small portion of the network would be shown in a way that is OVERLY RESTRICTIVE AND WRONG:

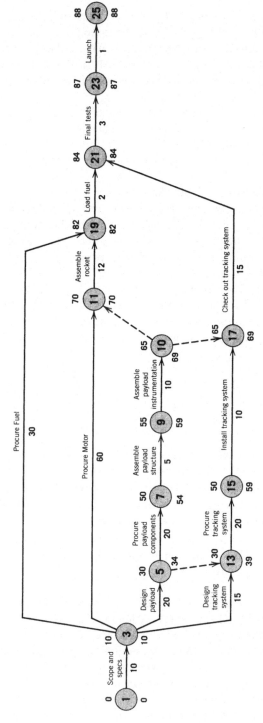

Figure 7.15. Activity-on-arrow schedule network for the rocket project.

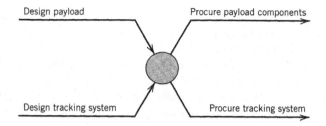

It is wrong because the start of "Procure payload components" does not depend on the completion of "Design tracking system." Note also the use of dummies 10–11 and 10–17. When we do the network calculations we treat dummy activities as regular activities of zero duration.

Calculations for the activity-on-arrow method focus initially on events, finding early and late times for each event, and then using these results to determine early and late start and finish times for each activity. On the network in Figure 7.15, we show early times above each respective event node and late times below. For events the following definitions apply:

early time = earliest possible starting time for any activity that begins at this event

late time = latest possible finishing time for any activity ending at this event

As before, we begin with a project start time of zero. The early time for event 1 is 0, and the early time for event 3, based on the duration of activity 1–3, is $0 + 10 = 10$. For event 5, the early time is $10 + 20 = 30$. When two activities enter an event, we must compare the possible early times and select the largest as being the earliest possible starting point for any activity that leaves this event. Thus,

early time = max (preceding event early time + activity duration) for all activities

entering the node

An example is found at event 11. The earliest that activity 11–19 ("Assemble rocket") can start is either $10 + 60 = 70$, based on activity 3–11 ("Procure motor") or $65 + 0 = 65$, based on activity 10–11, the dummy. We choose the larger, or 70, as the early time for event 11.

Another example is found at event 13, where the calculations are as follows:

$$30 + 0 = 30 \text{ (based on the dummy, 5–13)}$$

$$10 + 15 = 25 \text{ (based on activity 3–13)}$$

The choice is 30 for the early time for event 13.

Once we have completed the "forward pass" calculations and have determined, as before, that the project can be completed in 88 working days (i.e., "the early time for event 25 is 88 days"), we begin our "backward pass" calculations. Based on the rationale explained in the preceding section, we choose 88 as the late time for event 25, which will be our project's target completion time. All backward pass calculations will be based on this target.

The late time for event 23 will be $88 - 1 = 87$, and for event 21 it will be $87 - 3 = 84$. We proceed backward through the network, and when more than one activity leaves a node, we must make a comparison. See, for example, event 5, where activities 5–7 ("Procure payload components") and 5–13 (the dummy), originate. For the late time of event 5, we need to know how late activity 3–5 can be competed. The answer is 34, based on the late time of 54 for event 7 and the 20–day duration of activity 5–7. The other possibility is $39 - 0 = 39$, based on the late time for event 13 and the duration of 0 for the dummy. But, IF THE PROJECT IS NOT TO OVERRUN ITS OVERALL COMPLETION TARGET OF 88, activity 3–5 must be finished no later than time 34, so that activity 5–7 can start at that time. Thus,

late time = min (end event late time − activity duration) for all activities leaving

the node

Another example is found at event 3, where we compare four possible values before selecting 10 as the late event time. At last, we work back to the beginning, event 1, where the late time is 0, as expected. That is, the project must start no later than time zero if it is to be completed by time 88.

It is useful to develop a table that lists information about each activity. Table 7.5 contains the results of a series of calculations for this project, based on our event times shown in Figure 7.15. For each activity, we find the following:

early start time = early event time of the begin event

early finish time = early start time + duration

late finish time = late event time of end event

late start time = late finish time − duration

slack = late start time − early start time, or

late finish time − early finish time, or

late finish time − early start time − duration

Note that the table also contains the activity's begin- and end-event numbers, its description, and its duration. Activities with zero slack are denoted as critical (CR).

The results of this analysis are the same as those we found using the activity-on-node method. To the reader is left the choice of which approach to use for a particular application.

7.4.3.4   Precedence Method.   A variation of the activity-on-node method allows for "lags" and "leads" within the project network. Whereas the two methods just presented require that the relationships between activities be such that "activity A

TABLE 7.5 Activity Table for Activity-on-Arrow Schedule for Rocket Project, Indicating Critical (CR) Tasks

Begin	End	Description	(days) Duration	ES	EF	LS	LF	SL	CR
1	3	Scope and specifications	10	0	10	0	10	0	CR
3	5	Design payload	20	10	30	14	34	4	
3	11	Procure motor	60	10	70	10	70	0	CR
3	13	Design tracking system	15	10	25	24	39	14	
3	19	Procure fuel	30	10	40	52	82	42	
5	7	Procure payload components	20	30	50	34	54	4	
5	13	Dummy	0	30	30	39	39	9	
7	9	Assemble payload structure	5	50	55	54	59	4	
9	10	Assemble payload instrumentation	10	55	65	59	69	4	
10	11	Dummy	0	65	65	70	70	5	
10	17	Dummy	0	65	65	69	69	4	
11	19	Assemble rocket	12	70	82	70	82	0	CR
13	15	Procure tracking system	20	30	50	39	59	9	
15	17	Install tracking system	10	50	60	59	69	9	
17	21	Check out tracking system	15	65	80	69	84	4	
19	21	Load fuel	2	82	84	82	84	0	CR
21	23	Final tests	3	84	87	84	87	0	CR
23	25	Launch	1	87	88	87	88	0	CR

[a]ES, early start; EF, early finish; LS, late start; LF, late finish; SL, slack.

can start as soon as activity X has been completed," for example, a lag in a network might provide that "activity A can begin 5 days after activity X has been completed." This type of relationship might occur in a concrete-placing operation, where the stripping of forms can begin, say, 14 calendar days after a deck slab has been placed.

Another type of relationship might provide that a certain activity can start a given amount of time after another activity has started. Or, we might need to require that the completion of one activity precede the completion of another by at least 4 hours.

These relationships are incorporated into the precedence diagramming method, or PDM. There are four possible types: finish-to-start (the most common), start-to-start, finish-to-finish, and start-to-finish. Many network analysis computer programs provide for some or all of these conditions. The interested reader is referred to works by Bennett (1977), Meredith and Mantel (1995), and Moder, Phillips, and Davis (1983) for further details.

7.4.4 Resource and Cost Analysis and Control Based on a Network Schedule

A network schedule provides a convenient basis for planning and controlling both the resource needs and the cost factors that are always a part of a project. By resources, we mean the people, equipment, materials, and other physical assets that contribute to the project. By estimating the amount of each applicable resource required for each activity, we can develop an overall requirement estimate for each time period, based on completing all activities (1) as early as possible (the "early start schedule") and (2) as late as possible (the "late start schedule"). From there it may be possible to smooth or balance the resource requirements by taking advantage of the slack available in some of the activities. Furthermore, if resources are not available in sufficient quantities to carry out the proposed schedule, the network analysis can assist to determine which activities should be allowed to slip, again based on their amounts of slack.

Resource analysis can be extremely complicated, especially when several resources are involved. Authors such as Meredith and Mantel (1995) and Rosenau (1984) deal with the details, which are beyond the scope of this book. In Section 7.4.5, we show the results of computer analyses of resource requirements for our rocket project, including sample computer-generated curves of planned and actual cash flow.

Analysis of and accounting for project costs can also be accomplished with the network schedule as a basis. If we assign a dollar value to each activity, it is a straightforward matter to accumulate the totals for each time period. We can show these values for both early start and late start schedules and thus develop an envelope within which the actual cash flow can be expected to take place if the project remains on schedule. Then, as the project proceeds, we can determine how much each activity has earned, based on the proportion of its total that has been completed, and then add the earnings for each activity to find the total amount of earnings to date. In construction projects, for example, the contractor's bid price often is distributed among the project activities and periodic payments are based on the completion percentage of each project activity.

7.4.5 Computer Applications for Planning and Controlling the Schedule

As might be expected, the computer can play an important role in performing schedule calculations and generating useful output. The development of personal, laptop, and notebook computers has made possible the on-site planning and control of project schedules, with much more involvement of project personnel and faster analysis turnaround. The author remembers traveling to computer service bureaus in large cities in the 1960s, far from project sites, to have network schedules analyses performed. Often, the batch process would permit only one run per day. Times have changed!

The computer is useful in a number of ways, touched on briefly in the paragraphs that follow.

1. *Preparation of Input.* Many programs display blank nodes on the screen, into which the analyst places the required information, together with appropriate links between nodes; this permits a graphical display of a portion of the

network on the screen. Other software provides blank formats, which the analyst fills in.

2. *Error Finding.* Before processing the data, the typical program checks for loops and other input errors. Modern programs often perform these error-finding routines as the input data are being supplied.

3. *Basic Calculations; Analysis of Alternative Schedules.* The calculations, as explained in Section 7.4.3, are quite simple. The computer assists by performing them rapidly and without error (unless, of course, there are errors in input due to the analyst). Furthermore, it is a relatively simple matter to change some of the input and determine the effect of the changes on the project completion time, resource requirements, and the like. Many "what if" analyses can be conducted at one sitting.

4. *Generation of Tabular Output in a Number of Forms.* Once the calculations have been completed, the computer can be useful in providing listings such as that shown in Table 7.5. These lists of activities can be sorted by activity numbers, slack values, early start times, or other orders. Often it is especially helpful to request a sort by responsibility, in which case each activity is coded to indicate what trade or other group will be primarily responsible for its completion. For example, we could generate a list of "carpenter" activities, a list of "electrician" activities, and so on; within each of these, we could order the activities by, say, degree of criticality (i.e., by increasing value of slack).

5. *Resource Analysis.* Many programs provide analysis of the impact of resource requirements for the various activities on the overall project's resource needs. Given a limited supply of only 20 plumbers on each project day, for example, will our project needs exceed this amount? Can we shift some activities, because of their slack, to make sure the need does not exceed 20? Such analyses often include prediction of project cash flows, based on the cash flow requirement for each activity.

6. *Cost Reporting and Control.* The computer is also a good accounting tool. The amount earned by each activity and the total earned by the project can be computed, with this last amount being compared against the planned value. Some programs perform earned value analysis of the type to be described in Section 7.5.

7. *Monitoring and Updating the Project Schedule.* Once the basic input has been provided and the schedule has been analyzed, project progress can be monitored and compared with predicted progress. Some programs develop an initial "baseline" schedule, against which all future updates are compared. Often, the updating of the schedule and the reporting of cash flow earnings are performed concurrently—at the end of each month, for example.

8. *Generation of Graphical Output.* Computer-driven graphical output is a natural using modern plotters, laser printers, and other output devices. Types of output include the network itself, bar charts based on the schedule, cash flow curves, and resource requirements curves.

The availability of powerful, user-friendly scheduling software packages is changing so rapidly that an inventory of such packages would not be appropriate here. In 1993 such products as MacProject (Claris, Inc.), Microsoft Project for Windows (Microsoft Corporation), Primavera Project Planner (Primavera Systems, Inc.), Project Workbench (Applied Business Technologies, Inc.), and Texim Project (Texim Inc.) are but samples of well over a hundred software packages that can assist the project scheduler with the kinds of analyses listed earlier.

Instead, as examples, Figures 7.16 and Figure 7.17 and Table 7.6 present samples of output for our rocket project, generated by Microsoft Project for Windows. Figure 7.16 is a bar chart, with dark bars for critical activities; scheduled start and finish dates are tabulated, as well. Figure 7.17 is a special computer-generated graph that shows the requirements for assembler/installers, based on the assumed needs for this resource for certain activities and the schedules for those activities. Under the assumption that eight such persons are available, the graph shows potential shortages on five days in May. Table 7.6 indicates early and late start and finish dates, similar to Table 7.5, with values of slack for each activity.

7.5 PROJECT BUDGETING AND COST CONTROL

7.5.1 Budgeting

The development of a project budget requires that costs for the various elements be estimated by some rational method and at an appropriate level of detail. This step is followed by the preparation of a total budget that includes all direct costs plus overhead, contingencies, and profit, and this total amount will be "spread out" over the time the project is expected to take. Cost estimating and budget preparation are complicated and time-consuming responsibilities. We note some highlights.

7.5.1.1 Cost Estimating. There are many possible ways to estimate project costs. For example, it may be appropriate to estimate the cost of building your new house by multiplying the floor area by a cost per square foot that is based on experience tempered by such factors as inflation, special features, and market conditions. Builders also sometimes use volume measures, based on the total cubic footage of the structure, or linear measures, such as the cost per mile of a pipeline in a certain region. Such approaches typically are quite crude but often "good enough" at early stages of the design process.

Another approach is to determine the quantity of each physical element of the project (cubic yards of concrete, board feet of framing lumber, number of light fixtures, etc.) and multiply by historic unit costs for labor, materials, equipment, and other resources, again modified as appropriate for the particular conditions. Such a process is the basis for a large majority of construction cost estimating and is often performed by persons whose expertise in this field has been developed over many years.

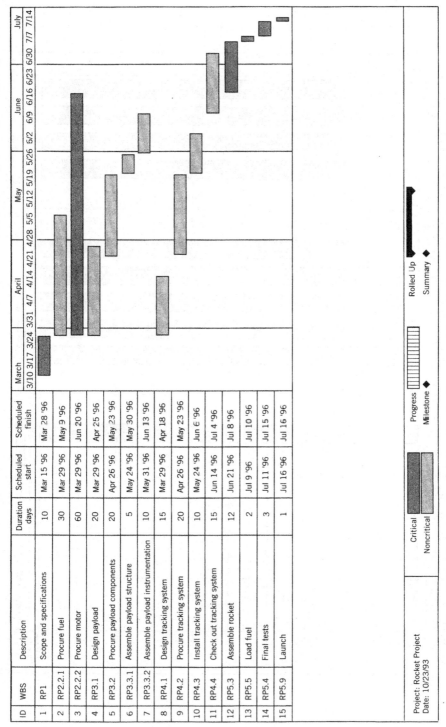

The table in the figure contains the following data:

ID	WBS	Description	Duration days	Scheduled start	Scheduled finish
1	RP1	Scope and specifications	10	Mar 15 '96	Mar 28 '96
2	RP2.2.1	Procure fuel	30	Mar 29 '96	May 9 '96
3	RP2.2.2	Procure motor	60	Mar 29 '96	Jun 20 '96
4	RP3.1	Design payload	20	Mar 29 '96	Apr 25 '96
5	RP3.2	Procure payload components	20	Apr 26 '96	May 23 '96
6	RP3.3.1	Assemble payload structure	5	May 24 '96	May 30 '96
7	RP3.3.2	Assemble payload instrumentation	10	May 31 '96	Jun 13 '96
8	RP4.1	Design tracking system	15	Mar 29 '96	Apr 18 '96
9	RP4.2	Procure tracking system	20	Apr 26 '96	May 23 '96
10	RP4.3	Install tracking system	10	May 24 '96	Jun 6 '96
11	RP4.4	Check out tracking system	15	Jun 14 '96	Jul 4 '96
12	RP5.3	Assemble rocket	12	Jun 21 '96	Jul 8 '96
13	RP5.5	Load fuel	2	Jul 9 '96	Jul 10 '96
14	RP5.4	Final tests	3	Jul 11 '96	Jul 15 '96
15	RP5.9	Launch	1	Jul 16 '96	Jul 16 '96

Project: Rocket Project
Date: 10/23/93

Critical   Noncritical   Progress   Milestone ◆   Rolled Up   Summary ◆

Figure 7.16. Computer-generated bar chart for the rocket project, showing critical and noncritical activities.

211

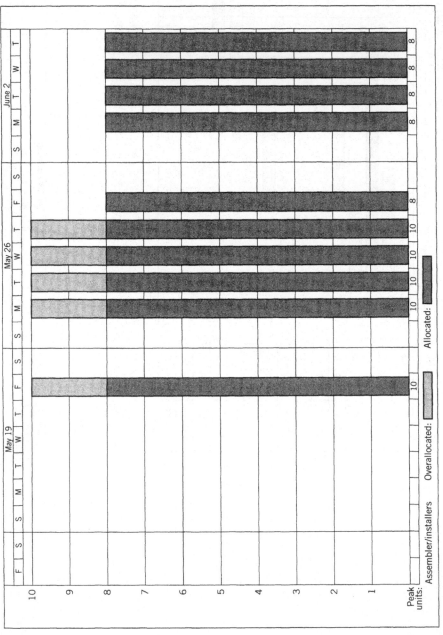

Figure 7.17. Computer output for the rocket project, showing resource usage graph with overallocation of assembler/installers on five days.

Still another method is to assemble a complete bill of materials for the project and determine the material cost from that listing. Labor costs are then roughed in based on estimates of the effort required by each individual operation, and equipment costs are figured from estimates of the time required for each operation. Every organization has its own approach, based on many factors, including the type of work, experience, and available support systems.

We advocate the use of the work breakdown structure in the development of project cost estimates. If the various elements of the WBS have been identified, at whatever level is appropriate for the project, an estimate of the cost of each element can be prepared. The total project cost is simply the sum of the cost of each element. The major advantages of this system are twofold: (1) it can be used in conjunction with the project schedule to develop relationships between cash flow and time over the life of the project schedule, and (2) the monitoring and control of costs as the project proceeds can be based directly on the WBS elements, each of which already has its own cost estimate. A disadvantage is the possibility that not all aspects of the project have been included in a WBS element; however, any cost estimating scheme carries with it the possibility of "leaving something out."

7.5.1.2 The Project Budget. There is a difference between the sum of the project's estimated costs and the total project budget! In the world of construction contracting, the contractor obtains projects with the hope of making a profit! Usually project budgets also include some amount, often called a "contingency," in recognition that the cost estimation process is not perfect and some risk is involved in conducting the project. Also, typically the initial cost estimate does not include such overhead costs as general supervision, project accounting and control, purchasing and expediting, and home office expenses. Every organization has its own way of handling these aspects of cost estimating and budgeting preparation. Indeed, the determination of the "markup" for a construction cost bid estimate, to include contingency and profit, is a responsibility usually reserved for those at the highest levels of the organization.

Even if the project is carried out in-house, with the organization's own forces, the project budget must reflect recognition that the estimated total cost of the project is greater than the sum of the direct costs for the various elements. There will still be general administrative costs, even if the project is not being done to make a profit.

A project budget may have more than one version, depending on its intended use. The owner of a building project who is paying for its construction will normally be interested in the total project budget, including the contractor's overhead and profit. The anticipated total cash flow requirements as a function of time are of primary importance. On the other hand, the contractor's field forces are expected to contain costs within their estimated amounts; comparisons of actual costs to date plus estimated future costs with those in the original estimate, at various points during the project's progress, are especially important. Obviously, if the actual total costs are high enough to equal the total income for the project, there will be no profit.

TABLE 7.6 Rocket Project Computer Output, Showing Calendar-Dated Early and Late Start and Finish Dates

ID	WBS	Description	Duration (days)	Early Start	Early Finish	Late Start	Late Finish	Slack (days)
1	RP1	Scope and specifications	10	3/15/96	3/28/96	3/15/96	3/28/96	0
2	RP2.2.1	Procure fuel	30	3/29/96	5/9/96	5/28/96	7/8/96	42
3	RP2.2.2	Procure motor	60	3/29/96	6/20/96	3/29/96	6/20/96	0
4	RP3.1	Design payload	20	3/29/96	4/25/96	4/4/96	5/1/96	4
5	RP3.2	Procure payload components	20	4/26/96	5/23/96	5/2/96	5/29/96	4
6	RP3.3.1	Assemble payload structure	5	5/24/96	5/30/96	5/30/96	6/5/96	4
7	RP3.3.2	Assemble payload instrumentation	10	5/31/96	6/13/96	6/6/96	6/19/96	4
8	RP4.1	Design tracking system	15	3/29/96	4/18/96	4/18/96	5/8/96	14
9	RP4.2	Procure tracking system	20	4/26/96	5/23/96	5/9/96	6/5/96	9
10	RP4.3	Install tracking system	10	5/24/96	6/6/96	6/6/96	6/19/96	9
11	RP4.4	Check out tracking system	15	6/14/96	7/4/96	6/20/96	7/10/96	4
12	RP5.3	Assemble rocket	12	6/21/96	7/8/96	6/21/96	7/8/96	0
13	RP5.5	Load fuel	2	7/9/96	7/10/96	7/9/96	7/10/96	0
14	RP5.4	Final tests	3	7/11/96	7/15/96	7/11/96	7/15/96	0
15	RP5.9	Launch	1	7/16/96	7/16/96	7/16/96	7/16/96	0

Whatever the type of budget, after it has been determined it is divided among the various time periods. In Section 7.1.3 in connection with the project life cycle, we noted that cash flow as a function of time is one oft-used measure of project progress. Figure 7.3 plotted cash flow to date versus time, in a typical S-curve shape indicating relatively slow progress near project start and end and more rapid progress in between.

Network scheduling, based on the WBS, can help us to determine cash flow curves as a function of time, both on an early start–early finish and a late start–late finish basis. Then, the project manager and the project team must establish an expected project schedule. Critical path activities don't afford any choice; they must occur at their early times (which are also their late times—that's why they are critical!). As a rule it is decided to schedule some critical activities at their early times, some at their late times, and some in between. After all the activities have been scheduled, we have a third curve, which should fall between the envelope of the other two curves. This third curve becomes our baseline; we shall measure our cost progress against it.

Figure 7.18 is a set of three such curves, for our rocket project, based on an estimated cost of each work breakdown structure element and the project schedule analysis presented earlier. We shall use the middle "as-scheduled" curve as the basis for our discussion of project cost control.

7.5.2 Cost Control

Suppose we come to the end of the fifth week of our project and our records indicate that we have spent $28,000 more than our "as-scheduled" curve of cash flow versus time had predicted. Are we in trouble? Is this situation good or bad?

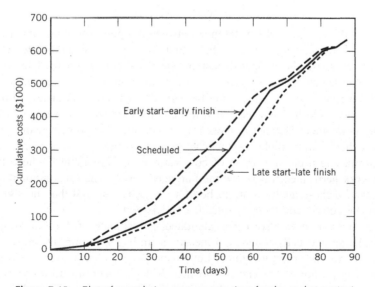

Figure 7.18. Plot of cumulative costs versus time for the rocket project.

The answer is, "We don't know!" More information is needed! Essentially, we need to know how much progress has been made during these 5 weeks. If we have made more progress than was originally projected, we may be in very good shape, even though we have "overspent" our budget to date.

In one approach to analyzing such situations, called "earned value analysis," we determine the value of the work that has been accomplished as well as the originally predicted costs to date and the actual costs to date. The U.S. Department of Defense, which requires such an approach on most of its projects, uses the system entitled cost/schedule control system criteria (C/SCSC). Although there are other variance systems for measuring cost and schedule progress, we shall confine our discussion to this one, which defines three measured quantities and three calculated variances. The quantities are:

- *Budgeted cost of work scheduled (BCWS)* = how much has been budgeted for the tasks we have scheduled as of a given date.
- *Budgeted cost of work performed (BCWP)* = the amount in the project budget for the tasks that have been completed or partially completed as of the given date.
- *Actual cost of work performed (ACWP)* = the amount spent to date on the tasks that have been completed or partially completed.

Given these three measurements, we define the three variances as follows:

$$\text{schedule variance} = \text{BCWS} - \text{BCWP}$$

$$\text{cost variance} = \text{BCWP} - \text{ACWP}$$

$$\text{total variance} = \text{BCWS} - \text{BCWP} + \text{BCWP} - \text{ACWP}$$

$$= \text{BCWS} - \text{ACWP}$$

The total variance is the difference between the point on our original progress curve as of the analysis date and the actual costs as of the same date. The other two variances use the budgeted cost of the work performed to date to explain this total variance. The cost variance compares how much the quantity of work should have cost with what it actually cost. The schedule variance compares the value of the work that was scheduled to be completed with the value of what was actually performed. Defined in these ways, a positive value of schedule variance can be called "unfavorable"—you have performed less than you were scheduled to perform, and a positive value of cost variance is "favorable"—the actual cost of the work performed was below the budgeted cost for the same amount of work. Obviously, if these analyses are to be of value, it is essential that the signs of the answers be correct and their significance understood.

Once the variances have been calculated, an estimate of the total project cost at completion is usually found, and measures are begun to improve areas in which cost problems have been identified. Let us apply these ideas to our rocket project.

Suppose we are at the end of the fiftieth day. According to our records, the total cost of project activities to date is $258,000, which is ACWP. The curve in

Figure 7.19, which is a small portion of the scheduled cumulative cost curve in Figure 7.18, shows that the amount we were scheduled to have completed by day 50 had a value, BCWS, of $270,000. So, we can find the total variance by simple arithmetic:

$$BCWS - ACWP = \$270,000 - \$258,000 = +\$12,000$$

Now, a further bit of information is that the value of the work we have put in place so far, the BCWP, is $239,000. We have found this amount by an analysis of each project activity. We determine its completion percentage, multiply this percentage by its respective value, and total these individual amounts. This BCWP of $239,000 is the project's "earned value" to date.

The calculations for the schedule and cost variances are as follows:

$$schedule\ variance = BCWS - BCWP = \$270,000 - \$239,000 = +\$31,000$$

$$cost\ variance = BCWP - ACWP = \$239,000 - \$258,000 = -\$19,000$$

Thus we conclude that we have completed to date a value of work $31,000 below what was scheduled (a positive, or unfavorable, schedule variance). Furthermore, the actual cost of the work completed was $19,000 greater than the budgeted cost of that work (a negative, or unfavorable, cost variance). Without the information about BCWP and the two variances calculated from it, we would know that actual costs were $12,000 less than scheduled costs; without a means of identifying the sources of that difference, however, the information is likely to prove misleading

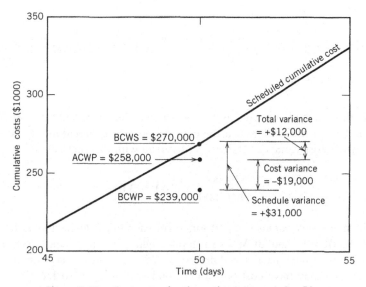

Figure 7.19.   Cost status for the rocket project at day 50.

We can attempt to estimate how much the project will have cost, in total, when it is completed. This estimate at completion (EAC) can be calculated in different ways, depending on our assumptions. For example, if we assume that the project will continue to perform at the same rate exhibited to date, we can calculate a completion percentage of

$$\frac{\$239{,}000}{\$636{,}000} \times 100 = \mathbf{37.6\%} \text{ as of day 50}$$

If we have spent an ACWP of $258,000 so far to do 37.6% of the work, *at the same rate* the estimate at completion, will be

$$EAC = \frac{\$258{,}000 \times 100}{37.6} = \mathbf{\$686{,}000}$$

which is a $50,000, or 7.9%, *overrun:*

$$\frac{\$50{,}000}{\$636{,}000} \times 100 = \mathbf{7.9\%}$$

Note that we could also say that the cost variance of $-$\$19,000 means an overrun to date of ($19,000/$239,000)(100) = 7.9%. Therefore, if this rate continues through the end of the project, the EAC will be $636,000 (1.079), which rounds to **$686,000**.

On the other hand, if we assume that the balance of the project after day 50 will match the original estimate, then the total cost at completion will be the sum of the actual cost to date and the originally estimated costs for the balance of the project, or

$$EAC = \$258{,}000 + \frac{100 - 37.6}{100} \times \$636{,}000 \approx \$258{,}000 + \$397{,}000$$

$$= \mathbf{\$655{,}000}$$

With this method, we are projecting an overrun of $655,000 $-$ $636,000 = $19,000. That is, we assume that the cost variance at completion will be confined to the current cost variance. The project overrun with this method is

$$\frac{\$19{,}000}{\$636{,}000} \times 100 \approx \mathbf{3.0\%}$$

Such an analysis can also be performed for each individual activity. The notions of ACWP, BCWP, and BCWS can apply equally to individual work elements. An activity may have been scheduled to be 75% complete, may actually be 85% complete, and may have cost 60% of its budgeted amount so far. (Now *that's* a well-managed, or lucky, work package!) The answers for the total project can be found as the sum of the individual activity results. Some computerized scheduling packages, such as Primavera, provide such capability.

It is most convenient to perform the type of cost control analysis just explained when the schedule is being updated. Some clients require satisfactory completion of such analyses of both cost and schedule before they will pay the contractor for work performed during the period being analyzed.

7.6 SOME FINAL THOUGHTS ON PROJECT CONTROL

We conclude this long chapter with a few final words on project control. First, if "the essence of successful project management consists of satisfying the triple constraint," as we suggested in Section 7.1.2, then it is necessary to control all three of those elements: technical achievement, elapsed time, and cost. This chapter has emphasized the planning and control of projects' time and cost dimensions. Planning and control of *technical performance* are just as important and cannot be ignored.

Second, some excellent references on project control are available. Meredith and Mantel (1995) devote an entire chapter to the subject, and they also have a section of another chapter on project management information systems (PMIS) that includes cautions against overreliance on completely automated computerized systems. A paper by Thamhain and Wilemon (1986), two highly respected researchers/practitioners, summarizes a survey of more than 400 technical project leaders that identified reasons for poor project control and suggested criteria for the effective control of projects.

Finally, it should be remembered that no real-life project control system can be completely automated. The project manager must maintain a considerable degree of flexibility and must use good judgment in interpreting the raw facts that constitute project status reports.

7.7 DISCUSSION QUESTIONS AND PROBLEMS

1. Suppose you are the project manager for a software development project. Discuss several tradeoffs that may be necessary or desirable in managing the triple constraint (Rosenau's term) or three-legged stool (Babcock's term) for this project.

2. You are involved in the planning, design, and construction of a major electric transmission line project. Identify several of the tasks that will be accomplished during each phase of the project's life cycle.

3. Consider a toy manufacturing company organized along traditional functional lines. What kind of organization structure do you recommend to accomplish the task of developing, test marketing, manufacturing, and distributing a radical new type of riding toy fire truck.

4. Suppose you are the sole mechanical engineer on a small shopping center design project. Your design firm is configured as a matrix organization. List five potential sources of role conflict in this position. To whom might you turn for help in resolving each conflict?

5. Your firm is preparing a proposal to dispose of spent uranium for the U.S. Nuclear Regulatory Commission, whose project representative will be a nuclear engineer with 20 years of experience. As part of the proposal, which must include a needs assessment and facility design, as well as oversight of construction and start-up, you are to submit the name and credentials of your proposed project manager. You have three possible candidates.

   - *Frank:* 45 years old, chemical engineer, MBA, extremely personable, trouble meeting deadlines, excellent relationships with clients; available immediately.

   - *Susan:* 32 years old, nuclear engineer, outstanding technical background and experience, likely to be your head of project management within 10 years, a little weak on people skills; in the middle of another project but could be moved.

   - *Michael:* 57 years old, Ph.D. in physics, research background with 20 years as a research project manager, highly respected for his technical expertise, fine writer and speaker; assisting with this proposal preparation but not available as project manager until 2 months after project start.

   As a company president, which name will you propose as project manager for this project, and why?

6. Draw an activity-on-node schedule network for a project whose activities are described in the accompanying tabulation.

Activity	Immediate Predecessor(s)
Establish scope	(None)
Appoint project team	(None)
Perform literature review	Establish scope
	Appoint project team
Conduct needs assessment	Establish scope
	Appoint project team
Identify alternatives	Conduct needs assessment
Conduct research	Perform literature review
	Identify alternatives
Carry out economic analysis	Identify alternatives
Identify recommendation	Conduct research
	Carry out economic analysis

7. Draw an activity-on-arrow network for the project in Question 6.

8. For the accompanying project schedule network, find the critical path and the number of weeks of slack in each noncritical activity. Assume that the early finish of the last activity is the project's target completion time.

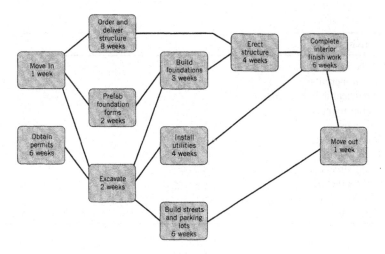

9. In Question 8, discuss the implications of a target completion time of 18 weeks.

10. Identify the critical path in the accompanying project schedule network. Give the early start, early finish, late start, late finish, and slack for each of the following activities: 2–6, 14–16, 20–22, and 22–24.

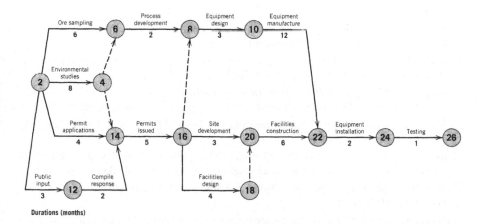

11. Now suppose the activities in Question 8 have the values shown below. Prepare a table of earnings to date at the end of week 0, 5, 10, 15, 20, and 22, on both an early start–early finish and late start–late finish basis. Graph this information, and describe the project status if $153,000 has been earned by the end of week 15. Assume that an activity's value is spread evenly over its duration.

Activity	Value
Move in	$2,000
Obtain permits	$18,000
Order and deliver structure	$40,000
Prefab foundation forms	$6,000
Excavate	$12,000
Build foundation	$24,000
Install utilities	$16,000
Build streets and parking lots	$18,000
Erect structure	$32,000
Complete interior finish work	$30,000
Move out	$2,000

12. Suppose a project has the following cost information at a certain point in time: ACWP = $270,000; BCWS = $270,000; BCWP = $239,000. Calculate all relevant variances and comment on the cost control status.

13. For ACWP = 100 kroner, BCWS = 90 kroner, and BCWP = 115 kroner, determine the status of the project's cost control. If 60% of the work has been completed, give the total estimated cost at completion.

7.8   REFERENCES

Anderson, D. R., D. J. Sweeney, and T. A. Williams. 1991. *An Introduction to Management Science.* St. Paul, MN: West Publishing Company.

Babcock, D. L. 1991. *Managing Engineering and Technology: An Introduction to Management for Engineers.* Englewood Cliffs, NJ: Prentice-Hall.

Bennett, F. L. 1977. *Critical Path Precedence Networks: A Handbook on Activity-on-Node Networking for the Construction Industry.* New York: Van Nostrand Reinhold.

Cable, D. P., and J. R. Adams. 1982. *Organizing for Project Management.* Drexel Hill, PA: Project Management Institute.

Cleland, D. I. 1982. "The Human Side of Project Management." In Kelley, A. J., Ed., *New Dimensions of Project Management.* Lexington, MA: Lexington Books, pp. 181–197.

Cleland, D. I., and W. R. King. 1983. *Systems Analysis and Project Management,* 3rd ed. New York: McGraw-Hill.

Goodman, R. A. 1976. "Ambiguous Authority Definition in Project Management." *Academy of Management Journal,* Vol. 19, No. 4, December.

Hogård, P. 1992. Personal interview. SKANSKA International, Stockholm, October 12.

Kelley, J. E., and M. R. Walker. 1959. "Critical Path Planning and Scheduling." *Proceedings, Eastern Joint Computer Conference,* Boston, December 1–3.

Kerzner, H.. 1989. *Project Management: A Systems Approach to Planning, Scheduling and Controlling,* 3rd ed. New York: Van Nostrand Reinhold.

Larson, E. W., and D. H. Gobeli. 1987. "Matrix Management: Contradictions and Insights," *California Management Review,* University of California, Vol. 29, No. 4.

Malcolm, D.G., J.H. Roseboom, C.G. Clark, and W. Fazar. 1959. "Applications of a Technique for Research and Development Program Evaluation." *Operations Research,* Vol. 7, No. 5, September–October, 646–669.

Meredith, J.R., and S.J. Mantel Jr. 1995. *Project Management: A Managerial Approach.* 3rd ed. New York: Wiley.

Moder, J.J., C.R. Phillips, and E.W. Davis. 1983. *Project Management with CPM, PERT and Precedence Diagramming,* 3rd ed. New York: Van Nostrand Reinhold.

Peters, T., and R. Waterman. 1982. *In Search of Excellence.* New York: Harper & Row.

Poirot, J.W. 1991. "Organizing for Quality: Matrix Organization." *Journal of Management in Engineering,* American Society of Civil Engineers, Vol. 7, No. 2, April, 178–186.

Poirot, J.W. 1993. Personal communication, June 11.

Posner, B.Z. 1987. "What It Takes to Be a Good Project Manager." *Project Management Journal,* Vol. 18, No. 1, March, 51–54.

Rosenau, M.D., Jr. 1984. *Project Management for Engineers.* New York: Van Nostrand Reinhold.

Silverman, M, 1987. *The Art of Managing Technical Projects.* Englewood Cliffo, NJ: Prentice-Hall.

Silverman, M. 1988. *Project Management: A Short-Course for Professionals,* 2nd ed. New York: Wiley.

SKANSKA. no date. "SKANSKA's Breakdown Structure for Construction Project Control." SKANSKA Project Management Systems, Stockholm.

Spinner, M. 1981. *Elements of Project Management: Plan Schedule and Control.* Englewood Cliffs, NJ: Prentice-Hall.

Stewart, J.M. 1965. "Making Project Management Work." *Business Horizons,* Indiana University School of Business, Vol. 8, No. 3, Fall, 54–68.

Stuckenbruck, L.C. 1981. *The Implementation of Project Management: The Professional's Handbook.* Reading, MA: Addison-Wesley.

Thamhain, H.J., and D.L. Wilemon. 1986. "Criteria for Controlling Projects According to Plan." *Project Management Journal,* Vol. 17, No. 2, June, 75–81.

─────8
Engineers and the Law

The United States system of jurisprudence is based on English law, which British colonists brought with them in the seventeenth century. Local law and custom had prevailed in England until the Norman conquest, but after 1066, national councils and, later, and a system of king's courts and a Parliament were established. The practice of early English judges of basing their opinions on previous similar cases formed a pattern that evolved into the process known today as common law, or the law of judicial precedent. Two ancient codes also influence the Western laws. In about 2000 B.C., the Babylonian ruler Hammurabi promulgated the code that bears his name. This code's best-known precept ("an eye for an eye and a tooth for a tooth") was often applied, figuratively if not literally, as justice was meted out in the early days in the United States. The Ten Commandments of Moses have also influenced our legislation and court decisions, and they have the distinction of being short and simple.

"Law," in the sense we use it here, has many definitions. The American Law Institute defined law as "the body of principles, standards and rules which the courts . . . apply in the decision of controversies brought before them." Justice Oliver Wendell Holmes wrote, "Law is a statement of the circumstances in which the public force will be brought to bear through the courts." Professor H. E. Willis suggested four characteristics of the law: (1) a scheme of social control, (2) for protection of social interests, (3) recognition of a capacity in persons to influence the conduct of others, and (4) the machinery of the courts and legal procedures to help the person with the capacity. (Definitions reported in Corley and Black, 1973.)

Law arises because individuals do not exist in isolation. Thus, society has established a set of rules and procedures that both restrict the conduct and protect the rights of its members.

One of the striking features discerned upon study of law, and one that is often frustrating to engineers, is the law's apparent vagueness. In a profession that is characterized by numbers representing the "exact" value of an impedance, a column stress, a pipe flow, or a rate of return (even though these exact values must always be interpreted and applied with some degree of judgment), we have trouble with what often seems to be a lack of logic or common sense. Lord MacMillan, Lord Chief Justice of England, wrote, "In almost every case except the very plainest, it would be possible to decide the issue either way with reasonable legal justification." And Justice Benjamin Cardozo, a famous jurist in the American system, said, "The problem is to find the least erroneous solution."

224

The legal process is devoted to resolving conflicts and is therefore inherently adversarial. The need to balance interests, to find some semblance of right, and to do so within a reasonable time limit, usually leads to *compromise,* an approach engineers and scientists might liken to deciding the thrust in a rocket engine by popular vote. Bockrath (1986) suggests the following attitude as you begin to study the world of legal principles and processes:

> Law should be thought of as a process rather than a set of rules: A process by which disputes are resolved. Oftentimes the dispute has no right or wrong but rather is a contest between competing interests, both of which are legitimate. Rules exist to be sure, but economic and political pressures and considerations are a part of the process, as are the personalities of the litigants, judges and attorneys. To expect perfect justice from a process of human intervention and administration is to seek disappointment.

Our purpose here is not to make you a lawyer. And we certainly do not suggest using the brief comments in these chapters in lieu of consulting a legal expert. It is common knowledge that a little knowledge is dangerous and may well be worse than no knowledge at all. With this caveat, we present an overview of some legal principles with which most engineers deal at least occasionally. We shall define some terms, give some examples, and suggest how these ideas apply to engineering design, construction, and operations. In the process, you should become aware of the circumstances in which you will need to consult an attorney.

In this chapter, we define several types of law and describe the federal and state court systems. Procedures used to prepare for and conduct trials are outlined. We then introduce three topics not covered in succeeding chapters that nevertheless can be especially important to practicing engineers—the laws of agency and such intellectual property rights as patents, and the importance of regulatory compliance. After a review of some forms of dispute resolution outside the more familiar legal setting, here we close the chapter by listing, with references, some other topics not covered for reasons of space.

Chapters 9, 10, and 11 deal with the basics of a topic of prime importance to engineers, the law of contracts (Chapter 9), the application of contract law principles to engineering and construction (Chapter 10), and the legal principles applicable to the various types of permanent and temporary business organizations (Chapter 11).

Closer to the end of the book, we discuss the law of torts, as a foundation for our consideration of professional liability (Chapter 12), and the role of the engineer as an expert witness (Chapter 14).

8.1 FOUR BASIC TYPES OF LAW

An understanding of the several categories of law will be useful when we discuss applications of legal principles to engineering management, which most often are based on common law. We adopt the fourfold classification presented by Schoumacher (1986).

8.1.1 Constitutional Law

Constitutional law sets up the basis for the operation of a government, including its powers and its limitations; a constitution establishes the framework for the government's political structure. Fundamental relationships between the individual citizen and the government are set forth, including rights that may not be infringed. A constitution may be unwritten, as it is in the United Kingdom, or written, which is the case for the United States and each of its 50 states. Constitutional law includes determining whether statutory laws are valid with respect to the governing document (e.g., "Is it constitutional?").

Constitutional law affects the engineer in a number of ways. For one, constitutions permit the federal and state governments to enact regulatory laws promoting safety, health, and welfare—an inherent power known as "police power." Laws governing the registration and practice of professional engineers and those that establish environmental standards derive from this constitutionally provided police power. Constitutions also set forth "due process" standards, which require that legal proceedings be conducted in accordance with certain rules.

8.1.2 Administrative Law

The ongoing operations of federal and state government depend to a large extent on the various agencies, commissions, and the like within the executive branch. Administrative law deals with establishing and enforcing the rules and regulations provided for in the police power provisions of the various constitutions. A regulatory agency may be required by a legislative statute to propose, accept comments on, revise, and implement a certain safety regulation; these actions constitute *establishing* the regulation. *Enforcing* the rules and regulations involves inspection, citations, hearings, reporting, decisions, and other proceedings related to routine operation and reported violations.

Examples at the federal level include the regulations promulgated by such agencies as the Interstate Commerce Commission, Federal Communications Commission, Department of Energy, Environmental Protection Agency, and Occupational Safety and Health Administration. At the state level, we have state utility commissions, licensing boards, and departments of the environment as examples. In Section 8.9, we discuss the engineering manager's role in several aspects of regulatory compliance.

8.1.3 Statute Law

A statute is a law enacted by a duly authorized legislative body. Bockrath (1986) writes that a statute "represents the express written will of the lawmaking power and is rendered authentic by promulgation in accordance with certain prescribed formalities." Although enactments by municipal legislative bodies are usually referred to as "ordinances," we can consider them to be "statutes," along with acts of Congress and laws passed by state legislatures.

An example of a statute that may affect the engineering manager is the statute of limitations; each state has determined through its legislative process a time after which a malpractice suit cannot be filed against an engineer or a product liability suit against a manufacturer. We deal with this topic in Chapter 12.

8.1.4   Common Law

The United States has primarily a common law system of jurisprudence, based on English common law. (Louisiana is a notable exception; most of the law in this state is based on the French Civil Code and some Spanish law.) Common law can be thought of as the law of judicial precedent, wherein decisions on prior similar cases are used to govern the suit under consideration.

> The common law is described by Blackstone as the unwritten law (lex non scripta) as distinguished from the written or statute law (lex scripta), i.e., enacted law.... Kent says it includes "those principles, usages and rules of action applicable to the government and security of persons and property, which do not rest for their authority upon any express or positive declaration of the will of the Legislature." (quoted in Bockrath, 1986)

The principle that prior decisions related to the same legal point are binding on the court is known as stare decisis, or "let the decision stand." Thus, the law is predictable (to some extent, at least!) because of the binding nature of earlier decisions. The court may override this binding effect, however, by deciding that the particular case before it is sufficiently different from the one on which the precedent was built. Furthermore, a particular case may have elements of several cases whose results taken together are contradictory. Also, a court may occasionally simply overrule an established precedent and in an unexpected way rule.

Some authors (e.g., Schoumacher, 1986) object to the term "common law" because English common law includes both decisions of English courts and statutes passed by Parliament. The term "case law" is often used to denote the law of judicial precedent which is the subject of this section.

Bockrath (1986) provides an example of the operation of judicial precedent and the role of common law as applied to an engineering situation, as follows:

> Assume that a dispute arises between an owner and a contractor regarding the meaning of the words "subsurface construction" in a contract. The question is whether this term refers to all construction below ground level or merely to whatever construction is to be finally covered with earth. The dispute is taken to court. The court decides that, under the particular wording of the contract and considering the intent of the parties, justice demands that the dispute be settled in a certain way and that the contractor and the owner fulfill their respective obligations accordingly. The official decision acquires a name and citation by which it may readily be located in the published reports of the cases. Henceforth, if other litigation arises in which all

significant circumstances are analogous to those in this previously settled dispute, the prior decision (unless it has been since reversed by a higher court or criticized in other cases) will be treated as a precedent of considerable weight.

8.2   CIVIL LAW: TWO MEANINGS

The term "civil law" has evolved to have two meanings. In one sense, it means codified law based on Roman codes; the countries of continental Europe use this kind of "civil" or written law. On the other hand, civil law is also defined as that which controls the action between persons or organizations rather than acts against the state. Thus, in the sense in which we shall use the term, this private law deals with such matters as contracts, real estate, personal injuries, sales, and other business matters (Lyden, Reitzel, and Roberts, 1985; Vaughn, 1983).

We distinguish civil law from *criminal law,* which defines crimes (e.g., felonies), establishes criminal trial procedures, and fixes limits of punishment. A crime is an act punishable by the state because it is an offense against the public order and well-being. It should be noted that someone who takes personal property unlawfully in a criminal act commits an offense against the state even though an individual was harmed. A fine, if any, will be paid to the state rather than to the victim. However, the victim has the right to bring a *civil* suit against the perpetrator for damages. Seldom is the engineer involved in criminal matters related to engineering activities; those rare cases involve such matters as fraud, perjury, and willful negligence (Firmage, 1980). The balance of the chapters on legal principles deals exclusively with private, or civil, law.

8.3   LAW VERSUS EQUITY

In early English law, only certain kinds of injustice were covered by the codified law of that time, and thus remedies were available only for a limited number of situations. For example, the king's courts provided no way to issue an order restraining someone from trespassing on another's land. Thus persons seeking relief in such matters would go directly to the king and his counsel. In time, the king began referring these cases to his spiritual adviser, the chancellor. Gradually, a separate system of courts "of chancery" arose, whose judges were called chancellors. The system continues today, although such "equity" matters, as well as law matters, are often handled by the same judge.

Common law remedies available today are limited primarily to money damages. If money is not a sufficient remedy, the case is not a common law case; that is, the common law has no remedy. For example common law cannot prevent a wrong from taking place, order persons to fulfill their obligations, or correct mistakes (Vaughn, 1983). It has been said that "as a remedy, law gives only money damages, whereas equity gives the plaintiff what he [or she] bargains for" (Dillavou and Howard, 1957). The remedies and courses of action that equity makes possible include injunction, specific performance, rescission of a contract for fraud, subrogation (substituting one person for another so that the second acquires the legal rights of the first), and reformation of instruments. Vaughn (1983) describes how an injunction works as follows:

> Black, a contractor, is hired to build a structure for White. During the excavation, Gray, next door, notes an impending separation of his house from its foundation. The walls start to crack and the ceiling begins to bulge. At this point, at common law, Gray has no remedy. At common law he would have to await whatever damage might be forthcoming and sue to get compensation for it. However, in equity he can do something about it *now*. He can obtain a *temporary injunction or restraining order* to prevent the excavation next door until something more appropriate can be done to prevent his house from sliding into the hole.

In a few states, such as Delaware, there are separate courts of equity. In others, the same courtrooms and judges are used but the procedures are different. In still others, as well as in the U.S. federal court system, law and equity are combined. Nonetheless, our courts still recognize a separate body of "equity law." For a suit to be considered an equity suit, a property right must be involved, and it must be established that a remedy at law would not be adequate. Equity also is distinguished from law by the generally more rapid proceedings in the former case, and by the lack of a jury, which results in more privacy (Vaughn, 1983).

8.4   THE UNITED STATES COURT SYSTEM

Laws are enforced in a number of ways: by example (i.e., general adherence by the public), by the court system, by administrative agencies, and by such alternate means as arbitration, which we shall discuss Section 8.10. We use the word *court* here to mean a tribunal established by a government to administer justice. A court's primary function is to decide controversies between two parties called *litigants*. The final decision is called a *judgment*. In the United States, we have both federal and state court systems.

8.4.1   Federal Courts

Most claims that are brought before federal courts could also be tried in state courts. Some matters, including bankruptcy, patent, copyright, and admiralty cases, actions involving the United States, and violation of federal criminal statutes, can be brought only in federal courts.

Figure 8.1 is a very simplified diagram of the federal court system. Article III of the United States Constitution provides that "the judicial power of the United States shall be vested in one Supreme Court, and in such inferior courts as the Congress may from time to time ordain and establish." At present, in addition to the Supreme Court, we have district courts, courts of appeals, and special courts. All federal judges are selected by the president and confirmed by the Senate; they serve for "good behavior," which usually means a lifetime appointment.

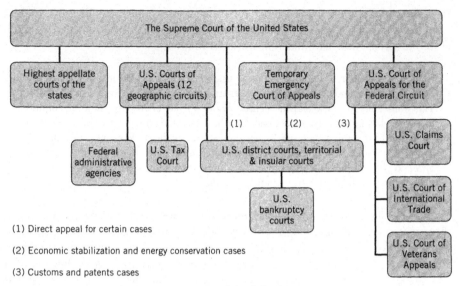

Figure 8.1. Diagram of the federal court system.

Only cases involving ambassadors, public ministers and counsels, or those in which a state is a party, may originate in the *Supreme Court of the United States* (i.e., the Supreme Court). All other cases go to the Supreme Court by means of the appeal process; and deciding these "other cases" is by far the court's major function. The constitution makes no provision for the number of Supreme Court justices. In practice the number has varied between four and ten; for a long while it has been nine.

A litigant who is not satisfied with the judgment of a district court may appeal the decision to one of the *courts of appeals*. The primary role of these bodies is to review final decisions of district courts and administrative agencies with the goal of discerning any errors in the law or its application. The three-judge panels that constitute courts at this level do not conduct trials; issues of fact have already been decided in a lower court.

The *district courts* are the trial courts of the U.S. federal judicial system. Every state has at least one federal district court. Most cases are tried before a single judge. In addition to controversies involving the federal constitution or federal statutes, these courts can hear civil disputes between citizens of different states. The amount in dispute in civil cases must be at least $50,000.

Special courts within the federal system include the Tax Court, the Claims Court, bankruptcy courts, the Court of Customs and Patent Appeals, the Court of International Trade, the Court of Military Appeals, and territorial courts. The Temporary Emergency Court of Appeals was created to handle appeals involving economic stabilization and energy conservation cases.

8.4.2   State Courts

State court systems in the United States exhibit considerable diversity in organization and titles. In every case, however, they have a minor judiciary, as well as higher trial courts and at least one level of appeals court. The *minor judiciary* includes courts of very limited jurisdiction, such as police courts, small claims courts, justices of the peace, and juvenile courts. Next, states have *trial courts of general and original jurisdiction,* with such names as circuit courts, superior courts, and courts of common pleas. Often, these courts may hear civil cases only when the amount in controversy exceeds a certain dollar limit. In some states, these courts also function as appeal courts for cases from the minor judiciary.

Finally, there is in each state at least one level of appeals court. Like the federal system, appeals courts hear no new testimony; they review the original trial procedure from the record of the trial and decide whether the law was applied properly. A state's highest court is usually called the "supreme court," although New York's highest appellate court is called the "court of appeals" (its trial court of original jurisdiction is called the "supreme court").

The State of Alaska offers an example of a relatively simple state court system (Figure 8.2). The routes of appeal for civil and criminal cases in Alaska are shown in Figure 8.3.

8.4.3   Jurisdiction

In what state should a lawsuit be filed, and should a given suit be filed in a state or a federal court? A court's *jurisdiction* is its authority, as specified by constitution or statute, to hear and decide cases presented to it. Whether a court has jurisdiction in a particular case depends on the type of case, the geographical region of both parties, the subject matter, and, in some cases, the amount of money involved.

If both parties reside in the same state, the case is generally filed there. When more than one state is involved, the question of location arises. The general guidelines, based on the U.S. Constitution as interpreted by the Supreme Court, are that "the courts of one state can exercise jurisdiction over a defendant located in another state only if the transaction in which the defendant is involved has sufficient minimum contacts with the plaintiff's state" (Schoumacher, 1980). For example, suppose two parties, one from Oregon and one from Massachusetts, enter into a contract to be executed in Boston; the contract goes sour, and the Oregon party decides to file suit for breach of contract. The Oregon party can always file the suit in Massachusetts, but could such a suit be filed in Oregon? The answer depends on what contacts the Massachusetts party has had in Oregon regarding the agreement and the contract performance.

Sweet (1989) describes a claim brought by a contractor in federal court in Alaska against three out-of-state parties—a supplier, a subsupplier, and a structural engineer who designed precast concrete work for the supplier. A structure built by the contractor in Alaska collapsed; the materials had been designed and shipped to Alaska by the out-of-state defendants. Since they were not in Alaska, the defendants were served papers out of state.

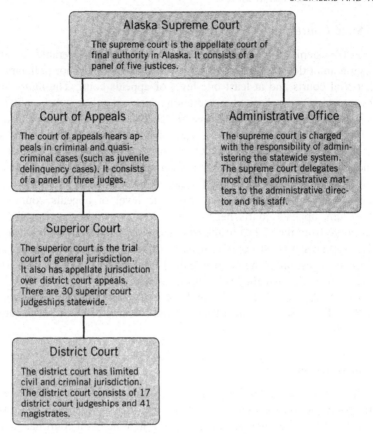

Figure 8.2. Organization of the Alaska court system. (From Alaska Court System, 1991.)

None of the out-of-state defendants were qualified to do business in Alaska, none had agents in Alaska, and all the work had been performed outside Alaska. However, the court held that all the defendants were subject to the jurisdiction of the Alaska federal court in which the action had been brought. Although the supplier and the sub-supplier to the supplier had had some minimal contact with Alaska, the structural engineer had *never* even been in Alaska in connection with the project. But because he knew that the building was to be constructed in Alaska and he sent his design into the stream of commerce aimed at Alaska, the court concluded that he, along with all of the other out-of-state defendants, was subject to the Alaska court. (Sweet, 1989)

The decision of whether to file a lawsuit in a federal or a state court is guided in part by federal laws restricting federal cases to those in which (1) residents of more than one state are involved *and* the amount of money involved exceeds $50,000, as we have seen, *or* (2) an interpretation of the U.S. Constitution or a federal statute is involved.

After deciding in which court system to file a suit, the plaintiff must select a specific court, based on the subject of the suit and the proper "venue," or

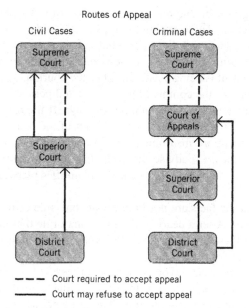

Figure 8.3.   Routes of appeal in the Alaska court system. (From Alaska Court System, 1991.)

geographical location. For example, a typical state venue law says that a suit must be filed in the defendant's county of residence or the county in which the actions leading to the suit occurred.

8.5   TRIAL PROCEDURE

Since the engineering manager is likely to become involved in a lawsuit at some time during his or her career, we list the steps through which a civil lawsuit typically passes. It is helpful to divide the process into pretrial activities and the trial itself. The party bringing the complaint ("filing the suit") is called the *plaintiff*, and the party being sued is the *defendant*. There can be more than one plaintiff and/or more than one defendant in any one lawsuit. Furthermore, the defendant is permitted to file a *counterclaim* against the plaintiff.

8.5.1   Pretrial

From the viewpoint of the party filing the suit, the first step, after identifying the problem and deciding to file suit, is to engage an attorney. Components of this action include identifying the issues, perhaps even seeking advice on whether to file suit, agreeing on a scope of work and a fee schedule, and establishing other details of the agreement between attorney and client. Sweet (1989) discusses some important issues with respect to the fee schedule.

The lawsuit begins with the *filing of a complaint* by the plaintiff in a court of appropriate jurisdiction. The complaint names the defendant(s), states the cause of the action, requests relief, and describes what is being sought. Next, a copy of the complaint, together with a *summons,* is delivered to the defendant. The summons states that the defendant is being sued, identifies the court, and directs the defendant to provide a response to the complaint within a stated period.

The defendant may respond in a number of ways. If the response consists of no answer at all, a default judgment is issued in favor of the plaintiff. Or, the response may (1) state that there are insufficient facts to support legal action (a demurrer), (2) deny the allegations, (3) advise the court that an affirmative defense is valid, including new information that would bar the plaintiff's recovery, and/or (4) file a counterclaim.

If it appears that the facts are not in question, one side or the other may request a *summary judgment.* After a hearing, the judge may rule that there are no material issues of fact and therefore no trial is necessary. In that case, judgment is then entered for the plaintiff or defendant.

Engineers often become involved in the process of pretrial *discovery.* The rules of discovery require each side to make available to its adversary as much information as possible about its position. The principal devices are as follows:

- Depositions (sworn testimony of witnesses, delivered at a hearing out of court).
- Interrogatories (written questions addressed by one party to the other).
- Requests to produce documents or other physical evidence.
- Requests for admissions (i.e., that one side admit to the genuineness of documents or facts: "Do you admit that Carolyn Jones was an electrician employed by FLB Contractors on the XYZ project?")
- Physical and mental examinations of persons in the controversy.

Before the trial begins, the judge will hold one or more *pretrial conferences* with the attorneys for both sides to determine the status of discovery, to frame and simplify the issues, to establish trial procedures and schedules, and, especially, to try to come to an out-of-court settlement.

8.5.2 Trial

For some trials, such as those in equity, no jury is allowed, and the judge alone decides the case. In others, either side may request a jury trial. If a jury trial is requested, the first item of business at the trial is to select the jury. After that, the steps described below proceed essentially identically regardless of whether there is a jury.

Opening statements by counsel for each side are followed by presentation of evidence. First, the attorney for the plaintiff introduces witnesses and elicits testimony to support the client's side through a process called *direct examination.* Any witness can be questioned by the opposing attorney under *cross-examination.*

One more round of questioning of the witness may take place, with the plaintiff's attorney conducting a *redirect examination,* followed, perhaps, by a *re-cross-examination* by the defendant's attorney.

After the plaintiff's side has presented all its witnesses and introduced its physical evidence, the plaintiff "rests." The defendant then may move for a *nonsuit.* If this so-called motion to dismiss is successful, it means that the judge believes that the plaintiff's case has not been proved. If such a motion is denied, the defendant then introduces witnesses and physical evidence to try to contradict the testimony and other evidence presented by the plaintiff, beginning with direct examination and proceeding as just outlined.

In a jury trial, after all the evidence has been presented, either side, or both, may argue that a verdict in its favor is indicated by the evidence and the applicable law. If the judge decides to issue such a *directed verdict,* he or she is essentially taking the case away from the jury. If there is no directed verdict, each party presents *closing arguments,* and the judge issues instructions to the jury (if there is a jury), which guide the panel in deciding the issue or issues.

The jury deliberates and reaches a *verdict.* In federal courts and many state courts, the decision must be unanimous. In other states, a minimum of three-quarters of the jurors voting on one side will be sufficient to return a verdict in favor of that side. If the decision is in favor of the plaintiff, the jury often must also decide the amount of damages to which the plaintiff is entitled.

After the verdict, the losing side may move for a *judgment n.o.v. (non obstante veredicto:* judgment notwithstanding the verdict). The thrust of such a motion is to argue that the decision of the jury is not supportable by the evidence presented at trial. When the judgment has been entered, either side may make a *motion for a new trial,* based on arguments such as the following: (1) the judge committed prejudicial error, (2) the damages were either excessive or insufficient, (3) the other side was guilty of prejudicial misconduct during the trial, or (4) new evidence was discovered after the trial.

If all the posttrial motions are denied, the judge enters a final judgment, which either party may appeal to a higher court. The party that files the appeal is the appellant; the other party is the appellee or the respondent. As pointed out in Sections 8.4.1 and 8.4.2, the appellate court does not retry the case. Its job is to determine whether any *error of law* was made during the trial court proceedings. After studying written "briefs" submitted by attorneys for both sides and listening to their oral arguments, the appellate court renders its decision in writing. If it finds no error of law, it will affirm the trial court's judgment. If prejudicial error is found, the original judgment may be reversed or modified. If the evidence does not clearly justify a decision for one party or the other, the case may be *remanded* (sent back to the trial court for a new trial).

Figure 8.4, from Lyden, Reitzel, and Roberts (1985), traces the path of civil litigation through the court system. Note that the flow is divided into the processes of issue identification and issue resolution.

8.5.3   Evidence

Evidence is the means by which questions of fact are proven in a trial. But what evidence is allowed, and which side has the burden of proving the facts if its case

Issue identification process   Issue resolution process

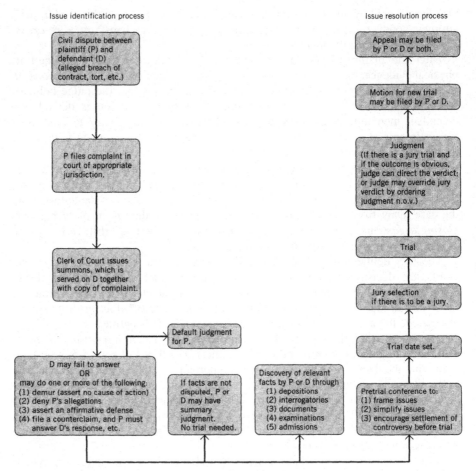

Figure 8.4. Flow of civil litigation through the court system. (From Lyden, Reitzel, and Roberts, *Business and the Law*, copyright © 1985. Reprinted by permission of McGraw-Hill, Inc.)

is to prevail? The rules of evidence are so numerous that we can only touch on some highlights here. In a criminal trial, the burden of proof to show that the defendant is guilty "beyond a reasonable doubt" always lies with the state. Without such evidence, the defendant is to be found innocent. In a civil case, the plaintiff generally has the burden of proof and the "burden" is to prove an issue by a "preponderance of the evidence," which means that the judge or the jury weighs the evidence from each side and decides which is "more correct."

Evidence can be classified as *real evidence,* which is a physical thing, such as a broken bolt or a signed contract, or *testimony,* which consists of statements by witnesses. Both these types of evidence are admissible under certain circumstances. The *best evidence* rule says that originals, and not copies, of written materials ("primary" evidence) must be used as evidence if they are available. If not, the court will admit copies of such documents or testimony about their existence and contents (secondary evidence).

Hearsay evidence is secondhand information; it is testimony that the witness claims to have heard from another person. Generally such evidence is not allowed if the other side objects in a timely fashion. As usual, there are exceptions to the hearsay evidence rule; testimony about dying declarations usually is admitted, for example.

The *parol evidence rule* states that oral testimony that alters the terms of a writing will not be admitted as evidence. The presumption is that the parties to a contract, for example, have reduced their most recent thoughts about the contract to writing, and therefore the written materials are the best evidence of the desired meaning. The rule does not force the exclusion of oral testimony that disputes an oral agreement or adds information about an aspect of the agreement that was not in writing.

Generally, witnesses are expected to limit their testimony to "facts" and to avoid expressing their opinions. However, two categories of *opinion evidence* are allowed. The *ordinary witness* is permitted to offer opinions that do not require any particular skill in arriving at, such as the weight of a person. If reasoning and conclusions are required on a complex issue, an *expert witness* from the special subject area is allowed to give his or her opinions as to the facts:

> The real distinction between the expert witness and the nonexpert is that the former gives the results of a process of reasoning which can be mastered only by those of special skills, while the nonexpert testifies about subject matter readily comprehended by the ordinary person and it gives the results of a reasoning process familiar to everyday life." (Bockrath, 1986)

In Chapter 14, we present some guidance for the engineer who serves as an expert witness.

8.6 CITATION OF CASES

We have discussed the essential position of case law in the American judicial system. Since cases must be accessible if they are to be used as precedents, we note here the common method of legal citation. The West Publishing Company of St. Paul, Minnesota, publishes reports of all state cases that reach the appeals courts and all federal cases. Volumes called *Reporters* are published for each state and for each of seven districts in the United States, in the case of state cases, and for each of the federal courts. Each entry consists of a summary of the facts of the case, a list of legal points, and the full text of the appeals court's opinion.

A rather standard means of citing cases has evolved with its own version of "shorthand." Consider the following citation:

South Burlington School District v. Calgagni-Frazier-Zajchowski Architects, Inc. 138 Vt. 33, 410 A. 2d 1359 (1980).

In this case, South Burlington School District was the plaintiff and Calgagni-Frazier-Zajchowski Architects, Inc., was the defendant in the original trial court proceedings. The case report may be found in Volume 138 of the Vermont case books beginning at page 33; it may also be found in Volume 410 of the second series of the *Atlantic Reporter* beginning at page 1359. The appeals court decision was rendered in 1980.

The following citation is similar; its interpretation is left to the reader:

United States Constructors & Consultants, Inc. v. Cuyahoga Metropolitan Housing Authority, 35 Ohio App. 2d 159, 163, 300 N.E. 2d 452, 454 (1973).

8.7 AGENCY

An engineer often acts as an agent for another party. The law of agency deals with the legal relationships involved when one person acts through another. The person who acts on behalf of another is called the *agent,* while the person or organization on whose behalf the agent is acting is called the *principal.* The principal authorizes the agent to deal with third parties in certain business undertakings, and actions of the agent who is acting within that authority will bind the principal. Difficulties arise primarily when the extent of the agent's authority is unclear.

The agency relationship differs from a contractual relationship, which is the subject of the next chapter. A contract is a two-sided proposition, in which the parties are dealing with each other "at arm's length" and looking out for their own interests. In an agency relationship, the parties are, as suggested by Nord (1956), "so to speak, 'locked in a fond embrace.'" This author goes on to say that an agency is a one-sided proposition.

> Both sides are the principal. Both the agent and the principal are looking out for a common interest—namely that of the principal. (If you have the choice, therefore, be the principal, rather than the agent.) Once this fact is grasped, most of the law of agency becomes immediately self-evident. (Nord, 1956)

At the risk of boring the reader with some "self-evident" principles, we spend the next several paragraphs describing how the agency relationship is created and terminated, and we present some examples of agency relationships in engineering management.

We distinguish between the principal/agent and the employer/employee relationship in that the agent is authorized to act for his or her principal in dealings with third parties, whereas the employee is typically hired to do a specific job under well-defined restrictions. Suppose your engineering firm puts a land surveyor on the payroll (rather than hiring a surveying consultant under a contract) to lay out subdivision lot corners, roadways, and watercourses. That person is an employee. If the surveyor also deals with outsiders on your behalf, such as property buyers or sellers or the drainage contractor, the relationship between you and this employee takes on the appearance of an agency relationship as well. Whether the surveyor's actions in dealing with those third parties are binding on you depend on the circumstances we describe in the following description of the creation of agency.

An agency relationship can be created in one of four ways (Vaughn, 1983). An *agency by agreement* arises when both parties intend to create the agency. This agreement, in most cases, does not need to be in writing. It grants to the agent actual authority to act for the principal in specified dealings with outside parties.

Suppose a person enters into a contract with a third party on behalf of a principal, even though he or she really is not authorized to act as this principal's agent. In that case, there was no agency relationship, and the "principal" is not bound to the contract. The same thing will happen if there is a true agency relationship but the agent exceeds the authority assigned by the principal. In either of these circumstances, the principal may choose to be bound by the contract under what is called *agency by ratification*. Thus, if a construction laborer who is sent to the hardware store to purchase nails also arranges for delivery of concrete blocks on his own initiative a contract between contractor and storekeeper has not been created, unless the contractor chooses to ratify the same, even if the blocks are in fact needed at the job site.

Most difficult agency cases arise when the principal's behavior reasonably leads a third party to believe that someone who claims to be an agent is acting on the principal's behalf, even though an agency relationship has not been created. Apparent authority is the key to resolving this situation. *Agency by estoppel* means that the principal is "estopped" from denying the existence of an agency relationship if that principal has clothed the agent with such apparent authority. The City of Delta Junction, Alaska, was unsuccessful in a lawsuit in which it claimed that Mack Truck, Inc., had granted to Alaska Mack sufficient authority to make Alaska Mack its agent. The city purchased from Alaska Mack a fire truck chassis that proved to be improper for its intended use. Claiming that Mack Trucks, Inc., should be at least partially responsible for the defect, since Alaska Mack appeared to be its agent, the city sued both Alaska Mack and Mack Trucks, Inc. If the city had been successful, the principle of agency by estoppel would have prevailed. In this particular case, the Alaska Supreme Court decided that such an agency relationship did not exist, and Delta Junction was left to deal with Alaska Mack alone. [City of Delta Junction v. Mack Trucks, Inc. 670 P 2d 1128 (1973)].

A fourth type of agency is *agency by necessity*. This case, which is rare, can arise in emergencies. For example, if an employee must deal with outsiders on behalf of his or her employer to avert a disaster, an agency is created even though the person did not have prior authorization from the employer.

The law with respect to negligent acts committed by an agent is somewhat unsettled. In general, an agent will be held liable to a third party for the agent's own negligent acts, even though they were committed while the agent was acting on behalf of the principal.

With respect to the design professional, Schoumacher (1986) cites three cases of interest. The first is an old Nebraska case [Erskine v. Johnson, 23 Neb 261, 36 N.W. 510 (1888)], in which an owner was held to be responsible for extra work performed by a third party because the architect who had ordered that work was found to be an agent of the owner. In another Nebraska case [Fuchs v. Parson Construction Co., 172 Neb. 719, 111 N. W. 2d 727 (1961)], a contractor followed plans and specifications for a building, and the architect certified final payment to the contractor, in essence certifying that the plans had been followed properly. After construction, the building settled. The court found that the architect was an agent of the owner; thus, by certifying completion, the owner's agent had acted with the authority of the owner and the contractor was not responsible for the settlement. In a Texas case [S. Blickman, Inc. v. Chilton, 114 S.W. 2d 646 (Tex. Civ. App. 1938)], the owner of a hotel was found to be liable to a hotel customer who was injured as a result of remodeling work. The injured customer sued the contractor, who made a successful defense by asserting that the architect was negligent. The court concluded that the hotel was at fault because the architect was an agent of the hotel.

When and how does an agency arrangement terminate? There are several possibilities. The agency agreement can specify that it will terminate (1) after a certain time period, (2) after the agent has completed a certain act, or (3) when a certain event occurs. Agency agreements usually terminate upon the death of the principal and almost always upon the death of the agent. Likewise, if some subject matter that is the reason for the agency is destroyed, or if some legislative act makes the purpose illegal ("supervening illegality"), the agency will terminate.

Certainly an agency relationship can be terminated by mutual consent of the two parties, and usually it can be terminated by one party or the other. Under the doctrine of apparent authority, if the principal acts in ways thay suggest to a third party that the agency still exists, the agent may retain the authority to bind the principal after the official termination (Bockrath, 1986; Sweet, 1989).

A *partnership* is a special kind of agency in which each of the partners is both principal and agent. The acts of any one of the partners bind the entire partnership. We take a closer look at partnerships in Chapter 11.

We conclude this section by referring to Figure 8.5, which depicts the "agency triangle," with its clear relationships between principal and agent and between agent and third party, and the question as to whether a relationship between the principal and the third party has been created. The ambiguity associated with this third leg is the primary difficulty with agency, as noted several times above. With respect to the three parties, Sweet (1989) reminds us that the design professional, whether in private practice or as a salaried employee, may be in any one of the three positions:

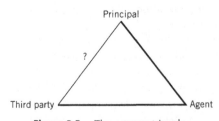

Figure 8.5.   The agency triangle.

Suppose the design professional is a principal (partner) of a large office. The office manager orders an expensive computer. In this illustration, the design professional, as a partner, is a principal who may be responsible for the acts of the agent office manager.

Suppose the design professional is retained by an owner to design a large structure. The design professional is also engaged to perform certain functions on behalf of the owner in the construction process itself. Suppose the design professional orders certain changes in the work that will increase the cost of the project. A dispute may arise between the owner and the contractor relating to the power of the design professional to bind the owner. Here the design professional falls into the agent category, and the issue is the extent of her authority.

Suppose X approaches a design professional in private practice regarding a commission to design a structure. X states that she is the vice-president of T Corporation. The design professional and X come to an agreement. Does the design professional have a contract with T Corporation? Here the design professional is in the position of the third party in the agency triangle.

8.8   INTELLECTUAL PROPERTY

One of the specialized areas of the law that may involve the engineer is the area of rights to and protection of "intellectual property." If you invent something, do you have exclusive use of it? How may whatever rights you have be protected? If you wish to use someone else's invention in your design, what restrictions may apply? The U.S. Constitution gives the Congress power to "promote the progress of science and the useful arts, by securing for limited times to authors and inventors the exclusive right to their respective writings and discoveries." We deal briefly with patents, trademarks, copyrights, and trade secrets. For greater detail, engineering law books by Schoumacher (1986), Sweet (1989), and Vaughn (1983), as well as *Understanding Patents and Other Protection of Intellectual Property,* coauthored by an engineer and an attorney (Foltz and Penn, 1990), are recommended.

8.8.1   Patents

A patent is a grant to an inventor by the government to "exclude others from making, using, or selling" the invention covered by the patent. The inventor has a lawful monopoly for a statutory period (17 years for most patents in the United States), in exchange for disclosing the invention to the public. Seven categories of inventions or discoveries may be patented: (1) an art or a process or method of doing something, (2) a machine, (3) a manufacture (anything made from raw material other than a machine), (4) a composition of matter, (5) a new or useful improvement of an item in any of categories 1–4, (6) a new variety of plant, and (7) a design. Design patents generally run for less than 17 years. Many examples probably come to mind. For example, a Swedish company obtained a U.S. patent for a rubberized asphalt pavement (a "composition of matter") and the process for producing the asphalt.

Note that an idea is not patentable, but the physical embodiment of the idea may be. Patent law requires that the invention be *novel* and *nonobvious*. It cannot be something that is already known and used. Publication or public use or sale of the invention more than a year before the patent application results in denial of patent rights. Nor can a patent be granted for an invention that would be obvious to someone with ordinary skill in the subject area. Furthermore, the invention must be both *useful* and *operative*.

After a preliminary search to determine tentatively that such an invention has not already been patented [not required, but strongly urged (Burgunder, 1995)], the inventor files an application and fee. The application has three main components: the petition, which requests that a "grant of letters patent" be issued; the specification, which describes the invention in words, pictures, drawings, and models, as appropriate; and the oath, in which the inventor certifies that he or she is the originator of the thing for which a patent is requested. The Patent Office then conducts an examination to ascertain whether the invention complies with the law and is truly something new, useful, and operative. The application may be approved in full, rejected, or approved in part. An application that has been rejected in full or in part may be amended and resubmitted. Rejected applications may be appealed.

Since holding a patent is a property right, the law provides procedures and punishments for infringements. Action is brought in any federal district court. Those found guilty of infringement may be subject to an injunction to bar their continued use of the invention, or to monetary damages to compensate the patent holder, or both. The plaintiff must prove that the defendant knew that the invention was patented; marking patented goods with the patent number is a common means of making sure the public knows about the patent.

An important issue among employees of research establishments, university schools of science and engineering, and manufacturing organizations is the right to ownership of a patent developed in the course of employment. Enlightened employers have clear written statements that set forth the organization's policies in this regard.

8.8.2 Trademarks

In the United States, trademarks are protected both by common law and by the Lanham Trademark Act of 1946. The act says, in part, "The term 'trademark' includes any word, name, symbol, or device or any combination thereof adopted and used by a manufacturer or merchant to identify his goods and distinguish them from those manufactured or sold by others." Part of the value of a trademark is its unique design, which identifies the products of a particular manufacturer.

To get a trademark registered in the United States, application is made to the Patent and Trademark Office. Only trademarks used in interstate commerce may be so registered. The office conducts a search, similar to a patent search, and, if no conflicts are identified, registration is issued. To protect the right to a trademark, such registration is not required, since some common law protection is available based on first use. But it does give national protection, and after 5 years with no protest, the mark becomes the property of the applicant. Renewal after 20 years is permissible.

8.8.3 Copyrights

U.S. copyright law is based on the Copyright Act of 1976, which provides for legal protection against copying for "original works of authorship." Categories of materials that can be copyrighted include literary works; musical works; dramatic works; pantomimes and choreographic works; pictorial, graphic, and sculptural works; motion pictures and other audiovisual works; and sound recordings. Copyright protection is provided for the life of the maker plus 50 years. Copyright registration is obtained by application to the Copyright Office of the Library of Congress; a fee is required, and usually two copies of the work must be filed in Washington. The law allows "fair use" of copyrighted materials by permitting copying "for purposes such as criticism, comment, news reporting, teaching (including multiple copies for classroom use), scholarship or research."

Sweet (1989) provides a helpful discussion of the issues related to copyright of the products of design professionals, beginning with the question of whether such protection is even desirable. In any event, neither ideas nor the completed project can be copyrighted. Thus,

> The principal problems relate to tangible manifestations of design solutions that are a step toward the project. These tangible manifestations vary considerably in the amount of time taken to create them and the amount of time and money saved by the infringer who copies them (Sweet, 1989).

Surely one way to start is to affix a notice of copyright to design documents. Methods are available to ease the burden of depositing all documents with the federal Copyright Office.

8.8.4 Trade Secrets

Many engineering companies make use in their business of proprietary information, the public disclosure of which would damage their competitive advantage. Intellectual property law provides that employees owe a duty of fidelity to their employers, even in the absence of a signed employment agreement. Thus, employees are enjoined from misusing trade secrets they have learned during their employment. In particular, anyone who leaves one employer and joins another will be barred from sharing with the new employer such trade secrets as specialized process information and mix designs. Customer lists are often considered to be within the trade secret category (Schoumacher, 1986).

8.9   REGULATORY COMPLIANCE

In Section 8.1, we identified administrative law as one of the four types of law. Administrative law establishes regulations and regulatory agencies, usually within the executive branch, under the police power provisions of the Constitution, to promote public health, safety, and welfare. Regardless of whether such regulation is excessive, too costly and unnecessary, or essential, well-controlled, and worth its cost, the engineering manager must be aware of, and comply with, a large number of federal, state, and local regulations. We discuss briefly some of the important regulatory areas and then give a sense of the kinds of regulation to which a hypothetical engineering organization may be subject.

Vaughn (1983) describes in some detail three regulatory areas of special importance in engineering—labor, workers' compensation, and safety. With respect to *labor,* the increasing influence of collective bargaining in the twentieth century led to federal legislation that both protected employee unions and limited their activities. The central federal law governing labor relations is the National Labor Relations Act of 1935 and its amendments, especially the Taft–Hartley amendment of 1947. The act established the National Labor Relations Board (NLRB), whose primary functions are to prevent or remedy unfair labor practices and to supervise employee secret ballot elections to determine bargaining representation. Employers are enjoined from interfering with collective bargaining elections or attempting to coerce employees into voting a certain way. Furthermore, they may not refuse to bargain with a duly voted-in union (just as the union may not refuse to bargain with the employer); and subject to limitations, they must not interfere with the employees' right to strike. The act protects employer rights as well, such as the right to merge two companies in order to reduce labor costs (or for other reasons), the right to subcontract work, the right to change production methods, and the right to institute protection against featherbedding (a labor practice resulting in payments for work not done or not to be done).

Every state has a *workers' compensation* law, as does the federal government. Although such acts differ in detail, all have the same overall purpose—to compensate a worker for injury arising "out of and in the course of" his or her employment. Employees never pay for workers' compensation coverage. Instead, depending on the state, the employer bears the cost (1) by buying insurance from an insurance company, (2) by carrying sufficient self-insurance, or (3) by buying insurance from a state agency. Basically, to comply with the regulation, the employer must do whatever is necessary to assure capability to pay an injured worker's medical costs and to compensate the person for lost wages. The amount of payment for lost wages depends on whether the injury resulted in temporary total disability, permanent partial disability, permanent total disability, or death. Many lawsuits have arisen over the determination of the appropriate category of injury. Often there are questions as to whether an employee is covered by workers' compensation while going to or coming from work (probably not), during lunch (probably so if the injured party was hurt on the premises; probably not otherwise), if intoxicated (probably not), or if injured by a coworker or a third party (probably so if the incident occurred in the course of employment).

A third regulatory area of special interest to the engineering manager is that of *workplace safety*. At the federal level, the Occupational Safety and Health Administration (OSHA) was created in 1970 to attend to the health and safety concerns of the American worker. OSHA issues rules limiting worker exposure to hazardous chemicals and requiring, for example, that temporary scaffolding and other structures on construction sites meet certain minimum standards. The agency regularly conducts investigations and inspections; it has the authority to issue citations and propose penalties if safety or health violations are noted.

We noted as examples in Section 8.1.2 a few of the federal regulatory agencies of interest to engineers. In addition, we might mention the Nuclear Regulatory Commission, which licenses and regulates nuclear energy; the Federal Aviation Administration, which issues and enforces standards for the manufacture, use, and maintenance of aircraft and the certification of pilots and airports, as well as operating the nation's air traffic control system; and the National Highway Traffic Safety Administration, which regulates the safety of motor vehicles and their components (Lyden, Reitzel, and Roberts, 1985).

Suppose Bell Construction, Inc., is formed to offer both on-site construction services (field operations and supervision) and construction project management services. The following is NOT an exhaustive listing of the regulations that will apply.*

- *Articles of incorporation* were filed with the lieutenant governor of the state prior to commencement of business as a corporation; the departments of commerce or business regulation or other appropriate agencies of other states in which Bell Construction planned to operate had to approve the company's qualifications.
- Before *corporate stock* was issued, the issue had to be registered under state law and federal securities law, and a detailed statement was filed with the federal Securities and Exchange Commission (SEC).
- A biennial *corporation report* is submitted to the state department of commerce.
- Bell Construction must file *corporate income tax returns* with the federal Internal Revenue Service (IRS) and the state department of revenue.
- The company must withhold *federal income taxes, social security taxes,* and *Medicare assessments* from each employee's earnings and forward these amounts to the IRS. It is also subject to *employee social security taxes* at the federal level and federal and state *unemployment insurance payments*. It must file *certified payrolls* with its state's department of labor.
- Bell Construction is required to file a periodic *Occupational Employment Survey for General Contractors* with the U.S. Department of Labor, reporting employment numbers, injury data, and other similar information. The report is filed via the state department of labor.

*I am grateful to Albert Bell, P.E., President of Ghemm Company Contractors, Inc., of Fairbanks, for helpful suggestions in the preparation of this hypothetical (but realistic!) example.

- Employees who seek *union representation* are protected by laws related to the National Labor Relations Board, which will also be involved in some employee *grievances*. Wage rates on any federally supported projects must meet the requirements of the Davis–Bacon Act.

- The company is subject to provisions of the *Equal Employment Opportunity* Act. It must set goals consistent with regulations for employing minimum percentages of certain protected classes and must file reports of actual employment percentages with the U.S. Department of Labor.

- Bell is subject to federal regulations with respect to carrying *workers' compensation insurance,* through the U.S. Department of Labor. The program is operated by the state's insurance commission.

- Several members of the firm are registered professional engineers, and others are currently seeking registration. The regulations concerning *professional registration,* with respect to determining qualifications for registration and the adherence to professional practice standards, are administered by the state's Board of Registration for Engineers and Architects.

- *Safety practices,* both on the job site and in the office, are regulated through federal regulations established through OSHA. The state department of labor has been designated the watchdog agency for this state, except on military projects, which are subject to the safety regulations of the U.S. Department of Defense.

- If the company undertakes design/build projects, the plans must meet the requirements of the *Uniform Building Code,* which has been adopted as a state standard and administered by the state department of labor. Plans are subject to approval by the city building official and/or the state fire marshal, depending on the project location.

- Project plans must meet the requirements of the local planning and zoning commission with respect to *zoning and land use.* A project involving removal or processing of hazardous fill requires a permit from the state department of environmental conservation. Projects near airports are subject to maximum height and minimum setback regulations set forth by the Federal Aviation Administration.

- A project necessitating the *discharge of water* into a river requires permits from the U.S. Army Corps of Engineers and the state departments of fish and game and environmental conservation.

- The company must secure permission to operate on lands designated as *wetlands.* Application will be made to the U.S. Army Corps of Engineers, which will coordinate the permitting process with the U.S. Fish and Wildlife Service and the state departments of fish and game and environmental conservation.

- Bell Construction is seriously considering offering its services in *Russia.* As a minimum, it will be subject to federal regulations under the departments of state and commerce.

• Unfortunately, Bell Construction has developed financial problems to the point of being obliged to declare *bankruptcy*. This process will be subject to regulation by the federal Bankruptcy Reform Act of 1978, together with appropriate state law regarding distribution of property.

8.10   ALTERNATE FORMS OF DISPUTE RESOLUTION

The high costs and delays of "going to court" have led many engineering managers to utilize alternative methods for resolving disputes. They range from negotiation, in which no outsiders are involved, to arbitration, a formalized process that involves a third party authorized to decide the outcome. We deal with four methods in order of decreasing control retained by the parties to the dispute. In this continuum, litigation would fall well beyond the end of the list, since the litigants seldom have much control over the legal process.

8.10.1   Negotiation

The idea behind negotiation is that two parties in a dispute get together and try to resolve their differences without assistance from an outsider. If this is possible, much time, effort, and money will be saved, compared with other methods. Indeed, Simon (1989) suggests that "proper representation has negotiation as its goal, not arbitration or litigation." In other words, the attorney should always encourage the parties to a dispute to work out their differences, with or without counsel but always without a third-party facilitator.

Although negotiation is not a viable method in many cases because the parties have honest differences of opinion resolvable only with assistance from a third party, it should not be overlooked as a possible first step. As starting points go, "Let's try to work it out ourselves" is always to be preferred to "I'll sue!"

8.10.2   Mediation

In mediation, the disputants meet with a respected neutral third party to discuss the case and try to reach a solution without formal procedures—in effect, they are negotiating with the help of an outsider. The mediator will help to clarify the issues and suggest what might be the result in a court or litigation proceeding, but he or she makes no decision on the case.

Mediation, a familiar method of attempting to resolve labor relations disputes, recently has gained recognition in the construction industry. The American Arbitration Association has established guidelines for the use of mediation; a suggested clause for a construction contract is the following:

> If a dispute arises out of or relates to this contract, or the breach thereof, and if said dispute cannot be settled through direct discussions, the parties agree to first endeavor to settle the dispute in an amicable manner by mediation under the

"Construction Industry Mediation Rules" of the American Arbitration Association, before having recourse to arbitration or a judicial forum. (quoted in Hughes, 1992a)

Mediation is both voluntary and nonbinding. Since it is voluntary, it can succeed only if both parties agree to the entire process. There is no obligation to continue to mediate if either party is dissatisfied with the proceeding. If no agreement is reached, moreover, the dispute can be taken to another forum, such as arbitration or litigation. Hughes (1992a) describes the mediator's role as follows:

A mediator provides assistance, is impartial, serves as a catalyst, is credible, is a symbol of authority...cannot dictate, has no power to make decisions, communicates and persuades the parties to reach agreements. The mediator's roles vary with every situation and range from translator or transmitter of information to provider of resources, sounding board and listener.

In addition to being simple, quick, inexpensive, and reliant on an expert third party, mediation is private and confidential. In exchange for this privacy, the parties give up the right to use information gained during the mediation in a later arbitration or litigation, and the decisions reached are not considered to be precedent-setting in the common law sense.

A variant of mediation is the alternative dispute review process established by the U.S. Army Corps of Engineers. This approach utilizes a three-person dispute review board, which makes nonbinding recommendations to the contractor and to the corps on construction claims or other disputes. The board reviews and recommends on such matters as interpretation of plans and specifications, delays and other schedule problems, changed conditions, and extra work (Hughes, 1992c).

8.10.3 Mini-Trial

The term "mini-trial" may be a misnomer, because the process is not a judicial proceeding. Rather, it may be described as a structured negotiation process that blends characteristics of negotiation, mediation, and arbitration (Hughes, 1992b). It is a streamlined, nonbinding counterpart of the court trial; it is voluntary, and it features the participation of top management representatives from each party, plus a neutral adviser. A common approach begins with the proposal to use a mini-trial, followed by the selection of management representatives and neutral adviser (who is often a retired judge, a law professor, or a technical expert), and the development of a mini-trial agreement. As in a court trial, there is a discovery period during which each side can learn details about the other's position. Unlike a court trial, the agreement will provide for a compressed discovery period, and there will be limitations on the amount of material gathered.

Then each side will prepare a position paper—"30 page summaries, not the discouragingly common 200 page megatomes—and in English, not legalese" ("Alternate Dispute Resolution," 1986). Following a preliminary meeting of the neutral adviser and the senior management people to clarify procedures, the "trial" (usually called a conference or hearing) is held. Attorneys present their cases to the senior management representatives during a short period (two or three days). The neutral adviser presides, keeps the proceedings running on schedule, asks questions to clarify each position, and, if the parties cannot reach a decision, may suggest a resolution. After the conference, the management representatives begin their private negotiations, sometimes with the neutral adviser present but never with attorneys in attendance. Hughes (1992b) reports that some settlements have been reached in principle within 30 minutes, while others took many negotiation sessions. In the event of an impasse, the case can be referred to arbitration or litigation. It has been reported that 90% of the cases that utilized mini-trials have been resolved successfully. The high success rate probably is attributable to the voluntary, nonbinding nature of the process ("Alternate Dispute Resolution," 1986).

The mini-trial is clearly more formal than pure negotiation or mediation, the process being agreed-upon under a contract, but it still may not result in closure because it is possible for the negotiation phase to reach an impasse. The last method, arbitration, contains features that assure a decision, and the parties can agree to be bound by that decision.

8.10.4 Arbitration

Arbitration is a process whereby the parties voluntarily agree to submit their disputes to an impartial third person or panel of persons, selected by the parties for the purpose of considering the evidence and arguments and reaching a decision. Whereas the other three dispute resolution methods provide for the parties themselves to reach the final decision, in an arbitration the outcome is decided by the third party. The parties make an agreement to submit disputes to arbitration, either in advance or after a dispute has arisen. Under such an agreement, arbitration can be either binding or nonbinding, meaning, respectively, that the parties agree to be bound by the arbitrator's decision or that they will not necessarily be bound by it.

Most states have statutes that recognize the right to arbitration for disputes of specified types, such as construction contracts. Many have adopted the Uniform Arbitration Act. Other legislation, the Federal Arbitration Act, applies to disputes related to interstate commerce contracts.

A leading organization in the arbitration field is the American Arbitration Association (AAA). This private organization promulgates guidelines and procedures, certifies arbitrators, and provides arbitration services upon request. Although the decision to place a dispute before arbitration can be made after the dispute has arisen, many construction companies include in their original contracts a clause such as this:

> Any controversy or claim arising out of or relating to this contract, or the breach thereof, shall be settled by arbitration in accordance with the Construction Industry Arbitration Rules of the American Arbitration Association, and judgment upon the award rendered by the Arbitrator(s) may be entered in any Court having jurisdiction thereof. (quoted in Firmage, 1980)

The arbitration process begins when one of the parties files a demand for arbitration on the other. Frequently the other party will file an answer and, sometimes, a counterclaim. The selection of the arbitrator or panel or arbitrators follows, in accordance with established rules. For example, the AAA provides lists of suggested arbitrators, and each party may strike unwanted choices. After sufficient time for preparation, the arbitration hearing is conducted. Rules of evidence are less formal than in a court trial. For example, hearsay evidence is allowed. Usually, there is no right to discovery before the hearing unless the parties agree otherwise. Under some state laws, the arbitrator may subpoena witnesses and documents (Schoumacher, 1986).

The arbitrators then reach a decision and make the award. In some cases, a complete explanation of the basis for the decision is given, while in others only the decision itself is announced. These decisions cannot be used as precedents in the common law sense, but the reasoning behind the decision, if it is given, can be cited in any subsequent court proceeding on the particular case. The right to appeal the decision is quite limited. Courts tend to uphold the findings of an arbitration unless (1) there was fraud in the process, (2) there was evident partiality on the part of the arbitrator(s), (3) the parties did not have a fair and full hearing, (4) the decision was outside the scope of the dispute as submitted, and/or (5) there was evident miscalculation in figures or mistake in the description of persons or objects referenced in the award (Firmage, 1980; Sweet, 1989).

The pros and cons of arbitration, as described by Schoumacher (1986), with assistance from Firmage (1980) and Simon (1989), may be summarized as follows.

Advantages

- The method provides for resolving disputes as they arise, rather than resorting to a lawsuit at the end of a project.
- An arbitrated case can be processed more quickly than a court trial, since a trial has to be fitted into crowded court schedules and typically involves more evidence and procedural sparring than are permitted in arbitration.
- Arbitration tends to be less costly than a court trial.
- A panel of technical experts, or a single expert, may be more capable of sorting out complex technical arguments than a judge whose primary training is in the law or a jury of lay persons.
- Court proceedings may preclude the admission of certain facts of evidence that are relevant to the case.
- While negotiating their contract, the parties have the opportunity to specify the procedure for dispute resolution to meet their exact needs.

Disadvantages

- Lack of discovery leads to inadequacies in the preparing presentations prior to the hearing.
- Hearsay evidence is allowed, and unjust decisions sometimes result.

- Arbitration decisions can be reached without concern for statute and common law.
- The right of appeal from an award is limited.
- Some important parties may not be subject to arbitration, such as subcontractors in a construction industry case.
- Decisions are not part of the public record and thus cannot be used as precedent.

Simon (1989) makes the bold statement that arbitration in the construction industry is no faster, no cheaper, no less cumbersome, and no easier than litigation. He says further:

> The selection is a philosophical and practical one, assuming a choice is available. The selection comes down to who you want to decide your claim: three independent peers—familiar with the construction process but very possibly without knowledge of the law, rules of procedure and evidence, legal precedent, or the technicalities of legal protocol—or a judge versed in the law but without knowledge of the construction process. The decision is yours.

8.11   TOPICS NOT COVERED

Many other legal topics of importance to the engineering manager could rightfully be presented. In a full course in engineering law, such would be given fuller coverage. Lack of space, however, forces us to omit considerations of sales law, real and personal property law, water law, insurance law, and environmental law; we have mentioned only briefly compliance with labor, workers' compensation, and safety regulations. Books by Bockrath (1986), Vaughn (1983), and Blinn (1989) present chapters on some of these topics, and entire books are available on every one of them individually.

8.12   DISCUSSION QUESTIONS

1. An engineering design firm located in Missouri has a contract to design a building complex located in, and owned by a party residing in, Colorado. If the designer decides to sue the owner for nonpayment of funds, under what circumstances might the designer bring suit in a state court in Missouri? In a state court in Colorado? In a federal district court?

2. Interpret the following court case citation: Roosevelt University v. Mayfair Construction Company, 28 Ill. App. 3d 1045, 331 N.E. 2d 835 (1975).

3. Black undertook to deliver a load of coal which White had ordered from Green. Green did not know initially that Black had delivered the coal. During the delivery, Black negligently broke a plate glass window in the front of White's building. When Green learned about the coal delivery (and the

broken window), he presented White a bill for the coal sold to him. Black was a member of Green's household but was not an employee or agent of Green. Is Green liable for Black's negligence? [Dempsey v. Chambers, 154 Mass. 330; 28 N.E. 279 (1891)]

4. Paragon was the manufacturer of a prefabricated home sold by Sewer, the manufacturer's local distributor, and erected by Romero, a contractor. The construction was faulty, and the homeowner sued Paragon. Paragon had supplied all materials and forms through Sewer, disbursed all construction payments and required that Sewer inspect the construction. Paragon, Sewer, and Romero were three separate legal entities. Paragon defended the suit on the basis that Romero and Sewer were not agents of Paragon. What should the result be? [Amritt v. Paragon Homes, Inc., 474 F. 2d 1251 (3d Cir. 1973)]

5. In 1966 Loretta Lynn signed an agreement with the Wil-Helm Agency in which the agency was made the entertainer's exclusive representative and adviser for a period of 20 years. In 1967 one of the partners of the agency began drinking excessively, in which condition he was abusive and ineffective. The partner was drunk on several occasions when he was acting as Lynn's agent. In 1971 Lynn notified the agency that it had breached the contract, that she would no longer be held to it, and that she would not pay a bill from the agency for more than $150,000. Was the agent's conduct sufficient to terminate the agency? [Wil-Helm Agency v. Lynn, 618 S.W. 2d 748 (1981)]

6. Farmer was an independent contractor hired by Teichert to haul asphalt to Teichert's plant. Farmer's truck struck a bicycle as the truck was turning into Teichert's plant, and the rider was killed. The deceased's father brought suit against Farmer and Teichert. A motion in trial court by Teichert for summary judgment was denied. On Teichert's appeal, what should be the result? [A. Teichert & Son, Inc. v. The Superior Court of Sacramento County, California, 179 Cal. App. 3d 657, 225 Cal. Rptr. 10 (1986)]

7. A went on a trip on behalf of her principal, B. Just before A, as agent, signed an agreement with C for the design of an industrial piping system, B died. Is there a binding contract?

8. An electrical contractor is awarded a contract to install a transmission line across 30 k of permanently frozen ground in a remote area of Alaska. Describe briefly all the government regulations you can identify that will apply to this project.

9. What advantages might be especially important in specifying that disputes in connection with a contract for the construction of an oil pipeline will be decided by arbitration rather than litigation? What are the primary disadvantages of arbitration in such a situation?

10. A contractor submitted a bid that contained a mistake. After the mistake was discovered, the contractor sought relief. Should the contractor's claim be resolved by arbitration? [Village of Turtle Lake v. Orvedahl Construction Company, 135 Wis. 2d 385, 400 N.W. 2d 475 (1986)]

8.13   REFERENCES

Alaska Court System. 1991. *Profile of the Alaska Court System.* Anchorage, January.

"Alternate Dispute Resolution." 1986. *Civil Engineering,* Vol. 56, No. 3, March, 56–58.

Blinn, K. W. 1989. *Legal and Ethical Concepts in Engineering.* Englewood Cliffs, NJ: Prentice-Hall.

Bockrath, J. T. 1986. *Dunham and Young's Contracts, Specifications and Law for Engineers.* New York: McGraw-Hill.

Burgunder, L. B. 1995. *Legal Aspects of Managing Technology.* Cincinnati, OH: South-Western College Publishing.

Corley, R. N., and R. L. Black. 1973. *The Legal Environment of Business.* 3rd ed. New York: McGraw-Hill.

Dillavou, E. R., and C. G. Howard. 1957. *Principles of Business Law,* 6th ed. Englewood Cliffs, NJ: Prentice-Hall.

Firmage, D. A. 1980. *Modern Engineering Practice.* New York: Garland STPM Press.

Foltz, R. D., and T. A. Penn, 1990 *Understanding Patents and Other Protection for Intellectual Property.* Cleveland, OH: Penn Institute.

Hughes, R. K. 1992a. "Alternative Disputes Resolution: Part 3—Mediation." *PEC Construction Reporter,* National Society of Professional Engineers, Vol. 14, No. 7, May, 2, 5.

Hughes, R. K. 1992b. "Alternative Disputes Resolution: Part 4—Mini-Trials." *PEC Construction Reporter,* National Society of Professional Engineers. Vol. 14, No. 8, August/September, 2–3.

Hughes, R. K. 1992c. "Alternative Disputes Resolution: Part 5—Alternative Disputes Review Boards/Partnering." *PEC Construction Reporter,* National Society of Professional Engineers. Vol. 14, No. 9, August/September, 2, 5.

Lyden, D. P., J. D. Reitzel, and N. J. Roberts. 1985. *Business and the Law.* New York: McGraw-Hill.

Nord, M. 1956. *Legal Problems in Engineering.* New York: Wiley.

Schoumacher, B. 1986. *Engineers and the Law: An Overview.* New York: Van Nostrand Rienhold.

Simon, M. S. 1989. *Construction Claims and Liability.* New York: Wiley.

Sweet, J. 1989. *Legal Aspects of Architecture, Engineering, and the Construction Process,* 4th ed. St. Paul, MN: West Publishing.

Vaughn, R. C. 1983. *Legal Aspects of Engineering,* 4th ed. Dubuque, IA: Kendall/Hunt Publishing.

___9
Contract Law Principles

A contract is the legal basis on which a predominant portion of engineering activities is conducted. Whether it is an arrangement to purchase a computer system, to provide building design or material testing services, to construct a power generating station, or to operate an airport parking concession, it is likely the work will proceed under some sort of contractual agreement. Because of the importance of contract law in engineering management, we devote an entire chapter to the topic. Even so, we are able to give only a survey of the basic principles.

We begin with some definitions, followed by a description of the elements required to form a valid contract. Then we turn to the sometimes difficult matter of interpreting the contract: What do the words or actions mean, and what did the parties intend to contract? The rights of third parties—those outside the contract itself—are considered next. We then discuss the ways in which a contract may be terminated and end with a short description of the remedies available in case of the breach of a contract. In Chapter 10, we apply some of these contract law principles to contracts for engineering services of various kinds.

9.1 DEFINITIONS

A *contract* is "a promise or set of promises for the breach of which the law gives a remedy or the performance of which the law in some way recognizes as a duty" (Lyden, Reitzel, and Roberts, 1985). Or, stripping away the legalese, Vaughn (1983) suggests that a contract is "an agreement enforceable at law." To be enforceable at law, a contract must have certain elements, which we discuss in Section 9.2. A promisor is a person who makes a promise, and a promisee is the person to whom the promise is made. In most two-party contracts, each party is both promisor and promisee, since each agrees to do something (or not do something) in exchange for the promise of the other.

Note that a promise can be framed as a negative: a party may promise to refrain from some act he or she would otherwise have the right to do. In a famous 1891 New York State case [Hamer v. Sidway 124 N.Y. 538], a man promised his nephew $5000 if the nephew would refrain from smoking, drinking, swearing, or playing cards or billiards until he was 21. The boy kept his promise, the uncle died, and the executor of estate refused to pay the $5000. The court, however, found that

254

the nephew's promise *not* to do those dirty things was sufficient consideration to make a valid contract and ordered the $5000 payment to be made.

Note also that a contract is a voluntary act, and the parties to the contract have the autonomy to choose its contents, provided it meets certain requirements. However, once a party has chosen to enter a contract and has agreed to its contents, the agreement is enforceable at law.

Most are *bilateral* contracts, meaning that each side promises to perform; the legal effects are reciprocal duties and obligations. In a *unilateral* contract, only one party promises performance, the other side already having done what was required. An airline ticket represents a unilateral contract: after you have purchased it, the airline promises to hold a seat for you on a particular flight. Your promise was fulfilled by paying for the ticket; the airline is the only party with a promise outstanding. A contract for services, such as painting your house, provides another example. Suppose you write to a house painter, saying that you will pay her $2200 if she paints your house. Then, without a formal reply, she paints the house while you are on vacation. If you refuse to make the payment, it is likely you will lose in court. By painting the house, the painter accepted your offer, and a unilateral contract was formed. The only unfilled promise was your obligation to pay. Even if it appears that $2200 is an unreasonably high price for these services, it is likely that you are obligated to pay in full because a unilateral contract for this amount was created.

An *express* contract is one whose agreement is manifested by oral or written words. An *implied-in-fact* contract is based on implications that can be drawn from persons' behaviors. If the parties, because of their acts or conduct (rather than oral or written words), make it reasonable to assume that a contract exists between them, then a contract does, in fact, exist. An auction is a good example. If you raise your hand when the auctioneer asks for a bid of $45,000 for a painting, and nobody offers a higher bid, it is likely that you have bought the painting without having written or spoken a word. In a construction-related case, an employee of a subcontractor rented equipment and brought it to a job site, where it was used on the project. The employer admitted that the equipment was there and was used but argued that he was not obligated to pay the employee for its use because there was no express agreement covering such a rental. The court decided otherwise:

> Even if there was no express oral agreement for equipment rental, the implied contract may be inferred from the conduct of the parties. Implied contracts arise...where there is circumstantial evidence showing that the parties intended to make a contract. Where one performs for another a useful service of a character that is usually charged for, and such service is rendered with the knowledge and approval of the recipient who either expresses no dissent or avails himself of the service rendered, the law raises an implied promise on the part of the recipient to pay the reasonable value of the service. [United States v. Young Lumber Co., 376 F. Supp. 1290 (D.S.C. 1974)]

Thus, an implied-in-fact contract is an actual contract, circumstantially proved. There is no difference in legal effect between such an agreement and an express contract.

Another type of implied contract is the *implied-in-law* contract, or *quasi-contract*. This "legal fiction" (Vaughn, 1983) is "a device created by the common law to avoid the unjust enrichment of a person who has received and retained valuable goods or services of another under circumstances in which it is socially reasonable to expect payment. The simple solution is to imply a promise on the part of the recipient to pay the reasonable value of what has been received and retained" (Fessler, 1989). In other words, the court will reason that even though there is no contractual obligation on the part of one of the parties, there ought to be; otherwise one party would benefit unjustly at the expense of the other. There was no contract, but if one party claims that the other is obligated, the court will agree. Bockrath (1986) distinguishes between a quasi-contract and a true contract by saying that "in the former, the duty defines the contract; in the latter the contract defines the duty." Fessler (1989) offers the following example of an implied-in-law contract:

> Rancher Smith notices a break in a fence holding Rancher Jones' cattle in a remote pasture. In order to prevent the escape of the animals, Smith undertakes to repair the break. Jones would be liable to pay Smith the reasonable value of such service as a matter of law. The liability rests on the policy goal of avoiding unjust enrichment of Jones.

Note that the equipment rental case cited earlier (United States v. Young Lumber Co.) involved a true contract rather than a quasi-contract. There *really was* a contract, which was implied from the actions of the parties. If the circumstances had been different—for example, if the employer had not known that the equipment was being used, meaning that no contract was implied in fact— it is still possible that the employee could receive payment by invoking the theory of quasi-contract.

If more than one person is involved as one side of a contract, the arrangement may be a joint or a several contract. In a *joint contract,* two or more parties unite in a combined action that will benefit them all. Firmage (1980) suggests the example of the construction of a swimming pool that will benefit a group of families. All the families will be individually responsible for payment to the contractor, but the pool cannot be divided into individual ownership. The contractor probably would insist on a joint contract, to ensure that if one of the families did not pay its portion, he could collect from the others.

In a *several contract,* two or more independent agreements are lumped into a single agreement, and each party is responsible only for his or her portion. An example might be the contract for land surveying services required by several property holders within a single subdivision. Although there might a single agreement, each property holder would be responsible only for the surveying of his or her portion of the survey. If one of the property holders defaulted on payment, the others would not be liable. (The surveyor might prefer a joint contract, but it is unlikely the property holders would agree!)

Some contracts are both *joint and several.* Suppose a contract reads, "We, Bill Jones and Sonja Smith, jointly and severally, promise to pay Margaret Bennett

$2750 upon satisfactory completion of computer programming services." If Bill and Sonja default in their payment, Margaret can bring separate actions against Bill and Sonja, or she can bring one suit against them jointly. But they cannot be sued both ways; the injured party must choose one or the other.

An important notion in construction contracts is the distinction between entire and severable contracts. "An *entire* contract is when full and complete performance by one party is a condition precedent to the right to require performance by the other party. In such contracts there is no liability for part performance, and failure of one party fully to perform relieves the other of any obligations" (Bockrath, 1986). An example is an old New Jersey case [Kelly Construction Co. v. Hackensack Brick Co., 91 N.J.L. 585, 103 A. 417, 418 (1918)] that involved furnishing brick for the construction of a high school. After delivering some of the brick, Hackensack refused to deliver any more because it had not been paid for the brick already delivered. The court decided in favor of Kelly Construction, stating:

> Where, as here, the sale of a specified quantity of brick (i.e., sufficient to complete a building according to stated specifications), the contract is entire, and a failure to pay when a part delivery has been made does not excuse the seller from completing delivery, no time for payment being stated in the contract. (quoted in Bockrath, 1986)

Even in the case of construction contracts that involve "progress payments" to the contractor after certain stages of completion, the agreement is often held to be entire; the reasoning is that although the payments were partial for the convenience of both owner and builder, the payment is for doing the whole work.

On the other hand, a *severable* contract (often called a "divisible" contract) is one that is capable of being divided into distinct parts, and the payment is to be made for each part. Suppose a single construction contract is divided into six phases, with a designated payment for each phase. If the contractor completes four phases but omits the remaining two, and if the contract were interpreted as being severable, or divisible, the contractor would be entitled to receive the designated payments for the four completed phases. In another case in which the contract was interpreted entire [LDA, Inc. v. Cross, 279 S.E. 2d 409 (W. Va. 1981)], a West Virginia court said:

> A contract is severable or divisible when the part to be performed by one party consists of a number of distinct and independent items and the price to be paid by the other party is apportioned, or is susceptible of apportionment, to each item. An agreement to pay a certain price for every bushel of wheat supplied which corresponds to a given sample is typical illustration of a severable contract. ... The primary criterion in determining whether a contract is entire or severable is the intention of the parties, regarding the various items as a whole or each contract item as a separate unit....

In an *executed* contract, the agreement has been fully performed by both parties and no obligations remain on the part of either. In an *executory* contract, on the other hand, neither party has carried out the promised performance. It is possible for a contract to be executory as to one party and executed as to the other, as in the case of a design engineer who has completed and delivered a design to a purchaser of services who has not yet paid.

A *void* contract is one that is invalid and produces no legal obligations. Strictly speaking, it is not a contract. A contract to commit a crime is a void contract. A painting contractor who was on a construction job for 10 months was denied payment for his work because his license had been suspended for part of that time. The court said he was operating under a void contract [Bierman v. Hagstrom Construction Company, 170 P. 2d 1138 (Cal. 1959)].

A *voidable* contract is one that can be set aside at the request of one of the parties (but not the other). Generally, a contract in which an underage person (a minor) is a party may be set aside at the minor's request; the same holds for a person who was induced into signing a contract by fraud, mistake, or duress and for a person suffering from mental infirmity. Bockrath (1986) says that such contracts "are fully effectual until affirmatively avoided by some act. Such contracts are prima facie valid but are subject to certain defects of which some party can take advantage. By electing to do so, that party avoids the legal relations which the contract creates or, conversely, by ratification of the contract may extinguish the power of avoidance." Fessler (1989) gives the following hypothetical example:

> Sean Riley, age 17, purchases a used 1980 Corvette from Byer's Motors. The agreed price is $11,000. Riley makes a $2000 down payment and agrees to make monthly payments of $300 to Byer's. Ten days after taking delivery, Riley is involved in an accident with the car. As a result of a misrepresentation which Riley made in his application for insurance, there is no coverage on the car on the date of the accident.
>
> As a minor, Sean Riley may disaffirm his contract with Byer's Motors. To do so he need only make an affirmative manifestation of his election and return the twisted remains of what had once been a proud automobile!

Some authorities suggest that Riley probably would get his $2000 down payment back. We shall consider the matter of competency of parties further in Section 9.2.1.

An *enforceable* contract is one that has been formed to meet all the basic requirements outlined in Section 9.2. It is one "for the breach of which the law gives a remedy" (Lyden, Reitzel, and Roberts, 1985); it may be enforced in a court of law, and the injured party may receive damages or, sometimes, specific performance. An *unenforceable* contract cannot be enforced by law. An obligation may have been created, but it cannot be enforced by legal proceedings. That is, the rights and duties have not been abrogated in the eyes of the law, but there is no legal means of enforcing them by bringing a lawsuit. One type of unenforceable contract is one which is required by the statute of frauds to be in writing (Section 9.2.5). Another type is one in which the object to be accomplished is illegal, such as wagering (unless modified by statute), charging usurious interest, or interfering with the administration of justice. The point is that a contract that is unenforceable may nevertheless be valid. If such a contract is performed, however, a party cannot later claim damages or other legal remedy on the grounds that the contract was unenforceable. If one party does not perform, the other cannot go to court to force performance or obtain damages. Nord (1956) uses the following example:

> A owed B $500 on a contract which was unenforceable because it was not in writing as required by the Statute of Frauds. A sent B a check for $500 to settle the account, not realizing that the contract was unenforceable. Later, when he discovered that the contract was unenforceable, he demanded his money back. B refused to return it. Was A entitled to a refund of his money? NO! He paid a legal obligation and is not entitled to a refund. While B could not have brought an action to obtain the money from A, A cannot bring an action seeking the return of the money. It was a valid, though unenforceable, contract.

Finally, we define some terms related to *negotiable instruments*, which are simply contracts for the payment of money. An important question here is whether such a piece of paper has any value to you if you happen to become the possessor of it. Negotiable instruments come in two basic types. A *promissory note* is a two-party instrument (e.g., sales note, mortgage note, bond coupon) in which a maker promises to pay money to a payee, and a *bill of exchange* (e.g., check, bank draft, time draft) involves three parties—the drawer orders a third party, the drawee (usually a bank), to pay money to the payee. To be negotiable, such an instrument must (1) be signed by the maker or drawer, (2) contain an unconditional promise to pay a sum certain of money, (3) be payable on demand, and (4) be payable to order or to the bearer (Bockrath, 1986).

The *holder in due course* doctrine makes negotiable instruments easy to transfer by securing the rights of the possessor. A holder in due course is a person who became the holder of the instrument before it was overdue, who took it in good faith and for value, and who had no notice of any defect. Thus, if you meet these qualifications and a check you just received meets the four requirements above, you have a valid claim to the money represented by the check even if some defect such as fraud had occurred between two of the parties involved initially in creating the check or passing it along.

9.2   FORMATION PRINCIPLES

A valid contract is made up of five elements, each of which must exist according to law. Sections 9.2.1 through 9.2.5 discuss these elements, which can be listed as follows:

1. The parties must be competent to enter into the contract.
2. The subject matter of the contract must be "proper."
3. There must be an agreement, consisting of an offer and an acceptance.

4. Consideration, something of value, must be given up by each side.

5. The agreement must be in proper form.

9.2.1 Competent Parties

All parties to a contract must have the legal capacity to contract; they must be "competent." We have already noted that a contract in which one of the parties is underage (a "minor" or "infant") may generally be voided at the minor's option. The limit usually is age 18 or 21, depending on the state. Most states make an exception for "necessities," such as food, lodging, education, and medical services, making the minor liable based on the theory of quasi-contract. Also, most courts hold that contracts made by intoxicated persons are voidable if a party can prove that he or she was too inebriated to understand the consequences of entering into the contract. Other courts take a harsher view. Believing that getting drunk is voluntary, they hold people responsible for contracts not soberly arrived at. Contracts made by insane persons are also voidable by the person when he or she recovers sanity. In this case, the insane person, or the guardian or conservator, must act promptly to have money, goods and/or property returned by both sides so as to return the parties to the position they were in prior to the contract. On the other hand, the sane person does not have the same privilege of cancellation; furthermore, if he or she took advantage of the insane person's mental condition, courts generally permit the insane person to avoid the contract (Lyden, Reitzel, and Roberts, 1985; Vaughn, 1983). As in the case of minors' contracts, courts generally hold intoxicated and insane persons responsible for the value of "necessities" for which they have contracted.

A corporation is a separate legal "person" that is treated quite differently from a sole proprietorship or a partnership (see Chapter 11). A corporation's charter limits the organization to certain specified activities. Within the terms of its charter, the corporation is free to enter into any otherwise proper contract. However, if a contract made by a corporation goes beyond its charter limits the contract is said to be *ultra vires* (beyond power), hence unenforceable. As we learned in Section 9.1, a party to an unenforceable contract cannot go to court to enforce its provisions, but neither can such a contract, once executed, be "undone."

In Chapter 8, we described the *principal/agent* relationship and noted the sometimes difficult problem of deciding whether an agent was authorized to act for his or her principal. In the language we employ in the context of contracts, we would say that an agent who is not clothed with the authority to act for a given principal is not "competent" to contract on the principal's behalf. Likewise in a partnership, each partner is generally an agent on behalf of all other partners; if, for some reason, a partner is held not to have this authority, we would say that the partner is not competent to enter into a contract for the partnership. In a rather complicated 1964 Minnesota case [Frank Sullivan Co. v. Midwest Sheet Metal Works, 335 F. 2d 433 (8th Cir. 1964)], an employee signed an agreement allowing another company to take over part of a contract to provide sheet metal and air conditioning installation for a post office project in Saint Paul. Part of the appeal in this case revolved around whether the employee had the authority to sign the agreement (whether he was "competent party" to the contract). The court held that he was, and therefore he had bound his company to the arrangement (Blinn, 1989).

Professional persons and others who must be *licensed* to provide services will be considered incompetent if they do not hold the proper certification. One who performs such services without being licensed may not resort to the courts to collect payment for the work (Vaughn, 1983). Recall Beirman v. Hagstrom Construction Company (Section 9.1), in which the painting contractor did not hold a valid licence.

9.2.2 Proper Subject Matter

The subject matter of a contract must be proper in two respects. It must be clearly defined, and its purpose must not be against common or statute law or against public policy. In a contract for engineering services, there must be sufficient words (written or oral) to permit the parties to understand their respective obligations and benefits:

> Courts can neither enforce, nor award substantial damages for the breach of, contractual agreements which are wanting in certainty. The terms of a contract must be complete, and the various obligations must be described with sufficient definiteness to enable a court to determine whether or not these obligations have been performed. (Vaughn, 1983)

A contract must have a lawful purpose. Thus, an agreement to commit a crime, to commit a tort (private injury, as explained in Chapter 12), or to defraud an individual or the public in general will be held to be invalid. Whether a particular contract is against public policy is a difficult question, and the answer seems to change over time. It is clear that tax evasion agreements are illegal, as are those that serve to perpetrate fraud or deception. An agreement that utilizes personal or political influence to secure work on a public project is also against public policy. The well-publicized scandals involving engineers and public officials in Baltimore County, Maryland, were a dark chapter in the annals of professional ethics; those kickback schemes were also clearly illegal, as being against public policy (Lewis, 1983). We deal more fully with the ethical side of this case in Chapter 13.

9.2.3 Meeting of the Minds

Phrases like *meeting of the minds, mutual assent to the terms,* and *offer and acceptance* are used to indicate that agreements of the parties are essential to contract formation. The idea is that both parties must agree to the same thing. A helpful way to grasp this concept is to look at the process of offer and acceptance.

No special words or formats are required to make a valid offer. All *offers* must contain three elements: (1) an expression of clear and sincere intent to enter into a contract, (2) sufficient articulation of the essential terms of the proposal, and (3) communication of the intent to the offeree (Fessler 1989). Sometimes it is difficult to ascertain whether there was serious contractual intent on the part of the offeror. Offers made in jest probably will not lead to valid contracts. In another automobile example, Fessler, (1989) presents two friends on a drive:

> Freda Fox and Paula Pence are taking a weekend pleasure drive when Fox's very expensive automobile breaks down. In obvious frustration Fox declares in a loud voice, "I will sell this piece of junk to the first person who offers me $5!" Should Pence attempt to form a bargain by promising or tendering the sum of $5, the evident frustration, the loud voice, and (perhaps most importantly) the preposterous price will all be factors tending to defeat the claim that an offer existed.

However, sometimes "joke contracts" are treated seriously by courts of law. In upholding the right of a plaintiff to $1000 plus a diamond ring which she had been promised if she performed certain business promotion services, the court said,

> Jokes are sometimes taken seriously by the young and inexperienced in the deceptive ways of the business world, and if such is the case, and thereby the person deceived is led to give valuable services in the full belief and expectation that the joker is in earnest, the law will also take the joker at his word, and give him good reason to smile. [Plate v. Durst, 42 W. Va. 63, 24 S.E. 580 (1896)]

A contract to purchase a farm was written on a napkin in a bar. The Virginia court upheld its validity even though the offer may have seemed informal and not serious [Lucy v. Zehmer, 196 Va. 493, 84 S.E. 2d 516 (1954)].

The actual offer must be distinguished from preliminary negotiations, a mere price quote, or an expression of desire to enter into a contract. Thus it is not likely that "I hope we can negotiate a $450 contract for these drafting supplies" will be held to be a valid offer. A public agency's request for proposals or invitation for bids is not an offer that can be accepted by submitting a proposal or a bid (Simon, 1989).

An offer may be *terminated* by several means. It can be revoked by the offeror at any time prior to its acceptance, provided the revocation is communicated properly. It can be simply rejected by the offeree as well, or rejected and followed by a counteroffer. The time limit specified in the offer, if any, may run out. If no time limit is specified, the limit will be interpreted as "reasonable." If the subject matter of the offer is destroyed before acceptance, such as a building that burns down before an offer to repair it has been accepted, the offer terminates.

Not only must there be an offer, there must be an *acceptance* of that offer: "It is a condition prerequisite that the offeree indicate a willingness to be bound by the terms of the offer" (Blinn, 1989). To be effective, an acceptance must express unequivocal, unconditional assent by the offeree to each and every term of the offer and must be communicated to the offeror at a time prior to revocation or termination of the offer.

If the "acceptance" modifies any part of the offer, the acceptance becomes ineffective. Thus, if you receive an offer for contract that would pay you $450 to perform certain specified computer programming services by next May 15, your reply that you "accept the offer but will not be able to complete the work until May 31" is not an acceptance at all. Instead, the law would likely treat this response as a *counteroffer,* which the original offeror could accept or reject. This "mirror-image acceptance rule" also means that a reply that is not a mirror image of the original offer is regarded as a rejection that terminates the outstanding offer.

In a North Carolina case [Howell Co. v. C. M. Allen & Co., 8 N.C. App. 287, 174 S. E. 2d 55 (1970)], a subcontractor sued to recover damages for an alleged breach of contract with a general contractor. The subcontractor had submitted a bid but failed to prove in court that an agreement had been reached as to the price of the work: "For a binding contract to exist, more than a mere proposal intended to open negotiations must exist. ... The parties must agree to the same thing in the same sense; their minds must meet regarding all the terms" (Simon, 1989).

Sometimes questions arise over the required form and timing of an acceptance. Generally no special form is needed unless the offer includes such a condition. An old English case [Adams v. Lindsell, 1B & Ald. 681, 106 Eng. Rep. 250 (K.B. 1818)] established the so-called mail box rule, which says that an acceptance is valid at the time it is put into the mail, if the offeror requested a reply by mail. With the advent of modern communication alternatives, the same rule is likely to be expanded to cover acceptances by fax and by electronic mail (Fessler, 1989).

If the offeree is silent after receiving the offer, there has been no acceptance, even if the offer stated, "If I do not hear from you by next Monday, I will proceed to paint your barn." An example is the case of Durick Insurance v. Andrus. The insurance company urged Andrus, an apartment building owner, to increase his coverage to $48,000, whereas Andrus wished only $24,000 in coverage. Durick sent Andrus a policy for the higher amount, stating that Andrus could cancel the policy by returning it. When Andrus did nothing, Durick sued for the unpaid premium, but the Vermont court held that Andrus' silence was not an acceptance [Durick Insurance v. Andrus, Sup. Ct. of Vt, 424 A. 2d 249 (1980); Fisher and Jennings, 1986]. Exceptions to this rule are sometimes allowed—for example, when the parties have used silence as acceptance in earlier dealings, or an offeree retains and uses goods that were provided under the terms of an explicit offer. (In the case of charitable institutions that send unsolicited items through the mail to prospective contributors, however, statute law has established that such items are gifts, and there is no obligation on the part of the recipient to either pay for them or return them.)

If an *error* is made in the preparation of an offer that is accepted by the offeree, but other contract elements are present, does a contract exist? Construction contract law continues to wrestle with the problem of mistakes in bids. Fessler (1989) describes such flawed offers as "unilateral blunders" and distinguishes between mechanical miscalculations, which are sometimes forgiven, and errors of business judgment, which will always find courts "deaf to any plea for relief." In reviewing a case in which a contractor mistakenly omitted $66,660 from a $699,500 bid [Powder Horn Constructors v. City of Florence, 754 P. 2d 356 (1988)], the Supreme Court of Colorado decided that the minds of the parties had never met because the bid accepted by the owner was not the bid intended by the bidder. Thus, there was no contract, and contractor was relieved from performance ("Bidder Gets Relief from Error," 1989). An even more interesting situation occurs when the low bidder turns out to have omitted an item or added wrong but would still be low bidder if the terms were corrected. Will this party be allowed to insist that the bid be accepted at the revised higher figure? We shall revisit this situation briefly in Chapter 10.

9.2.4 Consideration

A promise, to be enforceable, must be supported by consideration. Simply defined, consideration is something of value that is exchanged. In fact, Fessler (1989) calls consideration "the bargained-for incursion of legal detriment." Consideration is a bargained-for promise to do some act, or actually to do the act, which, absent the promise, the promisor is not legally obligated to perform. Or, consideration can be the promise not to do something, or actually not to do the act, which, absent the promise, the promisor is legally free to pursue. The promises, or the promise and the act, must have been consciously exchanged by the parties.

Several ideas emerge from this definition. First, if a certain act is adequate as consideration, then the promise to do that act is also valid consideration. If one party pays $15 to a hockey arena in exchange for a reserved seat at next Tuesday's game, the $15 payment is the consideration; a unilateral contract has been created, as discussed in Section 9.1. If another fan has promised to pay $15 for an equivalent seat reservation, the promise is the consideration, and a bilateral contract has been created.

Second, the promise not to do something is valid consideration. We earlier considered the case of a young man who promised not to smoke and do other bad things until he was 21. The court decided that the consideration was valid. In a Connecticut case [Osborne v. Locke Steel Chain Co., 218 A. 2d 526 (Conn. 1966)], a retired employee signed a contract with his former employer to be available for consulting assignments and not to compete with the company. This contract, which was held to be valid, is an example of a consideration that was a promise *not* to do something.

The law is very careful to assure that the promise puts the promisor in a position different from the one he or she would have been in if the promise had not been made. Thus, a promise by an employer to send an employee on a vacation trip because of past valuable service does not form a valid contract, because there is no bargained-for detriment on the part of the employee, whose service was in the past. Similarly, an agreement that adds extra work to a contract at no change in price, or increases the total payment without adding to the work scope, is not valid. In either case, there is no change in the position of one of the parties as a result of the contract change.

Finally, the law is quite settled on the matter of *adequacy of consideration*. According to Fessler (1989):

Bargained-for legal detriment, and not any economic factor, imparts "value" to valuable consideration. As long as any element of bargained-for legal detriment can be identified on both sides of the exchange, courts are totally disinterested in the distribution of economic gain or advantage resulting from the contract.

The belated recognition that a promisor made a stupid deal is no defense at all. Legal scholars refer to this concept as the "peppercorn theory," from an early Maine case [Whitney v. Stearns, 16 Me. 394, 397 (1839)] in which the court said, "A cent or a peppercorn, in legal estimation, would constitute a valuable consideration." So, if a party gets what has been bargained for, courts are unwilling to look into whether the consideration was of the same "value" on both sides. In the Osborne v. Locke Steel Chain case, the company sought to get out of the bargain by claiming that its promise to pay the retired employee Osborne $20,000 the first year and $15,000 each year thereafter for the balance of his lifetime was excessive. The court found in favor of Osborne, stating:

> The courts do not unmake bargains unwisely made. The contractual obligation of the defendant in the present case, whether wise or unwise, was supported by consideration, in the form of the plaintiff's promise to give advice and not to compete with the defendant. ... (quoted in Lyden, Reitzel, and Roberts, 1985)

9.2.5   Form

The final requirement for a valid contract is that it be in proper form. In the past, contracts of some kinds had to be sealed, but this practice is not at all common today. We have already described negotiable instruments, which are not considered to be negotiable unless they are in a certain form. Most other contracts can be formal or informal and written or oral, provided they meet the stipulations described above.

There is an important exception, however: contracts of some kinds are not enforceable unless they are in writing. The legislation that requires these contracts to be in writing is called the statute of frauds, which has been adopted in some form in every state. Statutes of frauds began more than three centuries ago in England during the reign of Charles II when Parliament wrote the Act for the Prevention of Fraud and Perjuries. The 1677 act contained 25 sections and required that certain kinds of contract be in writing. Nearly all English-speaking countries have adopted the statute in more or less its classical form. The rationale is that certain kinds of contract are important enough to need independent verification, rather than relying on witness testimony. The typical statute of frauds begins as follows:

> In the following cases and under the following conditions an agreement, promise or undertaking is unenforceable unless it or some note or memorandum of it is in writing and subscribed by the party charged [with liability for breach] or by the agent of that party: .... (*Alaska Statutes,* 1989)

The more important subjects addressed by the typical statute of frauds include the following:

- An agreement that by its terms is not to be performed within one year of the date of formation.
- An agreement that by its terms is not to be performed within the lifetime of the promisor (e.g., a will or a bequest of property).
- A promise to answer for the debt or default of another.
- An agreement made upon consideration of marriage other than mutual promises to marry.
- An agreement for the purchase or sale of real property or any fixture permanently attached to the land or for leasing real property for more than one year.
- Contracts for the sale of goods over a certain amount [typically $500], with certain exceptions. (Adopted from Fessler, 1989)

It is instructive for the engineering manager to note the many kinds of contract that do not have to be in writing to be enforceable. Of special importance are contracts for any kind of service. Thus, your contract to design a rocket, or build a sewage treatment plant, or manufacture a drilling rig does not have to be in writing to be enforceable, provided the work is to be completed within one year and the contract itself meets all the criteria listed above. Nevertheless, you probably will want to consider the obvious advantages of having even that kind of contract in writing.

9.3   INTERPRETATION OF WRITTEN CONTRACTS

Once a contract has been formed, the parties may disagree over its interpretation. Even though the agreement has been committed to writing, conflicts often arise. Sweet (1989) entitles his chapter on this subject "Contract Interpretation: Chronic Confusion." We list several generally accepted guidelines for the interpretation of written contracts, followed by three brief cases that illustrate some of the principles.

- Courts attempt to give a reasonable, "common sense," interpretation to contracts: "where one interpretation makes a contract unreasonable or such that a prudent person would not normally contract under such circumstances, but another interpretation equally consistent with the language would make it reasonable, fair and just, the latter interpretation would apply" [Elte, Inc. v. S. S. Mullen, Inc., 469 F. 2d 1127, 1131 (9th Cir. 1972)].
- The joint intent of the parties will prevail if it can be ascertained. That is, the court will be primarily concerned with trying to determine what the parties intended, based on the surrounding facts and circumstances and the oral or written words. "Words and other conduct are interpreted in the light of all the circumstances, and if the principal purpose of the parties is ascertainable it is given great weight" (Simon, 1989).

- A contract must be interpreted as a whole, with its meaning taken from the entire context rather than from particular portions or clauses. Related to this rule, "An interpretation which gives a reasonable, lawful and effective meaning to all the terms is preferred to an interpretation which leaves a part unreasonable, unlawful or of no effect" (Simon, 1989).

- If, when the contract is taken as a whole, there are two possible interpretations, one lawful and the other unlawful, the court will assume the lawful interpretation.

- Ordinary words are interpreted in their ordinary grammatical sense; technical and trade terms are given their technical and trade meanings when used in a transaction within their special field. Thus, the word "reservoir" has a certain meaning to a water supply engineer and a different meaning to a petroleum geologist; "Christmas tree" has a special meaning to an oil field engineer; the word "dry" means one thing to a cleaner and another to a bartender.

- Custom and usage may be used to determine the intent of the parties when the content of the contract is silent or ambiguous with respect to certain terms. If there is no provision to indicate which party pays shipping charges, the court might use common practice in the area to determine the outcome of a conflict. If a gravel aggregate specification gave a certain lower limit percentage for a certain fraction, such as "the percent passing the #40 sieve shall be at least 15%," and if common custom for contracts of that type was to allow 0.5% less than this limit, the court might determine that a contractor who provided 14.5% had met the intention of the specification.

- Where general words are followed by specific words, the specific words will prevail if there is a conflict. Thus, a sales contract for a microcomputer that reads "One XYZ computer system including microprocessor, disk drives, keyboard, monitor, and printer," without mentioning cables, software, or documentation, might be interpreted to deny that the purchaser was entitled to receive the last three items. (The custom and usage rule might produce the opposite result, however!)

- If there is an inconsistency between words and figures in a contract, the words will govern.

- As to inconsistency between printed, typed, and handwritten provisions, handwritten will be preferred over typed, and typed will be preferred over printed, on the grounds that the special effort to handwrite or type is probably a better indication of intent. [Sweet (1989) notes a problem with word-processed agreements, which appear to be prepared for the particular occasion but may be similar in use to printed forms: "The test in seeking to reconcile conflicting language ... should be whether some language was specially prepared for this transaction ... ."]

- As discussed in Section 8.5.3, the parol evidence rule deems oral evidence inadmissible if it contradicts what has been agreed to in writing. But such evidence is allowed if it clarifies the agreement, completes it by adding provisions that had been omitted, or confirms its very existence.

- Obvious clerical errors can be discounted in determining the intent of the parties: "In fact, the court will strike out an improper word or even supply an omitted word if from the entire context it can readily and definitely ascertain what terminology was intended and should have been used" (Bockrath, 1986). An example might be the replacement of the word "buyer" with "seller" when it was obvious such was the intention.
- An often-used rule of contract interpretation is that ambiguities will be construed against the drafter of the contract, since this party is responsible for the uncertainties. This rule is frequently invoked when a preprinted form prepared by one of the parties is used.
- When there are conflicts between documents, such as between drawings and specifications for a construction project, usually the written specifications will prevail (despite Bockrath, 1986). Note that a well-written construction contract will contain a "hierarchy of documents" clause that covers such cases.
- If the contract is silent as to the time of payment, the usual interpretation is that payment is due upon completion.
- If there is no provision for termination of a continuing contract, it will generally be terminable at will by either party.

Three construction contract cases (all filed against the United States!) will be used to illustrate some of the preceding guidelines. The first dispute [Edward R. Marden Corp. v. United States, 803 F. 2d 701 (Fed. Cir. 1986)] involved the existence of a requirement for a latex mastic floor covering in certain mechanical rooms at a Veterans Administration medical center. The contractor argued that the contract documents were ambiguous on this point. The court found that they were ambiguous but that the contractor had not been deterred by the inconsistencies and ambiguities in the original bidding process. The decision was based on the finding that the intention of the parties was clear:

> [We] base our decision on the cardinal rule of contract construction that the joint intent of the parties is dominant if it can be ascertained. It is uncontroverted that at the time the contract was awarded, both parties intended that composition flooring would be installed in the 18 mechanical rooms. Consequently, allowance of the contractor's claim would require us to ignore the clearly demonstrated intent of both parties. (quoted in Simon, 1989)

Another ambiguous contract documents case [Tibshraeny Bros. Construction, Inc. v. United States, 6 Cl. Ct. 463 (1984)] was decided in favor of the contractor. Here, on an Air Force contract, a dispute arose over the responsibility for the preparation of a necessary control wiring diagram. The U.S. Court of Claims held that the contract was clearly ambiguous with respect to that responsibility, stating, "In the absence of clear language shifting the responsibility to prepare the control wiring diagram to plaintiff as general contractor, the court finds that the responsibility remained with the defendant or perhaps its architect-engineer...." The ruling said that it is the obligation of the drafter to communicate effectively to the bidders what their respective obligations are (Simon, 1989).

Finally, a contractor and the U.S. General Services Agency disagreed on whether the surface layer of certain roadways and other areas was to be "bank run gravel," as shown on the drawings, or macadam, as indicated in the specifications. The contractor sought to be required to furnish only bank run gravel, claiming that the documents prepared by the government were ambiguous. However, the general conditions of the contract expressly provided that "in case of difference between drawings and specifications, the specifications shall govern," and the contractor was required to install macadam [William F. Klingensmith, Inc. v. United States, 205 Ct. Cl. 651, 505 F. 2d 1257 (1974); quoted in Sweet, 1989].

9.4   THIRD-PARTY RIGHTS

In early common law, only parties directly involved in a contract had enforceable rights. The relationship between the two parties is called *privity of contract,* and only a party in privity with another could claim to have enforceable rights. As business practices developed, the law began to recognize two major exceptions: (1) the assignment of rights to a third party by one of the parties, after the contract has been formed, and (2) the intention of the contracting parties to benefit a third party.

9.4.1   Assignment of Contract Rights

An assignment is the transfer to a third party, after a contract has been entered into, of the right to receive performance under the contract. For example, suppose Strong Steel Company makes a contract with Bell Construction, Inc., undertaking to furnish certain structural members for one of Bell's projects for a price of $183,000, with the provision that Bell will pay within 30 days after delivery. Ten days after delivery, Strong transfers its right to receive the payment to Second National Bank and notifies Bell of the transfer. Strong Steel Company thus has assigned its contract rights to Second National Bank, the assignee; and the obligor, Bell Construction, Inc., must pay the $183,000 to the bank. The relationships of Strong, the assignor, to the other parties are illustrated in Figure 9.1.

Most contract rights are assignable if there is no provision to the contrary in the contract. If the contract provides that any assignment shall be void, then none can be made. But sometimes a general provision against assignment is taken to prohibit assignment of duties but not of right. In Norton v. Whitehead [24 P. 154 (Cal. 1890)], the contract provided that "this contract shall not be assigned," but the contractor's right to assign the right to receive payment was upheld. The court decided that the responsibility to do the work was not assignable but the right to receive payment could be transferred.

Assignment can be made orally or in writing, provided prevailing statutory requirements are met. Assignment requires only an agreement between assignor and assignee. The obligor must be informed of its new obligation if it is to be binding, however. Common sense and the law dictate that the new obligation must be for no more than the original obligation.

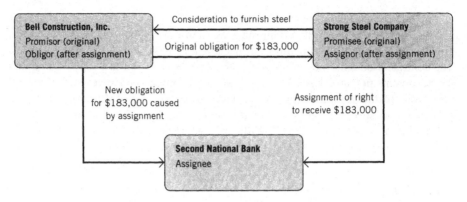

Figure 9.1. Assignment of contract rights.

Rights can be assigned, but duties cannot. The duties can be delegated, but the delegator remains responsible for their performance. Thus if you have a contract to develop a piece of computer software, you may delegate the writing of some of the subroutines to another programmer, but you remain responsible for the product—you are the guarantor of the performance. Exceptions to this rule involve specialized tasks that necessitate personal attention; Michelangelo probably would not have been permitted to delegate the painting of the Sistine Chapel even if he had guaranteed the performance of another artist.

An interesting case illustrating the validity of a contract right assignment was decided by the Supreme Court of Alabama in 1986 [R&A Construction Products, Inc. v. First National Bank of Russellville, 485 So. 2d 1119 (Ala. 1986)]. A contractor had been notified of its subcontractor's assignment to a bank of its right to receive payment; nonetheless the contractor made payment to the subcontractor's supplier. The court decided that the contractor's payment violated the subcontractor's assignment with the bank; therefore, the contractor had to pay the same amount again, this time to the bank, without receiving a "refund" from the supplier (Simon, 1989).

9.4.2 Third-Party Beneficiaries

The rights of so-called third-party beneficiaries have always been a controversial subject in contract law. Just as with assignment, early English common law insisted that to derive benefit from a contract, the parties had to be in privity of contract: "a person who did not give manifest assent or promise could not obtain a contractual right in an agreement of other parties" (Blinn, 1989). The modern view recognizes the rights of third parties whom the original parties to the contract intended to benefit from the contract. The key word is "intended," as we shall see. Whereas an *assignment* takes place after a contract has been entered into, a *third-party beneficiary contract* establishes the third party's interest as part of the original contract.

The effect, then, of a third-party contract is to give a legal right to a person who is not a party to the contract. In the past, courts have classified third-party beneficiaries into donee beneficiaries, creditor beneficiaries, and incidental beneficiaries, with only the first two categories having enforceable rights. Recently, however, the tendency is to use two classes: "intended" and "incidental" beneficiaries. Nonetheless, it is helpful to describe the three types briefly.

In the case of both donee and creditor beneficiaries, the original promisee intends that a third party will receive the benefits of the performance initially owed to the promisee. The difference is in the promisee's reasons for conferring those benefits on the third party. A third-party *donee beneficiary* contract is created when the purpose of the original promisee was to make a gift to the third party. Consider a life insurance contract in which a policyholder contracts with an insurance company, with the understanding that death benefits will be paid to a beneficiary. The beneficiary would be considered a donee. Or suppose that two persons enter a loan agreement: Mary lends George $500, with the understanding that George will repay the loan by sending $500 to the University Alumni Association as a gift from Mary. The Alumni Association is a third-party donee beneficiary of the contract between Mary and George.

In a third-party *creditor* beneficiary situation, the party who grants the right to receive performance does so not as a gift but to discharge a debt. If Mary had a $500 obligation to the Alumni Association (perhaps for the purchase of fifty nifty UAA T-shirts!), she could still enter the contract with George with the understanding that he would discharge his obligation to her by sending $500 to the Alumni Association. In this case the association would be a creditor beneficiary.

Why the distinction? The important difference is that creditor beneficiaries can sue either original party, promisee or promisor, whereas donee beneficiaries can sue only the original promisor. Thus, in the first Mary and George case, the Alumni Association can obtain remedy only from George, whereas it can sue either party when it has standing as creditor beneficiary.

Figure 9.2 diagrams a typical third-party beneficiary contract. Note that if a third party is a creditor beneficiary, he or she has the right to sue either party, whereas as a donee beneficiary the same person can sue only the promisor.

It is not required that a third-party beneficiary be identified by name or be told that he or she is a beneficiary. It is not even necessary that a particular person be known to the contracting parties at the time of contract formation. It is common that life insurance policies name unborn children or future spouses as beneficiaries, for example.

It often happens that a third party can demonstrate that satisfactory performance of a contract made by others would result in substantial advantage to that third party. Thus, the argument runs, unsatisfactory performance damages the third party, who should be able to obtain a remedy:

> Such a demonstration, no matter how compelling, is legally unavailing *unless the elements of intended beneficiary status can be established.* [emphasis added] Failing this, the third person is a mere "incidental beneficiary," which is to say that he has no legally protected interest in the performance of the contract duties. (Fessler, 1989)

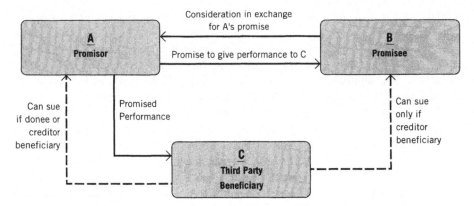

Figure 9.2. Third-party beneficiary contract.

That is, the third party is so far removed from the contract that there was no intention of including him as a beneficiary. The Supreme Court of California, on a vote of four to three, denied *intended* beneficiary status to a group who sought relief when a contract between a company and the U.S. Department of Labor, using more than a million dollars in antipoverty funds, failed to improve their hard-core unemployment status [Martinez v. Socoma Companies, Inc., 11 Cal. 3d 394, 521 P. 2d 841 (1974)]. The group sued the company, but the court concluded that their status was that of *incidental* beneficiaries (Fessler, 1989).

9.5   CONTRACT DISCHARGE

By *discharge,* we mean the termination of a contractual obligation. Contracts may be terminated in many different ways, ranging from satisfactory performance to the operation of law. We organize these types of discharge into seven categories and discuss each briefly.

9.5.1   Satisfactory Performance/Completion

The common mind-set that focuses on problems, difficulties, disappointments, and broken promises tends to obscure the realization that most contracts end happily, with satisfactory performance on both sides. Courts recognize three kinds of performance: complete and satisfactory performance, substantial performance, and material breach. In the realm of engineering and construction, it is probably quite impossible to achieve "complete" performance from contracts. Thus, the concept of *substantial performance* has evolved. Bockrath (1986) defines substantial performance as "the accomplishment of all things essential to the fulfillment of the purpose of the contract, although there may be insignificant deviations from certain contract terms or specifications." If the other party has received something less than full value, he or she will be compensated for the difference, but the party who has substantially performed will not be required to redo the performance. As might be expected, the line between substantial performance and material breach is often a thin one. In fact, Lyden, Reitzel, and Roberts (1985) use the term "nonmaterial breach" to refer to substantial performance that is nevertheless incomplete in one or more respects.

. The law recognizes a special circumstance described as "tender of performance," which indicates that one party has attempted to perform its contract duty completely but has been thwarted by the other party. Suppose you have a contract to perform subsurface soil sampling for a proposed residence but the owner denies you permission to bring your heavy drilling rig onto the premises, in accordance with standard practice. You are excused from further performance and should be entitled to recovery of your costs so far and, perhaps, to the profits you expected from this job, because you have effectively offered, or "tendered," performance.

It is important to reemphasize that most contracts end in complete or substantial performance, both of which we classify as satisfactory. Anything less can be considered as partial performance, which falls far short of the contractual requirements, even though it may confer partial benefit. Partial performance is tantamount to material breach, which will be discussed in Section 9.5.3.

Unless time is "of the essence" in a contract, the failure to finish performance by the stated completion date will not be held to be a violation of contract terms of the "material breach" variety. We deal further with this matter, especially as it relates to liquidated damages, in Chapter 10.

9.5.2 Contractual Provision

A typical construction contract will provide for at least two conditions under which the owner is allowed to terminate the contract. One is breach of contract by the contractor, which we discuss in Section 9.5.3. The other is "termination for convenience," which allows an owner to stop the work for a reason not related to the contractor's performance or lack thereof, such as absence of demand, lack of funding, or revised corporate emphasis. When the owner invokes such a provision, the contract documents typically provide a "fair" compensation for the terminated contractor, at least more compensation than would have been provided in the case of material breach.

9.5.3 Breach of Contract

Breach of contract is the failure, also called "material breach," by a contracting party to perform a duty imposed by the contract, absent sufficient excuse for nonperformance. If a party to a contract is found to have "substantially performed" his or her part, however, that party is not in breach of the contract, although some amount of compensation may be assessed to recognize the less-than-complete performance.

A contract may be broken by one party in ways other than "failure to perform," as stated earlier. The contract might be breached before the scheduled time of completion. Announcing beforehand that you cannot or will not perform your part of the bargain is known as *anticipatory breach,* or repudiation. In effect, you are admitting to breach, and the other party is relieved of further obligation. That party can begin immediately to sue you for damages, without waiting until the original time of performance. Such repudiation must be "positive and unequivocal" (Bockrath, 1986), rather than a mere threat to abandon the contract, and it must occur before the time for performance stated in the contract.

In addition, *voluntary disability,* whether by incapacitating oneself or by making proper performance impossible in some other way, leads to breach of contract, and the party that puts himself or herself in the position of being unable to perform is liable in damages. Interfering with the other party's performance is also a type of breach of contract. This sort of *prevention,* by an owner, of a construction contractor's performance, for example, discharges the contractor's further duty and makes the owner liable for breach of contract.

9.5.4 Mutual Agreement of the Parties

The parties can mutually agree to end a contract before performance is completed. Just as parties have the freedom to make a contract, they have the freedom to unmake it; they can call the deal off. If the contract is fully executory, such an act of mutual renunciation, a form of *rescission,* places the parties where they were before the contract was entered into. The consideration received by each party is the release by the other of the obligation to perform. If contract has been fully or partially executed on one side, there must be some arrangement for return of that performance or payment for it.

It is possible that the parties may then substitute a new contract for the original agreement. Or, they may agree to substitute a new party for one of the initial participants. Such an agreement among the three parties is called a *novation* agreement.

The parties can also agree that one will *waive* the performance that was expected of the other. Most often this means that one party accepts performance that was materially different from that which was specified in the contract. This voluntary surrender of a right must be based on the party's awareness of possession of the right and intent to give it up.

9.5.5 Impossibility of Performance

Suppose it becomes impossible for one of the parties to perform his or her part of the bargain. Is that impossibility sufficient excuse to terminate the contract? The answer, as in many cases, is "It depends!"

The law makes a clear distinction between subjective and objective impossibility. *Subjective impossibility* means that the promisor cannot perform, which is not a valid excuse, whereas *objective impossibility* means that the performance cannot be done, which is generally a valid excuse (Fessler, 1989). In the case of Ballou v. Basic Construction [407 F. 2d 1137 (4th Cir. 1969)], the subcontractor was to fabricate 200 columns; it fabricated 139, of which only 45 were accepted, and then claimed that the requirement was impossible. The court denied the claim, asserting that breach of contract is excused only in the case of objective impossibility; the fact that 45 columns had been fabricated satisfactorily indicated that it was possible to make columns that met the contract requirement (Simon, 1989).

Other excuses sometimes raised for failure to complete one's contract obligations are allied to the idea of impossibility of performance. One is *destruction of subject matter.* Although we might include this excuse under objective impossibility, it is usually listed separately. Bockrath (1986) explains the concept as follows: "where an obligated performance is such that accomplishment thereof absolutely depends upon the continued existence of a specific thing, that contract must be regarded as implying that, and should the aforementioned thing become unavailable (before time for performance and without fault of the promisor), the duty in question will be discharged."

In a Virginia case [Housing Authority, etc., v. East Tennessee L. & P. Co., 183 Va. 64, 31 S.E. 2d 273 (1944)], a public utility had contracted with a housing authority to supply natural gas. Both sides assumed that sufficient gas was available in the only known natural gas deposit in the area, although the written contract contained no mention of this assumption. When the natural source was found to be insufficient, the supplier was excused from further performance. The court ruled that the subject matter of the contract had been "destroyed" (Bockrath, 1986).

The *death or disabling illness* of a party to the contract may or may not be a valid excuse for nonperformance. Generally, if the contract is for personal services for which the individual was specially selected, failure to perform is not a breach of contract. However, in most other contracts, including the typical construction contract, a personal representative of the deceased or incapacitated party will be obliged to carry through if breach of contract is to be avoided.

Hardship or severe inconvenience is generally not a valid excuse and will result in breach of contract if performance is not completed. The author was involved in a case in Alaska in which a contractor claimed unsuccessfully that frozen ground underlying a rock material site was a valid excuse for nonperformance and extra compensation. The work was more difficult than anticipated, but it was not impossible. Similarly, unless there is a contract stipulation to the contrary, *labor strikes and other work stoppages* or slowdowns are not valid excuses for nonperformance.

9.5.6 Impracticability of Performance

The law seems to have moved away from the requirement for objective impossibility as a valid excuse in some cases to embrace a concept referred to as *impracticability*. The idea is that performance may be possible but, if carried out, would be burdensome enough to cause one of the parties to suffer large financial losses. According to Fessler (1989), "after-arising factors have rendered the cost of performing a contract promise grossly in excess of what had been assumed by the parties at the formation of the contract." One example is a case of Foster Wheeler Corporation v. United States [206 Ct. Cl. 533, 513 F. 2d 588 (1975)], in which the company agreed to design, fabricate, and deliver two boilers that would withstand shock up to certain intensities. After a computer model indicated that the specification was "impossible," however, Foster Wheeler ceased performance. The court found in favor of the contractor, indicating that the contract was both "commercially and absolutely impossible," explaining that the construction of such a shock-hard boiler, "even if ultimately possible, could not be accomplished without commercially unacceptable costs and time input far beyond that contemplated in the contract" (Sweet, 1989).

As early as 1966, the idea of impracticality was making its way into court decisions. The sudden closing of the Suez Canal in 1956 led to considerably increased costs for shippers already under contract. In writing the court's opinion on a relevant case [Transatlantic Financing Corp. v. United States, 363 F. 2d 312 (D.C. Cir. 1966)], Judge Skelly Wright gave a helpful example of the way the law changes over time:

> The doctrine of impossibility of performance has gradually been freed from the earlier fictional strictures of such tests as the "implied term" and the parties' "contemplation." ...It is now recognized that "A thing is impossible in legal contemplation when it is not practicable; and a thing is impracticable when it can only be done at an excessive and unreasonable cost." ...The doctrine ultimately represents the ever-shifting line, drawn by courts hopefully responsive to commercial practices and mores, at which the community's interest in having contracts enforced according to their terms is outweighed by the commercial senselessness of requiring performance .... (quoted in Fessler, 1989)

9.5.7   Operation of Law

Finally, contract performance may be discharged as a result of the operation of law. A *supervening illegality* is a statutory enactment occurring after the date of the contract that prevents performance of the contract. If Congress passes legislation that makes it illegal to use my ships to transport a certain product from Seattle to Singapore, I would be able to discharge a contract I had made to perform this task with no breach of contract. A *statute of limitations* may operate to discharge a contract. For example, suppose you and I have a contract: I am to build you a sidewalk of a certain description by a certain date for $1200. Suppose further that I finish the work, but you don't pay me. The law would provide a stated time limit, such as 7 years, within which I must file suit to try to recover my $1200. If I wait longer than the specified time, the contract will be considered to be completed, and the courts will not assist me in an attempt to obtain the $1200. The complicated process of bankruptcy is another means by which contract performance may be discharged with no breach of contract in some situations.

9.6   REMEDIES FOR BREACH OF CONTRACT

Finally, we outline the various remedies available to the nonbreaching party to a broken contract. The failure of one party to perform, unless there is some valid excuse such as impossibility, entitles the other party to affirmative relief as well as excusing it from the duty to perform. Justice Oliver Wendell Holmes Jr. declared that "the duty to keep a contract at Common Law means a prediction that you must pay damages if you do not keep it" (quoted in Blinn, 1989). In such cases, however, remedies other than damages may be available. We consider restitution and various equity remedies, as well as damages of several types.

9.6.1   Damages

Nominal damages are a very small sum (one dollar, six cents, ten dollars) fixed by the court more in acknowledgment that a technical default has occurred and the plaintiff is entitled to recovery than to compensate a bona fide loss. A court may impose this limitation because the plaintiff failed to prove actual damages or because the damages are too speculative to form a basis for recovery.

The most common remedy for breach of a contract is the recovery of a sum of money that compensates the innocent party for the amount of damages suffered. Thus, the term *compensatory damages,* or actual damages, is used. Bockrath (1986) quotes a court transcript that captures this basic idea:

> One who fails to perform his contract is justly bound to make good all damages that accrue naturally from the breach; and the other party is entitled to be put in as good a position pecuniarily as he would have been by performance of the contract.

To obtain compensatory damages, one must (1) show that a contract right existed and was not fulfilled, and (2) establish the amount of damages with reasonable certainty. The amount of compensation awarded to the plaintiff is decided by the jury, unless the judge believes the jury has weighed the evidence incorrectly. There can be no recovery for speculative damages, nor can the injured party be placed in a better position than would have been occupied if the breach had not occurred.

Also, the injured party has a duty to *mitigate the damages*—that is, do everything possible to keep the actual damages as small as can be arranged. If an apartment renter with a 6-month lease breaches the contract by moving out after 2 months, the landlord has the duty to try to rerent the apartment, rather than leaving it vacant and then seeking the entire 4 months' rent from the original tenant. At the expiration of the 4 months, however, provided an honest effort had been made to find another renter, the landlord would be entitled to recover from the original tenant the expected rent for those four months, minus what was collected from subsequent tenants, if any. In addition, there should be recovery of incidental costs for cleaning, advertising, and the like.

Punitive damages are damages awarded as punishment or to deter especially egregious conduct, rather than to compensate for actual injury or loss. The idea is to prevent the defendant from repeating the conduct or to make an example of the defendant in the hope of persuading others to desist. Such awards have been rare in engineering and construction contract disputes, although they seem to be increasing. Most courts award punitive damages in contract breach cases only if the breach included a tort such as negligence.

A fourth type of damages is often stipulated in construction contracts: *liquidated damages* are determined when the contract is formed and are used in lieu of actual damages in the event of breach of contract. If a breach occurs, no proof of actual damages is required. A typical application of this approach is the assessment of a predetermined amount for every calendar day a contractor is late in completing a project, if the time of completion is an essential condition of the contract.

Courts are consistent in insisting that a liquidated damages amount not be the same as a penalty. Two quotations are helpful:

> Damages for breach by either party may be liquidated in the agreement but only at an amount which is reasonable in the light of the anticipated or actual harm caused by the breach, the difficulties of proof of loss, and the inconvenience or non-feasibility of otherwise obtaining an adequate remedy. (quoted in Simon, 1989)

> *Liquidated damages* is the sum a party to a contract agrees to pay if he breaks some promise, and which, having been arrived at by a good faith effort to estimate in advance the actual damage that will probably ensue from the breach, is legally recoverable as agreed damages if the breach occurs. A *penalty* is the sum a party agrees to pay in the event of a breach, but which is fixed, not as a pre-estimate of possible actual damages, but as a punishment, the threat of which is designed to prevent the breach. [Westmount Country Club v. Kameny, 82 N.J. Super. 200, 197 A. 2d 379 (App. Div. 1964); Simon, 1989]

Generally a liquidated damages clause will be found to be enforceable if (1) the actual damages that would occur in the event of breach are difficult or impossible to estimate accurately, (2) the parties intended to provide for such damages by including appropriate wording in the contract, and (3) the amount specified is a reasonable preestimate of the actual loss in the event of breach and is not a penalty.

9.6.2   Restitution

Restitution is based on the verb "to restore." A plaintiff who desires restoration to the circumstances that existed before performance of the contract began, rather than being granted damages, may unilaterally "rescind" the agreement and demand the return of whatever consideration has been spent or otherwise contributed. We distinguish restitution from compensatory damages in that restitution seeks to place the aggrieved party to the position he or she was in before the contract, to the extent possible, whereas compensatory damages tries to place that party where he or she would have been if the contract had been performed satisfactorily. Generally the plaintiff is required to return what has been received, as well.

9.6.3   Equity Remedies

We discussed the distinctions between law and equity in Chapter 8. Equity remedies for breach of contract, of which specific performance and injunction are the most prevalent, will not be granted if damages provide an adequate remedy.

Specific performance is a requirement that the defendant live up to the promise made and perform in accordance with the contract, whereas an *injunction* is a court order to refrain from doing something. In both cases, the promisor is being required to live up to the bargain he or she has made. An example of specific performance might be a contract for the sale of land. If the seller reneges on the purchase agreement, a damages remedy would award some amount of money to the plaintiff. However, the buyer might be particularly interested in this particular piece of land rather than money in compensation. In that case, the buyer probably would file an equity suit asking for a decree to compel the seller to perform in specific accordance with the original sales contract.

Sometimes, when damages are deemed insufficient, both money damages and equitable remedies are awarded for breach of contract. For the most part, courts have been reluctant to require specific performance by engineering designers or contractors; in nearly every case, money damages are sought and awarded if breach of contract is proven (Sweet, 1989).

9.7   DISCUSSION QUESTIONS

1. Susan Smith is the president of Dirtmovers Construction Corporation, at an annual salary of $150,000. Midway through her 3-year written contract, a competitor company, Earthworks, offers her $200,000 annually to sign on as its president. Susan informs the Dirtmovers board of directors of the Earthworks offer, and the Dirtmovers board raises her salary to $200,000. Susan agrees to stay. Is this contract modification legally enforceable? Why or why not? If not, what could be done to make it enforceable?

2. You have a charge account at Southern Building Supply. You walk into the store, pick up three hammers and a roll of tie wire, and hold them in front of the bookkeeper, who nods affirmatively. You walk out of the store with the items, neither you nor the bookkeeper having spoken a word. Is there a contract between you and Southern, requiring you to pay for the four items?

3. *Haneman:* What is the lowest price for which you would sell your 1967 Chevy truck?
   Zarling: $450.
   Haneman: Sold!
   Zarling refused to turn his truck over to Haneman. Was a contract formed by this conversation? Use the words *offer* and *acceptance* in your answer [Harvey v. Facey, Judicial Committee of the Privy Council, based on A.C. 552 (1893)].

4. Company A sent Company B a letter asking for quotations on a certain type of lighting fixture. Company B responded with an offer to sell between 25 and 50 such fixtures for a specified price per fixture. Company A sent a fax and a letter ordering 15 of the lighting fixtures. Company B replied that they would not furnish this order. Company A then sent a fax ordering 25 of the fixtures, but Company B refused to furnish them. What is the legal situation?

Has a contract been formed? [Minneapolis & St. Louis Railway Company v. Columbus Rolling-Mill Company, 119 U.S. 149 (1886)]

5. *Brown:* I offer to sell you my used milling machine for $6500.
   Blue: I accept, provided you deliver the milling machine to my shop and have it here next Tuesday.
   Brown: Sorry, my delivery truck is tied up until the end of next week.
   Blue: Okay, I'll pick the machine up myself, since I need to start using it next Wednesday.
   Brown: I think I'll wait awhile before selling my milling machine.
   Blue: But I bought the milling machine! We have a contract!
   Who wins? Why?

6. On December 10, George sends Art the following offer: "I offer to sell you my transit and tripod for $625. This offer will remain open for your acceptance by mail until December 20." Art sent an acceptance by mail on December 15, but the Christmas rush delayed its delivery and George did not receive it until December 22. Was there a contract?

7. In Question 6, suppose Art received a withdrawal of George's offer at noon on December 16. Relying on the deadline of December 20, stated in the original offer, however, Art sent an acceptance at 4:30 P.M. on December 16. Was there a contract?

8. A patient was admitted to a hospital with an abdominal aneurysm. Her doctor ordered 24-hour nursing care, which was provided by an independent nursing care service. No contract, either express or implied in fact, existed between the patient and the nursing service. For 2 weeks of in-hospital care, the nursing service billed the patient about $3500, which the patient failed to pay. When the nursing service brought suit against the patient, the trial court denied recovery. What should be the result upon appeal? [Nursing Care Services, Inc. v. Dobos, 380 So. 2d 516 (1980)]

9. Suppose you contract orally with a manufacturing company to provide design services for $35,000 during a period of 2 years. Upon completion of your work, your client argues that the contract is not enforceable and therefore refuses to pay your bill. Upon what basis might you seek recovery? State your position in one sentence. How much are you likely to recover?

10. A 36-page construction contract provided, among other things, for a project completion date. The contractor introduced evidence that a prebid conference had established a different completion date. Should this evidence be admitted? [Lower Kuskokwim School District v. Alaska Diversified Contractors, Inc., 734 P. 2d 62 (1987)]

11. An engineering firm supplied a standard AIA contract (American Institute of Architects, 1987) for its design agreement with an investment company that was about to build a medical clinic. The owners were experienced business persons who read the document and made some revisions that were accepted by the engineer. If ambiguities were found in the contract, on what basis should a dispute about these points be resolved? [Durand Associates, Inc. v. Guardian Investment Company, 186 Neb. 349, 183 N.W. 2d 246 (1971)]

12. A contractor's bid included the following item: "400 linear feet of Type R cable at twenty-three dollars ($32.00) per linear foot." How will the contractor's price for this item be determined?

13. West owned a new home with a large front yard. She negotiated with North to have North provide landscaping of the yard, including seeding. North drafted the written contract. In part it read, "North will do a top class job of landscaping West's yard. (Back yard not included.)"

    A dispute arose as to whether North had provided a sufficient quantity of seed and whether the lawn had been raked and rolled an adequate number of times. Also, West contended that the side yard should have been seeded, but North disagreed. Give at least two bases on which this dispute might be resolved.

14. Manning contracted to deliver by a certain date "500 Draft-Fast drafting pens of assorted line widths" to Archibald Architectural Associates. Archibald intended to provide Manning with a list of the desired line widths and their quantities, but this intention was not made clear. Upon receiving the signed contract, Manning selected the mix of sizes and sent the order to Archibald, who refused to pay. If Manning brings a suit, will Archibald have to pay for the mixture of pens as provided?

15. A university contracted with Sammis to place tile work in a building for $27,344. In part, the contract read, "Sammis shall not sell, assign or transfer ... the contract ... unto any person ... without the consent of the architects. ... " Sammis did assign to Mueller $11,000 that was due from the university, with no consent from the architects. The university paid *Sammis* the $11,000, and Mueller sued the university for that amount. Will Mueller win? [Mueller v. Northwestern University, 195 Ill. 236, 63 N.E. 110 (1902)]

16. An owner of a New York City apartment building contracted with a security products company to install front door security locks. Some time after the installation was completed, a visitor entered the building and was accosted by a stranger, dragged to her host's apartment, and raped. The woman sued the security products company for breach of contract, alleging that the front door locks were negligently installed. What should the result be? [Einhorn v. Seeley and Rem Discount Security Products, Inc., 525 N.Y.S. 2d 212 (N.Y. Sup. 1988)]

17. A basketball official had a contract with the Big 10 Conference to referee various basketball games. The official called a foul against an Iowa player that resulted in a Purdue player's scoring the last-minute, winning point over Iowa, thus denying Iowa the league championship. Some Iowa fans blamed the official for their team's loss, claiming that the foul call was clearly in error. The owners of a store that sold Iowa memorabilia sued the official, claiming negligence and breach of contract and asking for $175,000 in damages because of loss of potential business. The official's motion for summary judgment was granted by the trial court. What should be the result of the store owners' appeal? [Bain v. Gillispie, 357 N.W. 2d 47 (1984)]

18. A building contractor and a school district entered into an agreement under which the contractor was to build an addition to the district's Public School No. 17. Before the construction was started, the school was destroyed by an earthquake. What is the contractor's obligation?

19. In Question 18, suppose the contractor had completed 40% of the work and had also stored $15,000 worth of materials at the site, which had been accepted by the school district. Then the earthquake destroyed the materials and the partially completed building. What are the obligations of the parties?

20. Now suppose the contract in Questions 18 and 19 had been for a new school rather than for an addition to an existing building. If the building was partially complete and the earthquake destroyed it, what are the obligations of the parties?

21. In 1969 Stoneway Concrete, Inc., leased a tract of land in the state of Washington from Weyerhauser Real Estate Company for the purpose of strip mining gravel and other aggregates. Both parties knew that several government permits would be required before mining could begin. Stoneway's application for a permit from the county was challenged by a public referendum and environmental lawsuits, which were eventually won by Stoneway. A new state environmental policy law was then enacted, but Stoneway successfully defended charges of violations of this act. Before any materials were extracted, state and county officials notified Stoneway of the need for environmental impact statements. Stoneway then abandoned the project and ceased further lease payments. Weyerhauser sued Stoneway for contract breach. Who won? [Weyerhauser Real Estate Company v. Stoneway Concrete, Inc., 637 P. 2d 647 (1981)]

22. Suppose you enter into a contract on March 1 to purchase a certain home in your city from Don Jones for $250,000, with the understanding that the sale is to be effective on September 15. Don sells the house to somebody else on June 30, and the new owners take possession immediately. You then purchase a comparable home in the same neighborhood for $325,000. When can you begin your lawsuit against Don, what will you seek to recover, and what are your chances of success?

23. For your new home, you engage a water well driller who, according to your signed contract, will "drill a 6" diameter cased water well for $2750 which will furnish water meeting local quality standards and have a flow rate of at least 200 gallons per hour." Your well driller abandons the job after spending $4000 without finding satisfactory water. How much are you obligated to pay the well driller?

24. Suppose an earthmoving contractor enters into a contract to furnish 750,000 cubic meters of a certain type of gravel to the Department of Highways for a road construction project. Should the DOH have to pay anything for partial performance if the contractor quits after 400,000 cubic meters has been removed?

25. A subcontract required the general contractor to make monthly progress payments to the subcontractor. After the project was started, the general contractor refused to make any payments to the subcontractor until the subcontractor's work was completed. Since it was not being paid, the subcontractor refused to do any more work. The general contractor sued the subcontractor on the basis that the refusal to perform further work was a breach of the subcontract. What was the result? [Silliman Company v. S. Ippolito & Sons, Inc., 467 A. 2d 1249 (1983)]

26. A short time after the completion of a $7000 swimming pool, defects were discovered that rendered the pool unusable. The cost of correcting the defects was estimated at $11,000. Assuming that the defects were the fault of the contractor, describe briefly two possible remedies that might be provided to the owner. Which of these do you think the court chose? [Mayfield v. Swafford, 106 Ill. App. 3d 610, 435 N.E. 2d 953 (1982)]

27. A contractor agreed to construct a 1200-foot tunnel. As protection against tunnel cave-ins, the other party to the contract, a water company, furnished timbers and specifications for their installation. The contractor protested that the timbers were defective and the specifications inadequate. After approximately 750 feet of tunneling, a cave-in occurred. The water company refused to allow the contractor to continue work. The court found that the contractor had performed satisfactorily and in accordance with the specifications. What damages do you think the contractor was awarded? [McConnell v. Corona City Water Company, 149 Cal. 60, 85 P. 929 (1906)]

9.8   REFERENCES

Alaska Statutes. 1989. Code of Civil Procedure. Sec. 09.25.010. As amended in 1989.

American Institute of Architects. 1987. Standard Form of Agreement Between Owner and Architect, Document B141, 1987 edition. New York: AIA.

"Bidder Gets Relief from Error." 1989. *Civil Engineering,* Vol. 59, No. 3, March, 24.

Blinn, K. W. 1989. *Legal and Ethical Concepts in Engineering.* Englewood Cliffs, NJ: Prentice-Hall.

Bockrath, J. T. 1986. *Dunham and Young's Contracts, Specifications and Law for Engineers.* New York: McGraw-Hill.

Fessler, D. W. 1989. *Contracts: Casenote Law Outlines.* Beverly Hills, CA: Casenotes Publishing.

Fisher, B. D., and M. M. Jennings. 1986. *Law for Business.* St. Paul, MN: West Publishing Company.

Firmage, D. A. 1980. *Modern Engineering Practice.* New York: Garland STPM Press.

Lewis, B. J. 1983. "The Story Behind the Recent National Scandals Involving Engineers." In Schaub, J. H., and K. Pavlovic, eds., *Engineering Professionalism and Ethics.* New York: Wiley, pp. 239–245.

Lyden, D. P., J. D. Reitzel, and N. J. Roberts. 1985. *Business and the Law.* New York: McGraw-Hill.

Nord, M. 1956. *Legal Problems in Engineering.* New York: Wiley.

Simon, M.S. 1989. *Construction Claims and Liability.* New York: Wiley.

Sweet, J. 1989. *Legal Aspects of Architecture, Engineering and the Construction Process,* 4th ed. St. Paul, MN: West Publishing.

Vaughn, R.C. 1983. *Legal Aspects of Engineering,* 4th ed. Dubuque, IA: Kendall/Hunt Publishing.

10

Engineering and Construction Contracts

This chapter applies many of the concepts in Chapter 9 to contracts of the kinds engineering managers often deal with. We begin with some information about contracts for engineering and architectural services. Then, because preparation and execution of construction contracts are such important parts of the activities of many engineers and architects, and because such contracts exhibit the characteristics common to contracts in general, we devote a majority of the chapter to contracts for construction services. In Section 10.2 we discuss the five contract elements as they are present in a valid construction contract and review several types of contract as to the basis for payment and the relationships among the various parties. We then summarize the process by which the contract steps proceed, from advertising for bids to project completion, and describe the several construction documents. There are many contract provisions of special interest, including bonding, time of completion, delays, and different site conditions; we select eight such provisions and discuss them in Section 10.6.

10.1 ENGINEERING AND ARCHITECTURAL SERVICES CONTRACTS

In Section 9.2, we noted five elements that must be present to make a contract valid. These elements are essential in the formation of contracts between engineers, architects, and other design professionals and their clients, whether in the public or the private arena. In writing about the formation and interpretation of design services contracts, Sweet (1989) points out two attributes of such contracts that set them apart from ordinary commercial contracts. First, it is often the case that clients are relatively inexperienced in such contractual matters. While the petroleum refining company or the highway department will surely have contracts departments staffed by experienced personnel, the school board, the church building committee, and the individual homeowner take on such projects so infrequently that they have little knowledge or experience in contracting with engineers and architects. The second attribute is that the initial phase of contract formation often is rather vague and informal. There will normally be some preliminary discussions of content and price; after that, there may be a written contract, using preprinted forms supplied by the design professional or a letter written by one of the parties. Thus, even when the five elements are present, these two characteristics of inexperience and imprecision can lead to honest, and interesting, misunderstandings between the parties.

10.1.1   Contract Elements

Both parties must be *competent,* in the sense that they are capable of carrying out their respective obligations. The design professional must have the expertise he or she claims to have, must be licensed as required in the particular jurisdiction and, if a corporation, must operate within the scope of the corporate charter. The owner/client must have the financial capability required and be capable of making decisions during the design process. Principles relating to the avoidance of contracts by those who are under age, insane, or intoxicated apply here as well.

The *subject matter* must be proper in two respects. The agreement must define things sufficiently to ensure that the intent is clear: What is the scope of each party's obligations? What are the fee arrangements? What are the relationships among the varoius parties? Also, the purpose and provisions of the contract must be lawful; agreements to set up kickback scemes clearly violate this requirement.

We have alluded to the informal manner in which contracts with design professionals are often formed. Nonetheless, there must be a *meeting of the minds.* Whatever the process, eventually one party must make an offer that the other party accepts without change. "I accept your proposal to design my back porch provided you add four chartreuse flower boxes" will not form a valid contract unless the designer says, "Okay." Valid contracts also require proper *consideration.* Although courts generally do not inquire into the value of such contributions (the so-called adequacy of consideration rule), each side must give or promise to give something of value. When the promises of each—certain design services in exchange for a certain payment—are clear, this aspect should cause little problem.

Finally, with respect to *form,* we note that the typical statute of frauds does not require that design service contracts be in writing for them to be enforceable. This statute applies only if the contracts are not to be performed within the statutory time frame or if they somehow involve real estate transactions. (The dollar limit above which contracts must be in writing applies to sales of *goods* but not *services.*) The author once made a very satisfactory $600 oral contract with an architect for the design of a house addition. Both parties performed to the satisfaction of the other, and all was well; if such had not been the case, the agreement could have been enforced in a court of law.

All the preceding good words notwithstanding, it is excellent practice to make sure that all contracts for design services are in writing. (Do as I *say,* not as I *do!*) And certainly it is the case that most such contracts, excepting only the smallest and shortest, are indeed reduced to writing.

10.1.2   Scope of Services

Here, we assume that the contract for professional services will be in writing. The agreement should state precisely what the design professional is expected to do. It is important, also, to have an understanding of what the designer is not obligated to do under the contract. While much variety in the scope of services is provided in such contracts, it is helpful to divide the work into planning, design, and construction phases.

During the *planning phase,* the design professional typically assists the owner in determining general project feasibility. Should the project be built at all? What sites are available that might be considered? What are some alternative physical arrangements? What are the costs of the various alternatives? How much time is available, and what schedule impacts can be predicted for the alternatives? Even though the information developed at this stage is tentative and must be understood to be such, the owner must have it to be able to (1) decide whether to proceed and (2) seek any necessary funding, if it is decided to proceed.

The *design phase* is the work the public most commonly associates with the engineer and architect. A common means of subdividing this work—as, for example, in the American Institute of Architects (AIA) standard contract (1987)—is to designate schematic design, design development, and construction documents subphases. During the first subphase, general layouts are developed to show relationships of spaces, flow of materials, and other aspects of how the elements of the project will "fit together," depending on project type. Design development includes the detailed calculations—pipe sizes, electrical circuit design, sizing of structural members, and a host of other similar details—as well as rough drawings and specifications that begin to put these results into graphical and written form. The output from the construction documents phase is a set of documents that will form the basis for entering into a contract with a construction contractor. It includes final drawings and technical specifications, as well as general and special conditions and other written materials that will be part of the construction contract. In Section 10.5, we discuss these documents in more detail.

During the design phase, the designer will often engage the services of outside consultants to be responsible for certain specialized aspects of the design, such as electrical instrumentation or acoustical design. The choice of these experts is usually left to the lead designer. The owner/client will need to participate in some decisions prior to the design phase, however. The contract should state whether it includes responsibility for obtaining outside data such as subsurface conditions or landownership status, acquisition of a variety of permits, purchase of land, utility arrangements, and estimation of construction costs. If this last item is included, the degree to which the estimate is "guaranteed" must be made clear. In a 1974 Minnesota case [Griswold and Rauma, Architects, Inc. v. Aesculapius Corp. 310 Minn. 121, 221 N.W. 2d 556], a project budget for a medical clinic was established at $300,000. Bids were opened, and after some items had been deleted, the low bid was $413,037. The owner decided not to build the project and refused to pay the architects for services rendered. The outcome of the case depended in large part on whether the $300,000 "maximum cost figure," as stated in the contract, was an approximate estimate or a guarantee. If a guarantee, then the architect had breached the contract by designing a project that could not be built for the stated price. In this case, the figure was held to be an estimate, and the Supreme Court of Minnesota directed that the architects be paid.

The scope of the design professional's work during the *construction phase* is often the most difficult to define and understand. Usually the engineer/architect has some responsibility for assisting with contractor selection, including advertising the project, receiving and evaluating proposals, and awarding the contract. Furthermore, usually the design professional has some responsibility to represent the owner on site during the actual construction work. It is essential that the scope of this work be defined clearly.

Some of the matters to be clarified entail the following questions (Manzi, 1984; Schoumacher 1986).

1. Does review of shop drawings constitute "approval"?
2. Does inspection of the contractor's work make the inspector responsible for its fitness?
3. Will the designer be responsible for certifying work completion for payment purposes and will such certification constitute "acceptance"?
4. To what extent is the design engineer responsible for job site safety?
5. To what degree do any of these on-site "observation of work" activities mean that the designer has become the de facto supervisor/coordinator of the work and is thus liable for difficulties that arise?

A further recent question involves whether the design professional is an agent of the owner when he or she carries out certain project obligations. If so, a disenchanted party such as the contractor may have some chance of suing the owner successfully for an action of the designer. In Prichard Brothers, Inc. v. Grady Co. [Minn. App. 1989, 436 N.W. 2d 460], the contractor sued both the designer and the owner for damages due to delays caused by the designer. The court decided that Grady, the architect, was acting independently and not as an agent of the owner when he did the things that caused the delays; thus the court allowed Prichard to recover from Grady but not from the owner (Loulakis and Thompson, 1990).

The subject of professional liability of engineers and others is discussed in some detail in Chapter 12. At this point, we simply state the importance of clarifying the various aspects of the design contract scope, in writing, in as much detail as is needed to avoid misunderstanding.

10.1.3   Fee Arrangements

On what basis shall the design professional be paid for services rendered? The contract for those services ought to provide sufficient information to answer that question. There are several possibilities, ranging from a fixed fee to one that is based on the cost of the construction contract. It is often quite difficult to estimate in advance the amount of effort, and thus a fair level of compensation, for such services, for the following reasons:

- The design professional does not control the basic project requirements presented by the owner, thus making it virtually impossible to foresee all the necessary work.
- The basic requirements are vague and uncertain at the beginning of the project.
- The effort required for engineering work during the planning phase— feasibility studies, development of alternatives, site investigations, and the like—is especially difficult to predict, even though such work has potential for large savings later in the project.
- The delay and expense associated with the often frequent changes in the owner's perceptions of project needs and wants cannot be predicted with any degree of certainty. (Bockrath, 1986)

There are several bases on which the design professional's fee can be determined.

- *Fixed Fee.* The designer is paid a fixed amount, which is decided on prior to commencement of work. The obvious difficulties of predicting the effort required limit the use of this arrangement to cases in which the scope of the services is clearly defined. A fixed fee will include direct costs, overhead, profit, and provision for contingencies. The contract must state whether the scope includes only basic services or whether such activities as site surveys, financing arrangements, and/or permit acquisition are included.
- *Percentage of Construction Cost.* This popular method simply multiplies the construction cost by a preagreed-on percentage. It is important to define "construction cost." For example, usually it does not include the cost of land, financing, or other owner's costs. A disadvantage from the owner's standpoint is the lack of incentive for the designer to produce a low cost project.
- *Expenses Plus Professional Fee.* The designer is paid his or her actual expenses plus a fee that is either a fixed amount or a percentage of the expenses. It is essential that "expenses" be defined: Do they include travel, telephone, data processing, copying and other reproduction of documents, overtime work, and the like, or are those costs to be taken out of the fee?
- *Multiple of Direct Hourly Expense.* The cost of supporting an employee, including salary and fringe benefits, is increased by a multiplier that will then cover administrative costs and profit. The rate can be either hourly or daily, although hourly probably will be preferred owing to the vagaries of defining a "day." If an employee's salary is $22.50 per hour, and an additional $8 per hour is required for insurance, pension, annual sick leave, and the like, the $30.50 would be multiplied by a factor such as 2.7, in which case the client would be billed $82.35 for every hour the employee worked on the project.

10.1.4 Relationships with Other Parties

The contract for design services must make clear any relationships that are to be established with parties other than the owner and the design professional. For example, will the designer have some contractual relationship with the construction contractor or, perhaps, with the lender? There should also be a clear understanding about the designer's use of consultants and the means for paying for their services; if a cost-plus contract, are those services "costs"? (The answer is likely to be "yes," but the contract should make this clear.)

As we shall learn in Chapter 12, the designer may be held to have some liability even toward parties with whom it has no direct contract. This recent trend in legal thinking, the so-called loss of the privity defense, is having a major impact on the way designers do business.

10.1.5 Standard Documents

The National Society of Professional Engineers (NSPE), the American Institute of Architects (AIA), and other professional organizations publish suggested documents that can be used in preparing contracts for design services. These templates have stood the test of time, having served, with modifications, over many years. They require filling in the blanks and inserting any special provisions. They also require careful reading prior to signing, however.

One contract interpretation principle says that contract ambiguities are likely to be interpreted against the party who prepared the contract. In the case of standard documents, the designer normally selects the papers and presents them to the owner for review, negotiation, and signing. Thus, the designer is likely to be held accountable for any discrepancies between sections of these "standard" documents that develop as changes are made during the negotiation process, just as would have been true if he or she had prepared them personally.

We now turn from contracts for design services to contracts for the actual construction of the project, beginning with a discussion of the elements necessary to make a valid construction contract.

10.2 CONSTRUCTION CONTRACT ELEMENTS

No matter what method is used to select the contractor, or what basis is used to pay the contractor, a construction contract must include the five basic elements outlined in Chapter 9 and discussed here in the context of design contracts. We now apply those elements to the typical contract for construction services.

10.2.1 Competent Parties

In the typical two-party contract for construction services, the principal parties are the owner and the contractor. Both must be capable of fulfilling their respective roles. The owner must be capable of providing information for use by the contractor and paying for the work. There may be other obligations of the owner, such as review of proposed materials, inspection of the work, and the like, which the owner must be competent to perform.

Likewise, the contractor must have the expertise, resources, and time to conduct the work in accordance with the contract documents. Determining whether the contractor is qualified is often difficult. Some owners prequalify potential bidders by evaluating the sufficient experience, personnel, equipment, financial, and other resources of interested contractors. We discuss prequalification further in Section 10.4.2

10.2.2   Proper Subject Matter

The subject matter must be definite, so that both parties know what is to be done in sufficient detail to be able to assemble the desired product. Indeed, the purpose of the planning and design phases is to develop the various documents to accomplish this purpose. The drawings, technical specifications, and general and special conditions described in Section 10.5 provide this definition.

The construction contract cannot be against public policy. A New York case [S. T. Grand, Inc. v. City of New York, 32 N.Y. 2d 300, 289 N.E. 2d 105, 344 N.Y.S. 938 (1973)] in which a contract for cleaning a reservoir was fully performed by the contractor offers a sad example. After contract completion, the company and one of its officers were convicted of violating state bribery laws because the work had been obtained through payment of a kickback to a public official. The contractor filed suit to recover the $148,735 unpaid balance, and the city countersued to recoup the $689,500 already paid to the contractor. In finding in favor of the city, the court said,

> The rule is that where work is done pursuant to an illegal municipal contract, no recovery may be had by the vendor.... We have also declared that the municipality can recover from the vendor all amounts paid under the illegal contract.... The reason for this harsh rule, which works a complete forfeiture of the vendor's interest, is to deter a violation of the bidding statutes. (quoted in Simon, 1989)

10.2.3   Meeting of the Minds

As we noted in Chapter 9, for a contract to exist, there must be an *offer* and an *acceptance* of that offer. Depending on local statute and case law, there need not necessarily be a strict form to the offer. As long as it is clear, it should be valid. In some cases, bidding requirements stipulate that an offer to do construction work must contain certain items; one that does not is deemed to be unresponsive.

The point at which a contract offer has been accepted by a contractor is sometimes difficult to determine. Clearly, there must be acceptance of the exact thing that was offered. That is, "the minds must meet." The confusion often present at this point is illustrated by a North Carolina case involving a contract between a contractor and a subcontractor [Howell Co. v. C. M. Allen & Co., 8 N.C. App. 287, 174 S.E. 2d 55 (1970)]. The subcontractor submitted a bid but failed to prove that an agreement had been reached as to the price of the work to be done.

For a binding contract to exist, more than a mere proposal intended to open negotiations must exist. The court indicated that for a valid contract to exist the parties must agree to the same thing in the same sense; their minds must meet regarding all the terms. (quoted in Simon, 1989)

It is sometimes claimed that an owner's invitation for bids constitutes an offer, and the submission of the low bid is therefore the acceptance of that offer, thus binding low bidder and owner into a contract if the other contract elements are present. However, this view is very much in the minority, with most legal minds asserting that the invitation for bids is just that—an announcement of an opportunity to submit offers, one of which may be accepted by the owner.

We should also note that an oral acceptance may be sufficient to bind the parties. In many cases, formal execution is simply a formality, the actual contract having existed ever since the successful bidder was notified, perhaps by telephone, that his or her bid had been accepted.

10.2.4 Consideration

Something of value must be promised or done on each side. In nearly all construction contracts, the contractor promises to build the specified project in accordance with the contract documents, and the owner promises financial consideration. No money need change hands at the contract signing. There is no need for such statements as "for ten dollars and other valuable consideration" or "this agreement is executed in consideration of the mutual promises herein." Both are considered to be self-evident.

As we emphasized in Chapter 9, courts almost never take on the task of determining whether the consideration was "fair." Rather, it is assumed that the parties have looked out for their own interests before entering voluntarily into a contractual arrangement:

Commercial enterprises are presumed to be familiar with the terms of the agreements they sign, and they will be held to those agreements. If hindsight proves the agreement to be foolhardy, that is the risk the party assumed when signing the agreement. (Jervis and Levin, 1988)

The matter of consideration is sometimes raised when dealing with changes and extras to a construction contract. It is well settled that a change order that provides for additional compensation for doing what one was already obligated to do under the contract is unenforceable. There is consideration only on one side of the new part of the contract, and that does not make for a valid contract [Bergstedt, Wahlberg, Berquist Associates v. Rothchild, 302 Minn. 476, 225 N.W. 2d 261 (1975)]. But, a change order providing for an extra $40,000 to add two more birdhouses ought to be considered valid, no matter how small the structures. The court will not be dragged into determining whether the consideration is "adequate."

10.2.5 Form

The statute of frauds requires that certain kinds of contract be in writing if they are to be enforceable. Generally, contracts for construction services are not caught by the statute unless they are not to be performed relatively promptly, such as within one year. Stated another way to avoid the double negative, an oral construction contract is generally enforceable if it is to be performed within one year. An exception might occur if the provision of goods rather than services were in fact at issue or if a real estate transaction was involved.

The term "to be performed within one year" can lead to interpretation difficulties. The Supreme Court of Arkansas dealt with such a dispute in Township Builders, Inc. v. Krause Construction Co., Inc. [286 Ark. 487, 696 S.W. 2d 308 (1985)]. The guideline was whether the contract was "capable of performance within one year" (Simon, 1988). The chance, or even the likelihood, that more than 365 calendar days would be required was not the issue. In this case, the Arkansas court decided that whether the contract could or could not be performed within a year was a question for the jury, and the case was remanded to the trial court for that determination.

10.3 TYPES OF CONTRACT

In this section, we deal with three traditional bases for compensating the contractor; in each instance, we assume that the contract is "traditional" in that it involves only the owner and the contractor. We then present five more recent approaches to the relationships among the various parties, each of which utilizes one or more of the three payment methods.

10.3.1 Fixed Price

Under the fixed price, or lump sum, payment arrangement (the terms are used interchangeably in American practice), the contractor is paid a single, preagreed amount for the entire project. This amount must include all direct costs, overhead, and profit. Without provisions to the contrary, all risks of unanticipated costs fall on the contractor (Blinn, 1989). Thus, the contractor, in preparing a bid, will include contingencies for such risks, with due recognition that too large a contingency will mean that someone else comes out the low bidder. In a Wisconsin case involving a fixed price contract [American Casualty Company v. Memorial Hospital Association, 233 F. Supp. 539 (E.D. Wis. 1963)], a contractor whose costs were twice those anticipated was allowed no extra payment (Sweet, 1989). And, when Blount Brothers Corporation sought additional compensation from the U.S. Government because of an increase in the cost of union labor, the request was denied because the fixed price contract provided for no such escalation (Appeal of Blount Brothers Corp., 1964 Board of Contract Appeals 4422, 1964).

Some fixed price contracts do provide for changes in the price. One example is a provision under which payments will be reduced if the contractor's performance fails to meet the specifications (e.g., by not completing until after a specified date). Or, if a target performance level has been set, certain percentage adjustments in the "fixed" price may be made for variations above or below that target. Stokes (1977) describes a nonconstruction case relating to the U.S. government's first contract for a "heavier than air flying machine." Specifications required that the vehicle fly at 40 miles per hour, carry two passengers with a combined weight of 350 pounds, and carry enough fuel for 125 miles. For every mile per hour above 40, the lump sum would be increased by 10%, with a concomitant decrease for speeds below 40 mph. The Wright brothers obtained this contract in 1908 and built the aircraft, which flew at 42 miles per hour. The plane was accepted and paid for, and then it crashed.

Yet another important possibility for changes in a fixed price contract arises when site conditions turn out to be different from those expected by the parties. These "differing site conditions" are discussed in Section 10.6.8

Lump sum contracts are best suited for projects with a minimum number of uncertainties. Examples are buildings that are completely designed before bids are received and for which the quantities of materials to be incorporated are completely defined in advance. From the contractor's standpoint, the advantages of this approach are the potential profit from a well-managed job, the need to keep fewer detailed records to "prove" costs than are required by other methods, and the potential for profit from changes to the work after the initial contract is signed. On the other hand, a poorly managed or unlucky fixed price project can be a financial disaster for the contractor. Furthermore, a major effort is required to develop the fixed price bid, which is costly to the contractor and is "wasted" by all but the successful bidder. In the end, the owner pays for the cost of such efforts.

Probably the major advantage of the fixed price contract from the owner's viewpoint is that the price is known in advance. Financial arrangements can be made with a high degree of assurance that they will not need to change. Another advantage is the lower cost of tracking the contractor's costs. The owner pays for the product, without needing to verify how many carpenter hours were spent in January. Disadvantages include the possibility of barely acceptable work by a contractor seeking to maximize profit by "squeaking by" with minimal quality. The inherently antagonistic interests of owner and contractor must be recognized by the owner considering this contract type. Other disadvantages to the owner, as cited above, include the high cost and extended time needed for contractors to prepare proposals and the potentially high costs of change orders.

Fixed price contracts are not appropriate where uncertainties exist, where it is important to start the work without delay (e.g., in an emergency), or where routine work on the site is likely to interrupt the contractor's operations (e.g., in an operating factory or educational facility) (Vaughn, 1983). The arrangements discussed in Sections 10.3.2 and 10.3.3 can help to overcome these objections.

10.3.2   Unit Price

Under a unit price contract, for each item of work the contractor is paid the actual quantity of work put in place times a preagreed price per unit. Bids are based on estimated quantities times a contractor's unit prices, respectively, and the low bidder is the one whose total estimate is the lowest. Thus, the two steps are

- *Prior to the Work:* the contractor is selected based on his or her unit prices times the estimated quantities.
- *During and After the Work:* the contractor is paid based on the actual quantities times unit prices.

The unit prices developed by contractors must include not only their direct costs but also indirect costs, overhead, contingencies, and profit. Thus, considerable study is required, just as in the case of a fixed price contract. The only way for the contractor to recover costs is through the unit prices.

Table 10.1 illustrates the use of unit pricing on a four-bid-item contract. Note that clearing and grubbing is a lump sum item; presumably the amount of work in this item was well defined prior to bidding. The contractor was selected based on the low total bid of $317,750. During the work, the actual quantities varied in such a way that the total payment was $329,750.

The unit price contract is most appropriate when the required quantities cannot be determined precisely in advance. Examples include highway construction (the embankment quantities depend on the elevations of the ground below, and the pavement quantities can vary significantly with small variations in thickness) and building projects (the quantities of certain finish work such as painting and floor covering often are not precisely determined until after the contract has been signed). From the owner's standpoint, the method has the advantage of removing some of the risk from the contractor, and this tends to reduce the amounts included for contingencies. Also, it may be possible to prepare prebidding contract documents less detailed than are necessary in the fixed price method. There are disadvantages, however: the final cost is not known in advance; the contractor's operations must be closely "tracked" by measuring the actual quantities incorporated into the project; contractors must spend time and money to prepare their bids.

On the contractor's side, the unit price tends to reduce some of the gamble in the bidding process, compared to the fixed price method, since payment is based on actual quantities. Also, there is the potential for a larger than expected profit if the actual quantities are greater than estimated and the fixed costs have been covered by the original quantities. For example, if a concrete item was bid at $150 per cubic yard for 1000 cy, and the actual quantity turned out to be 1100 cy, the extra $15,000 payment ought to mean some unanticipated profit. On the other hand, the opposite will be true if actual quantities are less than estimated. Also, considerable bookkeeping is required on the contractor's side as well to keep track of quantities.

If the actual quantities vary significantly from the estimates, the contract may provide for an adjustment in the unit prices. In our concrete example above, if the actual total quantity were only 500 cy, it would be unfair to pay the contractor only $150 per cubic yard, since there will be some fixed costs that do not depend on the quantity. The parties may negotiate a higher unit price, or they may determine the

TABLE 10.1 Unit Price Bid and Actual Payment

Bid Item	Description	Estimated Quantity	A		B		C: Low Bidder		Actual Quantity	Actual Payment
			Unit Price	Total Price	Unit Price	Total Price	Unit Price	Total Price		
1	Clearing and grubbing	1 job	$3700	$3,700	$5000	$5,000	$4000	$4,000	1 job	$4,000
2	Excavation	6,000 cy	$10	$60,000	$9	$54,000	$10	$60,000	5,800 cy	$58,000
3	Embankment	15,000 cy	$15	$225,000	$15.50	$232,500	$12.75	$191,250	16,000 cy	$204,000
4	Pavement	2,500 sy	$22.50	$56,250	$24	$60,000	$25	$62,500	2,550 sy	$63,750
				$344,950		$351,500		$317,750		$329,750

value by using a specific formula contained in the contract. If the actual quantity is 2000 cy instead of 1000, it seems fair to reduce the unit price somewhat, which the parties would likely do. Variations in quantities more than on the order of 10–15% are often handled this way in unit price construction contracts.

If the owner deletes some of the work, an adjustment in the unit price may be appropriate. In two interesting cases involving the U.S. government, the contractors were required to bid a single unit price for work involving tasks of varying difficulty. In one case, the work involved a combination of easy and difficult flood cleanup work, and payment was based on the volume of debris removed; in the other, an excavation project, the work was of varying difficulty. In both cases, the owner deleted some of the easy work but kept the difficult work without change. Ultimately, both contractors were allowed upward adjustments in their unit prices (Loulakis, 1985; Sweet, 1989).

10.3.3 Cost Plus

The two basic types of cost-plus contract are *cost plus a percentage of costs* and *cost plus a fixed fee.* In each case, the contractor is paid for the costs incurred, provided they are of the type agreed upon, plus a fee to cover nonreimbursable costs and profit. With the first type, the fee is figured as a percentage of the costs, whereas the second type gives the contractor a fixed fee regardless of the costs.

The major challenge in writing a contract of this type is to define precisely the reimbursable costs. It would be difficult to interpret a payment agreement that says "The contractor shall be paid the cost of the building plus 12%." What is the "cost of the building"? Does it include all expenses associated with the contractor's field supervisor, the home office manager, the field office, and the temporary on-site utilities? What about the interest on the money the contractor borrowed to finance the early phases of the job? Stokes (1977) offers the following list of costs that are typically reimbursed:

> Wages, payroll taxes and fringe benefits; cost of all materials, supplies and equipment incorporated into the work, including transportation charges therefore; payments to subcontractors; rental and maintenance charges for all necessary equipment, trucks and hand tools; cost of salaries for contractor's employees stationed at the field office or while they are expediting production or transportation of materials or equipment; proration of reasonable travel, meals and hotel expenses of the officers or employees of the contractor directly incurred with the work; premiums of all bonds and insurance, seals and use taxes related to the work; permit fees; minor expenses, such as telephone and telegraph costs; cost of temporary site facilities and removal of debris; and losses and expenses not compensated by insurance which result from causes other than the fault or negligence of the contractor.

The same author notes that the following costs typically are not reimbursed and therefore must be covered by the fee:

Salaries or other compensation for the contractor's officers and employees while working at the contractor's main office or branch offices; expenses of the contractor's principal or branch offices other than the field office; any capital expenses including interest on the contractor's capital and any additional capital required to perform the work; all general overhead expenses; and cost due to negligence of the contractor, subcontractors or anyone directly employed by the contractor.

An interesting case arose in Louisiana over whether a contractor should be reimbursed for the costs of his personal labor. In this particular case, the oral contract called for payment of the cost of labor and materials plus 12%, but apparently there had been no discussions about whether the contractor would be reimbursed for labor he himself might expend. The court said that "for a contractor to charge for his own labor, there must be an agreement between the parties allowing this charge." Since such an agreement was missing, the charge was denied ("Contractor Must Outline All Charges in Contract," 1992).

The point here is that reimbursable costs should be defined when the contract is being written. If this is done, there should be little controversy.

From the owner's standpoint, there are two primary concerns with the cost-plus method of contracting. First, there is no way of knowing what the total cost of the project will be; and second, there is little incentive for the contractor to control costs. The cost-plus-percentage contract, especially, carries a disincentive to control costs, since a larger "cost" will lead to a proportionately larger percentage fee. Advantages to the owner include the ability to begin a project before the design is complete or in an emergency situation, simply with an agreement that defines reimbursable costs and stipulates the fee arrangement. From the contractor's side, the method shifts the risk of cost increases to the owner and assures that reimbursable costs will always be covered. However, there is not as much reward for innovative construction management as there might be with a fixed price contract; if the contractor can keep costs down, that savings is realized by the owner.

Some cost-plus contracts include provisions for target prices or sharing of savings. With a target price, the contractor may guarantee a maximum price for the job, with a penalty to the contractor if the price exceeds that amount; the contractor may pay all the overage, or there may be a sharing between contractor and owner. If the costs are less than some preagreed amount, there can be a sharing of the savings (Sweet, 1989).

10.3.4   Design/Build Contracts

A design/build contract is a project organization method under which the owner contracts with a single party to design and build a facility. In a typical arrangement, there is a design/build contractor who signs a contract with the owner. That contractor arranges for the design, either by an independent design professional or by in-house staff, and then signs contracts for various parts of the construction with contractors and subcontractors, often performing some of the work itself. The owner's only official, contractual relationship is with the design/build contractor.

The planning phase is carried out, with conceptual design and a preliminary cost estimate. Then, the parties usually negotiate a contract price; a typical arrangement is a cost plus fixed fee contract with a guaranteed maximum price. As the work progresses, the owner pays all allowable costs up to the guaranteed maximum (Jervis and Levin, 1988).

The owner must be very clear about his or her desired project scope before turning the project over to the design/build contractor. A panel of experts responding to a questionnaire in *Civil Engineering* magazine offered some guidance:

> A major problem is the question of scope of the project.... For design/build to be effective, the owner must be clear about what is—and is not—included in the scope of the work. For the inexperienced owner, design/build can be very traumatic...the owner's requirements and expectations must be clearly articulated at the outset. (Tarricone, 1993b)

The owner enjoys the advantage of working with a single party who holds responsibility and authority for a successful project. Such an arrangement allows the possibility of so-called fast-tracking, or phased construction (Section 10.3.7), which could provide major time savings. For heavy, engineered projects that are capital intensive, such as industrial facilities and utility plants, the approach can greatly benefit the owner, who relies on a single designer/builder with special expertise in this type of project. On the other hand, the owner tends to have less control over both the project's definition and its execution than in a more traditional project that is designed completely by one party before being bid to another.

10.3.5 Turnkey

Although sometimes used synonymously with design/build, the *turnkey* method is most often considered to include an even larger scope of responsibility for the party that contracts with the owner. "The core element... is, at least in theory, that the owner simply tells the contractor what it wishes and does not appear on the scene again until the contractor says the project is completed and hands the owner a key and the owner 'turns it'." (Sweet, 1989). Thus, the contractor is responsible not only for the design and construction but also for land purchase, construction financing, and any other arrangements necessary for a complete project. When the project is complete, the buyer purchases the land and facilities, based on the contract that defined the project and set the price.

The contractor may be selected based on a priced proposal and/or other factors. The price may be a lump sum or some other arrangement. In a public housing project, for example, developers are invited to submit priced proposals for housing. The winning proposer obtains the land and designs and builds the project, in accordance with the owner's requirements. When complete, the project is turned over to the local public housing authority. As a rule in such cases, a fixed price is agreed on at an early point in the project (Sweet, 1989).

10.3.6 Construction Manager

The terms "construction manager" and "construction management" have several uses when it comes to contracts for construction services. Central to the whole concept is that the owner uses the services of an expert in the management of construction, and there is opportunity for beginning the actual construction before all the design work has been completed.

With one version, the owner retains the services of a construction manager as an independent consultant, who may be considered to be an agent of the owner. The owner also engages an engineer/architect to do the design. Furthermore, the owner contracts with several prime contractors, who are responsible for various phases and aspects of the work. Thus, the owner has prime responsibility for scheduling, coordination, and financial management of the project. The construction manager is the owner's adviser in all these matters.

Under another approach, the construction manager is responsible for contracting with the various contractors and/or subcontractors. Thus, he or she has a role similar to that of the traditional general contractor. The major difference is that the construction work begins before the completion of the entire design. Normally this construction manager guarantees a maximum project price based on conceptual drawings and the owner's stated requirements (Jervis and Levin, 1988).

A third approach has the architect/engineer functioning as construction manager. Here, the design professional, who is called the "project manager," has responsibility for design and then advises the owner about construction. Usually, however, the project manager neither contracts with the various prime contractors nor guarantees a maximum project price. Once again, a fundamental feature is the ability to begin construction when design documents are incomplete.

The construction management approach, in whatever form, has the advantage of permitting a project to be fast-tracked, with lower owner financing costs and earlier occupancy. Also, if the owner contracts directly with various contractors, the markup that would be added to a general contractor's price is avoided. Of course, this saving is somewhat offset by the need to pay the construction manager for services rendered. The owner probably has less control over the final project price and over its nature than under a fixed price contract. Furthermore, there is really no single point of responsibility, since the construction manager is generally just an adviser. Thus, there may be "reciprocal finger pointing" (Jervis and Levin, 1988) among the various prime contractors, with the owner in the middle. Also, there tends to be more legal liability exposure for the owner, who now has responsibilities for scheduling and coordination.

Is the construction manager an agent of the owner? A 1981 U.S. District Court case in Louisiana involved damages suffered by a concrete contractor: the construction manager had allowed several prime contractors to discharge water through partially completed storm drains, and the resulting flooding caused delays. The owner was responsible for coordination, including allowing contractors timely access to the site, but the construction manager was held to be an agent of the owner. Thus the concrete contractor was awarded a $700,000 judgment against the owner and the construction manager jointly [R. S. Noonan Inc. v. Morrison-Knudsen Co., Inc. 522 F. Supp. 1186 (D. C. La. 1981)].

10.3.7   Fast Track

Fast tracking, or phased construction, is not really a different payment method, nor is it an organizational variation. Rather, it is "a process where construction of a facility begins prior to the completion of design" (Tarricone, 1993b). The basis for paying the contractor, and the contractual relationships among the parties, may be any of several already discussed. An example might be the design and construction of an office building, in which the designer furnishes preliminary design information to the owner and the contractor, who thereupon enter into an agreement, perhaps on a cost-plus or a fixed price basis. As design proceeds, working drawings of the foundations are provided, and construction begins while the balance of the design is being worked on (Schoumacher, 1986). The challenge is for the designer to stay far enough ahead of the construction process for the field work to proceed effectively.

There may be a design/build arrangement, or a construction management contract, but these are not essential elements of the fast track process. Both designer and contractor may have separate contracts with the owner, as in more traditional construction contracting, with the possibility of some on-site inspection responsibilities going to the designer.

The prime advantage is a shortened project delivery process. Some have claimed savings of as much as two-thirds of the time required, compared to the conventional design–review–advertise–bid–construct process (Tarricone, 1993b). Also, the owner may have access to construction loan money earlier than in a more traditional project, since those funds usually become available at the start of field construction. The incomplete design is itself a disadvantage, however: many changes may be needed as the project proceeds, or on the contrary, the designer and owner may be constrained from making changes after part of the project has already been built. Also, there is a greater likelihood of design omissions when a project is designed piecemeal. Furthermore, the likelihood of delays is increased if all parties do not or cannot perform as scheduled.

10.3.8   End-Result Contracts

Finally, we mention an emerging approach to construction contracting that focuses on the end result rather than the method of getting to it. For example, in road construction, various statistical measures may describe the concrete or asphalt strength, thickness, smoothness, and riding quality. Target values are specified, and the contractor may be paid extra amounts for exceeding statistical specifications (Tarricone, 1993a). The approach seems to be consistent with the recent emphasis on total quality management and statistical quality control.

Such an approach has been tried with some degree of success in Sweden and other European countries. Designated the "function contract," the agreement provides that the contractor will meet certain descriptive performance specifications, such as rut depth, roughness, and cross slope on a highway surface. The contractor guarantees the performance of this "product" for a period between 5 and 10 years (Grennberg, no date).

10.4   THE CONTRACTING PROCESS

Now we are ready to consider the steps in the traditional contracting process. We assume that the planning and design phases have been completed, as described in Section 10.1.2. Thus, the project is now ready to be advertised for bids.

10.4.1   Advertising

The owner, often with the assistance of the design professional, now wishes to make the contracting community aware that project will be built. In the public sector, where there will be a competitive bidding process, it is important and often required by law that public notices be made. For privately funded projects, less formal means may be used, but the private owner still is interested in attracting competent, financially attractive proposals.

What means do owners use to get the word out to prospective contractors? The methods are many; several are listed.

- Classified advertisements in local newspapers
- Notices in regional or national trade magazines, such as *Engineering News Record* or *Pacific Builder and Engineer*
- Publications of contractor organizations, such as the newsletter of the local chapter of the Associated General Contractors
- Listings with local or regional construction plan services
- Distribution by mail to a list of contractors who have expressed interest in this type of project
- Informal contacts in writing, by telephone, or in person, if the project is privately financed

Although the advertisement for bids must be somewhat restricted in length, certain kinds of information should be included. A typical advertisement might contain the following:

- A brief description of the type of job and its magnitude
- Project location
- Identity of owner and design professional
- Date and place for receipt of bids and opening of bids
- Type of required bid deposit or bond, if any
- Whether bids may be withdrawn and whether the owner reserves the right to reject bids
- Time for starting and completing the work
- Place and cost to obtain contract documents
- Special conditions or features

With this information, prospective contractors can determine quickly whether they are interested in obtaining the contract documents and pursuing the preparation of a proposal.

10.4.2 Proposals

At the time and place set by the advertisement, proposals are received for constructing the project. Construction cost estimating and bid preparation is a highly specialized endeavor and one that is handled in different ways by different contractors. It is a combination of hard work in determining quantities and applying appropriate cost information, plus judgment and strategy. The proposer wants to be the low bidder but still wants to turn a profit. If you are too high, you won't get the job, and if you are too low you may get the job but lose your shirt. It is not our purpose to discuss the techniques and strategies for preparing proposals.

Most owners provide special proposal forms on which bidders fill in the blanks. This format gives some assurance that all bidders are providing the same information, which can clearly be an offer capable of being accepted by the owner. The simplest will be the fixed price and the cost-plus proposals. In the most straightforward case, the bidder gives a single figure, plus assurances that all bidding information has been utilized, required licenses obtained, and bonding requirements met. Proper signatures must be affixed. Often, a fixed price bid will ask for additional proposal data for alternate ways of doing the work or for adding or omitting certain aspects. Unit price contracts require considerably more data from the contractor; basically, the contractor submits a unit price for each bid item, multiplies each by its respective quantity, and finds the total bid price.

Most advertisements stipulate that the contract will be awarded to the *lowest responsible bidder,* if it is awarded at all. How can the owner be assured that the bidder is "responsible"? One way is to check the low bidder's qualifications after the person has been identified. The contractor may be asked to provide evidence that he or she has sufficient finances, personnel, and equipment to do the job, that the company has had enough experience in that line of work, and that it is not so burdened with other work as to be unable to prosecute this new job adequately. Providing such evidence is a difficult and inexact task! After the bidding, however, it is almost impossible to declare a bidder not responsible and make such a declaration stick.

Another approach is to require that any interested bidders be *prequalified.* Through questionnaires and other means, the owner determines whether each prospective contractor has the resources and experience to conduct the work successfully. Then, only those who are determined to be qualified are invited to submit proposals. This approach saves unqualified contractors the expense of preparing proposals, and it assures the owner that proposals come only from firms capable of doing the work. But it does invite controversy because a contractor's qualifications may change over time, may fit one line of work and not others, may be sufficient for small jobs but not big ones; also, determining qualifications is an inexact process, as noted, and the preclusion of a potential contractor seems to violate the notion that any interested contractor should be able at least to submit a proposal (Bockrath, 1986; Vaughn, 1983).

A large volume of case law has arisen over the matter of mistakes in bids. If Ben Bidder makes a mistake by adding incorrectly, omitting the cost of an element, or committing some other clerical error, should he be allowed to withdraw the bid (1) prior to the bid opening or (2) after the bid opening when he has been declared the low bidder? Furthermore, suppose (3) he was low bidder but wants to change his price (upward, no doubt) and still be low bidder, or (4) he was not low bidder but a correction would make him low bidder. The answers to 3 and 4 are most always "No!" The answer to 1 is usually "Yes." The answer to 2 seems to depend on the circumstances.

Jervis and Levin (1988) suggest that a bidder is generally entitled to withdraw a mistaken bid after the bid opening if (1) the mistake is both factual (e.g., an error in mathematics or a transposition as opposed to an error in judgment) and "material," (2) there is not gross negligence on the part of the bidder, (3) it would be unconscionable for the owner to enforce the bid, (4) the bidder promptly notifies the owner of the mistake, and (5) the owner can be returned to its prior position. In Liebherr Crane Corporation v. United States [810 F. 2d 1153 (Fed. Cir. 1987)], the plaintiff proposed to supply a crane for a price of less than $4 million, when all other bids were at least $7 million. When questioned, the supplier assured the government that the $4 million bid was correct. Unfortunately, the supplier had read only a few of the 98 pages of specifications and was in fact proposing to supply a crane that did not meet the government's requirements. The federal court held that such a mistake was one of judgment, and the plaintiff was required to fulfill the contract requirement at the original bid price (Simon, 1989).

When an Alaska bidder on a unit price contract committed a mathematical error that made the "bottom line" total not the low bid, the court allowed the bid to be reformed by using the individual unit prices and correct multiplication and addition. The result was that this bidder became the low bidder and was awarded the contract. (Vintage Construction, Inc. v. Alaska Department of Transportation and Public Facilities, 713 P. 2d 1213, 1986; "Total Price Error: Unit Price Upheld," 1986) Various authors, such as Jervis and Levin (1988), Simon (1989), and Sweet (1989), present a myriad of interesting cases having to do with mistakes in bids.

Some owners are changing the construction proposal process in interesting and apparently effective ways. Among those approaches reported recently are

- *Cost-Plus-Time:* bidders on highway projects provide a bid price and a "societal cost," which is the number of days to complete the project times the daily cost to society of having the roadway closed.
- *Lane Rental:* similar to cost-plus-time in that the contractor is assessed a lane rental fee for having the highway out of commission.
- *Technical Merit:* contractors submit both a technical proposal and a price proposal, which are evaluated by allocating points to the various elements of the proposals.

• *Competitive Negotiation:* several bidders are invited to revise their technical proposals and change their prices if appropriate, after the initial bid opening.

Although these newer methods have been criticized for being confusing and for introducing nonquantifiable factors inappropriately on public projects, such alternative contracting may find increasing use as agencies and private organizations seek to utilize decreasing funds more effectively (Tarricone, 1993a).

10.4.3   Bid Opening

Sealed bids in single copy, for performing all work for each of the above named projects described herein, will be received until 10:00 am prevailing time, July 2, 1993, at the Department of Transportation and Public Facilities, 2301 Peger Road, Fairbanks, at which time bids will be publicly opened and read. (48538 Legal, 1993)

This statement, or one like it, is a very common part of the typical invitation for bids for a public project. Among the things we learn are that bids will be sealed and that they will be opened publicly and read aloud. In the private sector, bids do not have to be opened in public, but often they are. Usually, a representative of each bidder is present. The public official opens each bid in turn and reads appropriate information from it, especially price information. The information is recorded and checked at the time or soon thereafter. Often, the public official will declare the "apparent low bidder," although such a statement must make it clear that the low bid has not yet been accepted.

The owner often reserves the right to reject any or all bids; this need may arise if one or more bidders are not qualified or if all bid prices exceed the available funds. Sometimes there is a stipulation that the owner may waive informalities in the procedures—for example, a bid that was a few minutes late or did not fit exactly the required form (but see below for examples of rejected bids). Occasionally the owner will reserve the right to award the contract to other than the low bidder; in the public arena this action is difficult to sustain, but sometimes an extremely well qualified bidder submits a price only slightly higher than that of the less qualified low bidder, and the owner is convinced that a better job will result from selecting the higher priced proposal.

The right of an owner to reject a bid, provided such right has been made clear in the bidding documents, was established in a hundred-year-old Missouri case whose description from Jessup and Jessup (1963) we quote:

In Anderson v. Board of Directors of Public Schools, 122 Mo. 61, 27 S.W. 610, 26 L.R.A. 707 (1894), the plaintiff was a building contractor, and defendant was a corporate entity having charge of public schools in St. Louis, Missouri. Defendant advertised on August 28, 1891, for sealed proposals for the erection of a high school building. Bids were to be opened 10 days later. The cause of action arose when defendant refused to accept plaintiff's low bid of $196,965 and, instead, accepted another bid of $197,000. Plaintiff sued for loss of profits in the amount of $15,000. The trial court found for the defendant school board, whereupon the plaintiff appealed to the Supreme Court of Missouri. The court agreed with the trial court, noting that the advertisement included the statement "The Board reserves the right to reject any and all bids," and said:

1. That language demonstrates the nature of the advertisement as a mere invitation for offers for a contract.
2. The [defendant's] right to reject the bid was unconditional...even without any assignable cause.

Before the bid can be accepted, the owner must ascertain whether the proposal is responsive to the invitation for bids. Nonresponsive bids include (1) those whose price information contains incorrect arithmetic, (2) those submitted after the deadline, (3) those containing qualifications or changes in such areas as completion date or material specifications, and (4) those that fail to include a required bid bond, assurances that workers from disadvantaged and protected classes will be employed, and/or a list of subcontractors, if required (Firmage, 1980). If the owner has reserved the right to waive informalities, however, the rejected low bidder may have a case for reinstatement (although even here it seems that the owner has the right to decide whether informalities will be waived).

Recently the city of Atlanta accepted a bid that was three minutes late; it turned out to be the low bid, and the contract was awarded to the very slightly tardy bidder. The second bidder filed a protest and eventually had the city's initial award overturned. The Georgia appellate court noted that both the contract specifications and the invitations for bids required bids to be submitted prior to the 2:00 P.M. closing time. Because it had failed to observe this condition, the low bidder was deemed nonresponsive (City of Atlanta v. J. A. Jones Construction Company; Loulakis and Cregger, 1991a).

There are other recent examples of court rulings of nonresponsiveness as a result of bidders' failure to comply with invitations to bids. In a Louisiana case, the low bidder neglected to present along with the bid a list of subcontractors who would be used on the project, as required in the contract documents (C. R. Kirby Contractor v. Lake Charles, Court of Appeals of Louisiana 1992; "Contractor Names Subs Late, Loses Bid," 1993), and in a Pennsylvania case a deposit amounting to 10% of the bid price was required, but the low bidder provided a smaller percentage [Karp v. Redevelopment Authority, Commonwealth Court of Pennsylvania, 566 A. 2d 649 (1989); "Courts: Follow Bid Instructions," 1990].

10.4.4 Evaluation and Award

After the proposals have been received, they must be evaluated. We have discussed (Section 10.4.3) the need for the owner or owner's representative to determine whether the bid is responsive and the bidder responsible. In addition, it is necessary to make sure that sufficient funds are available to pay for the low bidder's price. For a fixed price project, this task should be straightforward. For a unit price contract, if it is assumed that the owner-supplied quantities for the bid items are correct, the task also should be straightforward. In the case of a cost-plus contract, the owner should have arrived at estimates of the reimbursable costs, and thus an estimate of the total project price can be determined.

When bidders have been asked to supply proposed prices for alternatives in addition to the basic bid, then the owner will need to decide which if any of the alternatives to select. Occasionally, when affordable alternatives are included, the contractor with the low bid on the basic proposal will not be low on the total package. The design professional is usually involved in assisting the owner with all these evaluations.

The next step is to award the contract, if it is to be awarded. It is good practice for the owner to notify all bidders that a low bidder has been selected, both as a courtesy and to let them know that their bids will be considered open until all contractual arrangements have been made. This way, if the original low bidder fails to sign the contract, the owner may still contract with another firm.

Normally, the contract documents will include a copy of the agreement form, so that prospective contractors will know in advance what they will be asked to sign if they are selected. After the formal signing of the agreement, the contractor usually must wait for a *notice to proceed;* the owner will make such notification fairly soon after contract signing, within time limits stated in the contract. When the contract stipulates a certain number of days from the beginning of the project until project completion, the clock is started on the date of the notice to proceed. The bidding process is now complete, and the actual construction can begin.

10.4.5 The Work Proceeds

It is not possible here to give an exhaustive description of all the various contractual events that occur during the construction process. We shall mention some of the highlights.

Shop drawings are detail drawings furnished by the contractor that provide sufficient information to permit completion of the field work. Often these drawings are prepared by subcontractors or material suppliers; examples are reinforcing steel drawings that show fabrication and installation details beyond those provided by the design professional. The term "shop drawing" also includes material samples as required in the contract, such as paint chips, floor covering samples, and other finish materials. The contract ought to specify that until these drawings have been approved by the owner or a representative, no work based on them is to proceed. However, it is important that the contractor check the drawings as well. Even after approval, the contractor is still accountable for the work. "The important point to be emphasized about this procedure, and to be set forth explicitly in the contract, is that although the shop drawings are to be approved by the engineer, the contractor is responsible for the accuracy of the work" (Bockrath, 1986).

In most every construction project except the purest turnkey type, the owner will arrange for some sort of *inspection* of the work as it proceeds. Often, the design professional who prepared the plans will inspect the work to assure compliance with the contract documents. Such inspection is usually periodic rather than continuous, but a representative is present for major events: a large concrete placement, the installation of a primary piece of equipment, and so on. An important general caveat in the inspection process is that even if work is "passed" by the inspector, the contractor remains responsible. Sometimes exceptions arise— in one Massachusetts case, a housing project contractor defended successfully a claim of hidden defects discovered after the expiration of the statutory 6-year time limit. The court decided that the presence of an inspector on the site during construction was sufficient to relieve the contractor of responsibility [Kingston Housing v. Sandonato & Bogue, Appeals Court of Massachusetts, 577 N.E. 2d 1 (1991); "Monitored Construction Protects Contractor," 1992].

Problems with overzealous inspectors sometimes arise. In one case, the inspector adopted his own standards for some gypsum drywall construction, requiring reworking of many walls that were more than an eighth of an inch out of tolerance. The Armed Services Board of Contract Appeals found that such a requirement was unwarranted, especially since the specifications did not provide for a "skim coat" of paint as is the usual practice for such a finish (Loulakis, 1986b).

As the work proceeds, it is common practice for the owner to make *periodic progress payments* to the contractor. Most contracts set forth a procedure under which the contractor makes a formal request for payment, based on work completed during the preceding period, and the owner or a representative certifies that the work has been accomplished (or has not!). Monthly progress payments are probably most common. A fixed price contract often requires the contractor, prior to commencing work, to prepare a "schedule of values" that indicates the breakdown of the bid price into several work items; requests for payment will then be based on the proportion of each item that has been completed at the time of the request. With unit price contracts, payment will be based on the quantities put in place times the contracted-for unit prices. Cost-plus contract payment requests require the contractor to submit detailed documentation verifying the costs.

In Chapter 9 we discussed the concept of severability, or divisibility. If a contract is severable, rather than entire, it is capable of being divided into parts, and the completion of one of those parts entitles payment for that part. The law is quite settled that periodic progress payments do not imply a severable contract, even though payments are made in several stages (Sweet, 1989). Most construction contracts are entire; the contractor is required to complete the whole job or be charged with breach of contract.

A somewhat controversial practice is the matter of *retainage,* in which the owner is entitled to keep a certain percentage of the amount due as security to cover claims it may develop against the contractor. As much as 10% may be retained until the end of the project, although the percentage is often reduced after, say, half the project has been completed. Retainage has two negative aspects (1) the owner may already be protected against risks by payment and performance bonds, and (2) retainage can place an undue hardship on the contractor's cash flow, the interest cost for which will ultimately be covered in the bid price and therefore paid by the owner.

Most construction contracts require the contractor to submit a *project schedule* showing when the various elements of the work will be done. A bar chart or network schedule may be required. As the work proceeds, periodic *schedule updates* will often be required, indicating progress to date and the current expected completion time. In some cases, the submittal of the updated schedule is made in concert with the periodic request for payment, and one of the conditions for payment is an acceptable schedule.

A major topic in construction contract law has to do with *changes in the work.* Simon (1989) writes:

> The probability that a construction project will go from the design through plan, bid, award, performance, final completion, and acceptance without a change in the design or construction work is highly improbable. At some stage during the construction process a light bulb, room, wing or building will either be added, relocated, modified or in some other aspect changed.

There are at least two aspects of importance. First, there ought to be a clear contractual provision for dealing with changes the owner desires to make after construction has begun. Generally the contractor will be asked to propose an adjustment in the contract price (upward or downward) for the requested change, which will then be considered by the owner. If the chance is accepted, a *change order* is issued. Usually, the contract will be clear that the contractor is entitled to the adjusted price only if the approved change order was received before the work was performed.

A second aspect of changes in the work involves the basis of payment for extra work for which a formal change order has not been issued. There are many examples. One would be a unit price contract, in which it might seem straightforward to pay for additional units at the agreed unit price. However, suppose the extra work is considerably more difficult than anticipated? On a project for construction of a reservoir for the Metropolitan Water Reclamation District of Greater Chicago, the contractor was required to excavate roughly 7% more soil than the quantity set forth in the contract, using a method more costly than that for the basic contract. The parties agreed that additional compensation was due. However, the owner claimed that payment should be based on the contractually agreed-on rates, while the contractor sought payment at a higher rate for the more difficult work. A U.S. circuit court of appeals found for the owner [Brant Construction v. Metropolitan Water Reclamation District, U.S. Court of Appeals for the Seventh Circuit 967 F. 2d 244 (1992); "Court Upholds Contract Price Despite Added Costs," 1993].

The doctrine of *substantial performance* reflects acknowledgment that even though most construction contracts are not performed completely and exactly as specified, the contractor usually is entitled to payment. Again, Simon (1989) explains it well:

> Completion, like every other alleged breach, is a question of degree and fact. A contractor who has failed in some minute detail to comply in strict accord with the contract is not necessarily denied recovery of substantial moneys. This principle of

law ... acknowledges the contractor's right to payment when the contractor's error is neither fraudulent nor willful, and when it has substantially performed its work.

A leading and oft-quoted case is Jacob & Youngs, Inc. v. Kent [230 N.Y. 239, 129 N.E. 889 (1921)], which involved a contractor's use of some pipe of a brand different from that specified. Instead of Reading pipe, as specified, Cahoes pipe was installed. The change was inadvertent, and the substituted brand was identical in every respect except for the name. When the owner refused to make final payment, citing breach of contract, the contractor brought a successful suit. Writing for the court, Justice Cardozo said,

> An omission, both trivial and innocent, will sometimes be atoned for by allowance of the resulting damage, and will not always be the breach of a condition to be followed by a forfeiture.... Nowhere will change be tolerated, however, if it is so dominant or pervasive as in any real or substantial measure to frustrate the purpose of the contract.... There is no general license to install whatever, in the builder's judgment, may be regarded as "just as good" .... The question is one of degree, to be answered, if there is doubt, by the triers of the facts [jury] ... and, if the inferences are certain, by the judges of the law. (quoted in Simon, 1989)

If performance is to be "substantial," the deviation must be minor in relation to the overall project, the mistake must be inadvertent and neither willful nor fraudulent, and compensation must be made for variation in performance, if any (Jessup and Jessup, 1963).

10.4.6 Completion

As the construction project nears its final stages, a significant milestone is reached at *substantial completion*. At this point, the project can be occupied by the owner and be put to use for its intended purpose. There are analogies between substantial completion and the doctrine of substantial performance; in each, the work is "almost but not quite" perfect or final. Here, all details are not yet in place but the work is functionally complete. After this point, the contractor cannot be held in breach of contract, and the assessment of any liquidated damages ceases.

When the contractor believes that substantial completion has been reached, he or she notifies the owner and requests an inspection and certification. In conducting the inspection, the owner's representative makes a list of incorrect, incomplete, and missing items, traditionally called a *punch list*. When the parties agree that substantial completion has arrived and the punch list is accurate, a certificate is issued. Usually, at this point, the moneys retained by the owner are reduced significantly.

The contractor works on remedying punch list items and, when he or she believes they are complete, once again notifies the owner. A *final inspection* is held, including visual checks and testing of equipment and systems; on large projects, this process may last for several days. If the project "passes," the owner formally accepts it by issuing a *certificate of completion*. At this point, any money owed to the contractor is paid. Legally, the owner, upon final acceptance, releases any claims against the contractor except for warranty claims and hidden defects. On the other hand, the contractor loses the right to bring claims against the owner for extra compensation.

Since both the contractor and the owner may be occupying the facility between substantial and final completion, the contract should make clear (1) the point at which the guarantee period starts, and (2) the point when responsibility for operation and maintenance of the facility shifts to the owner.

10.5 CONSTRUCTION DOCUMENTS

We give here a quick review of the parts of a typical set of construction documents. All these materials are prepared by the design professional, acting on behalf of the owner, and are made available to prospective bidders as a package. Normally the drawings are bound separately, with all other documents bound together in one or more volumes as a "project manual" or other designation.

10.5.1 Agreement

The "agreement" is no doubt the shortest of all the construction documents; it defines the work to be performed and the rights and duties of the parties. One important purpose is to incorporate other documents by reference. The actual agreement states the names of the owner and contractor, gives a brief description of the work, states the time of commencement and completion, gives the contract price and the manner of payment, and names the design professional. Any special agreement pertaining to the project would be included, and sometimes such items as the hierarchy of documents (see Section 10.6.3) are also a part of the agreement. Finally, the document is signed by authorized persons on behalf of both parties.

10.5.2 Drawings

The drawings, also called "plans," are prepared by the design professional to depict the project details in a graphical way. Certainly they are a primary communication medium between the designer and the builder. Firmage (1980) points out that if "a picture is worth a thousand words," a drawing may be worth several thousand. The drawings as prepared by the engineer/architect are known as the *design drawings;* after the contract has been signed and the drawings incorporated by reference, they are referred to as *contract drawings.*

Depending on the size and type of job, the number of design drawings can vary between just a few and several hundred. A typical set of drawings for the construction of a building is divided into sets for each of the following: (1) general layout and "civil" works such as roads and parking areas, drainage, and landscaping, (2) architectural, showing all dimensions and locations of all features in the building, (3) structural, with details of all such elements including connections and fastenings, (4) mechanical, including plumbing, heating, ventilation, air conditioning, and special mechanical equipment, and (5) electrical, with light fixtures, motors, conduit and cable, instrumentation, junction boxes, and other such details.

Bockrath (1986) devotes an entire chapter to the subject of construction drawings.

10.5.3 General Conditions

The general conditions are general in that they could apply to every project of the type under consideration. That is, they do not contain information specific to the particular job. This "boilerplate" material addresses "everything from the mundane (such as site cleanup or maintaining order among the crew) to the crucial (such as change order procedures or compensation for unanticipated subsurface conditions" (Jervis and Levin, 1988). In the past, general conditions were often preprinted and simply inserted into the manual of project documents; this practice has been made less prevalent by word processing, but its past use helps to underscore the general nature of the material.

A recent set of general conditions (West Ridge Natural Science Facility, 1992) contained 65 general conditions, including such topics as time for completion and liquidated damages (including the declaration that the time for beginning and completion are essential conditions of the contract, but no specific dates or amounts for this project), progress meetings, inspection and correction of the work, periodic payments to the contractor, substantial completion, suspension of work and delays, changes, disputes, apprentices, and minimum wages. The important points to grasp are the variety of topics covered, their nontechnical nature, and their "general" usefulness on any similar project, at least in the same area and for the same owner.

10.5.4 Special Conditions

In contrast to the general conditions, the special conditions (sometimes called "supplementary general conditions" or just "supplementary conditions") are drawn for the particular project. These also are nontechnical conditions, but they include, for example, the particular project start and completion dates and the amount of liquidated damages that would be assessed against the contractor for each late day if the completion date were overrun. Other topics that might be covered include requirements for the project sign, toilet facilities and parking for workers, protection of existing facilities, temporary offices for owner and/or designer, dust control, disposal of waste materials, and traffic control.

10.5.5 Technical Specifications

The "specs" describe the technical requirements in words and numbers. Often they are quite standard, and in the old days many were reproduced and bound into the project manual with no editing. Of course, some will be specific to the project. The technical specifications state the requirements for materials and components, as well as installation. Despite the introduction of prescriptive, or functional, specifications into some projects, as discussed in Section 10.3.8, descriptive specifications ("tell them what to do and how to do it") are still by far the dominant type.

A popular means of organizing technical specifications for the construction of buildings is that originally developed by the Construction Specifications Institute (CSI). The 16 categories comprising the Uniform Construction Index (Jervis and Levin, 1988) are generally organized to reflect the traditional building trades, as follows:

1. General requirements
2. Site work
3. Concrete
4. Masonry
5. Metals
6. Wood and plastics
7. Thermal and moisture protection
8. Doors and windows
9. Finishes
10. Specialties
11. Equipment
12. Furnishings
13. Special construction
14. Conveying systems
15. Mechanical
16. Electrical

Bockrath (1986) provides a helpful review of technical specifications for both materials and installation.

10.5.6 Other Documents

Sometimes the project manual contains other documents of interest and use to bidders and, especially, the eventual successful contractor. These materials may include (1) instructions to bidders (an expanded version of the advertisement), (2) bid form, (3) bid, performance, and payment bond forms, (4) insurance forms, and (5) minimum rates of pay for laborers and mechanics, if applicable to this project.

10.6 CONTRACT PROVISIONS OF SPECIAL INTEREST

It is impossible here to list all the contract provisions that apply to a typical construction project, just as we could describe only briefly some of the highlights of the contracting process. We have selected eight topics of special interest.

10.6.1   Bonds

The bid bond, the performance bond, and the payment bond are important in construction contracting. A bond is a guarantee to the owner that a third party, the bonding company or "surety," will protect the owner in case the contractor defaults. Bonds are contracts between the contractor and the surety, written for the benefit of the owner. The contractor buys the piece of paper, the bond, from the surety and includes the cost of the bond in its estimate of the cost of the job.

A *bid bond* is furnished at the time of bidding. It guarantees that the bidder will sign the contract at the bid price, if it is accepted by the owner, and will furnish any required performance and payment bonds. If the selected bidder fails to honor its bid, the surety will pay to the owner the difference between the price of the contractor it eventually contracts with (usually the next to low bidder) and the low bid, up to a specified limit, such as 10% of the bid price. A 10% bid bond on a $200,000 bid means that the surety guarantees that if this bid is accepted and the contractor fails to sign the contract, the surety will pay the owner up to $20,000. Thus if the owner eventually signs a contract with a bidder originally deemed too high, the net cost to the owner will not exceed $200,000. Note that this arrangement will work only if that second price is $220,000 or less.

A *performance bond* is furnished by the successful bidder at the time of contract signing or shortly thereafter. It is a contract between the contractor and the surety under which the bonding company guarantees the contractor's performance. The "obligee" is the owner, just as in other construction bonds. Here, the surety's obligation is activated only if the contractor defaults on the contract by failing to complete the work as agreed. If this happens, the surety must complete the work or pay the owner the cost of having it completed. Usually, at least on public projects, performance bonds are written for the full contract price. Thus for our $200,000 project, the bonding company would be promising to pay up to $200,000 to have the work completed in the event of default by the contractor.

A *payment bond,* like a performance bond, is furnished by the contractor to the owner after the contractor has been selected. It guarantees that material suppliers and subcontractors will be paid in full in the event that the contractor fails to make the necessary payments. On private projects, the purpose is to protect the owner from liens that unpaid parties might otherwise place.

Public property cannot be subject to liens; here the purpose of the payment bond is make some recourse available to unpaid subcontractors and suppliers, as a matter of public policy. Like other bonds, the surety's obligation comes into play only if the contractor defaults. Often, the payment bond will guarantee coverage of up to 50% of the contract price.

For a project with a total bid price of up to $100,000, a bonding company charges on the order of 2.5% of the bid price for 100% performance and 100% payment bonds, with no extra charge for the bid bond; the percentage cost of those bonds declines with increasing project price, so that the jobs exceeding $2.5 million, the cost is typically 0.75% of the project price. These charges vary with different bonding companies and especially with different contractors.

More detailed information on construction bonds can be found in Bockrath (1986), Jervis and Levin (1988), and Sweet (1989).

10.6.2 Incorporation by Reference

When a construction contract is created, the "agreement" form tends to be quite short, perhaps four pages or less. Section 10.5.1 gives the information that may be included. In addition, the agreement will most likely refer to other materials and make them a part of the contract by reference. Simon (1989) writes:

> It is strongly urged that each and every contracting party attempt to list, in detail, each and every document it intends to include as part of its contract. This "incorporation by reference" means listing by date, description, title, revision, addenda, and any other available information and designation that it has, so that the parties can have a true meeting of the minds and more readily avoid litigation.

Certainly the drawings, specifications, and conditions discussed in Section 10.5 will be made a part of the contract by reference. Furthermore, it is common practice to incorporate applicable government regulations and such industry standards as those by the American Iron and Steel Institute and the National Electrical Manufacturers' Association and thus make them fully as binding as if they had been physically attached to the agreement.

10.6.3 Hierarchy of Documents

Since so many different documents are involved in a typical construction project, there is some chance that one might conflict with another, despite every sort of good faith checking during the planning and design phases. The technical specifications might specify a certain quality of illumination in a certain area, while the drawings indicate something different. The special conditions and the agreement might indicate different completion dates. If such discrepancies are not found until the contract has been signed and construction is under way, how shall the conflict be resolved? One effective way is to include a "hierarchy of documents" section in the general or special conditions; sometimes owners even place such a section in the agreement itself. Here is a typical hierarchy section:

> In the event that any provision of one contract document conflicts with a provision of any other contract document, the provision of that contract document first listed shall govern, except as otherwise specifically stated:
> a. This Agreement
> b. Exhibits, attachments, etc. incorporated herein by reference
> c. Instructions to Bidders
> d. The Special Conditions
> e. Special written instructions to the contractor, if any
> f. The General Conditions
> g. Manufacturer's instructions with reference to approved materials
> h. The Technical Specifications
> i. The Contract Drawings (West Ridge Natural Sciences Facility, 1992)

10.6.4   Indemnity

Indemnification, or "hold harmless," clauses come in many forms. The basic idea is that one party agrees to protect another party, or hold it harmless, against claims brought by third parties as a consequence of the first party's negligent acts. (Even that sentence is confusing; read it again!) Loulakis and Ingberg (1988) call indemnification provisions "one of the most feared—and least understood—clauses in construction contracts." And Simon (1989) reports that the law surrounding this clause is in a state of flux. He observes several guidelines: "(1) [the clause] must be written in clear, precise and unambiguous language; (2) a court will enforce it with great disfavor; and (3) it will be interpreted in a strict and narrow manner."

Table 10.2 presents in lay language a very helpful set of interpretations of this provision, from the viewpoints of the owner, the contractor, and the architect.

As might be expected, insurance plays a major role in the indemnification process. A promise to indemnify may be worthless unless it is backed by a solvent insurer.

10.6.5   Time of Completion and Liquidated Damages

One of the most important parts of any construction contract is the provision for time of completion and the consequences of the contractor's not meeting that deadline. The general rule is that the work must be *substantially complete,* in the sense used in Section 10.4.6, by the date specified in the contract. Extensions are permitted for any *excusable delays* enumerated in the contract or if it is impossible to meet the deadline because of an act of God, some operation of law, or the fault of the other party to the contract (Stokes, 1977).

Many contracts provide for the assessment of *liquidated damages* against the contractor if this party fails to meet the specified time of completion. Three points must be made with respect to such damages. First, they will be enforced only if the contract declares explicitly that *time is of the essence.* Courts have held consistently that time is not "of the essence" in a construction contract unless there is such a declaration in the contract. Second, damages must not be intended as a penalty, to punish the contractor for being late; rather, they are to compensate the owner. See the discussion in Section 9.6.1 for a useful discussion from an actual court case.

Third, the damages must be a "reasonable" approximation of the actual damages the other party has suffered. In this regard, a liquidated damages dispute between a contractor and subcontractor is interesting. The subcontract was to be completed on November 15 and the prime contract on December 1, but the prime contract was not completed until the following July 19. After negotiation that established that some of the delay was not its fault, the prime contractor was assessed $17\frac{2}{3}$ days of liquidated damages. The prime contractor sought to collect liquidated damages from its subcontractor, as provided in the contract, for the period between November 15 and July 19, but the appeals court decided that a more reasonable amount would be based on the 15 days between November 15 and December 1, plus the $17\frac{2}{3}$ days assessed the prime contractor [Mattingly Bridge Company, Inc. v. Holloway & Son

TABLE 10.2   Three Interpretations of Indemnity Clauses in General Contracts, in Lay Language

From the Viewpoint of the Owner

"I have turned over the site to you. It is your responsibility to see to it that the building is constructed properly. You must not expose others to unreasonable risk of harm. The increasing likelihood that I will be sued for what you do makes it fair that you relieve me of ultimate responsibility for these claims by your agreeing to hold me harmless or to indemnify me."

From the Viewpoint of the Contractor

"I know that you may be concerned about the possibility that a claim will be made against you by a third party during the course of my performance and that you will have to defend against that claim and either negotiate a settlement or even pay a court judgment. I always conduct my work in accordance with the best construction practices, and I have promised in my contract to do the work in a proper manner. I am so confident that I will do this that I am willing to relieve your anxiety by holding you harmless or indemnifying you if any claim is made against you by third parties relating to my work. You have nothing to worry about as I will stand behind my work. If you are concerned about my ability to pay you I will agree to back it up by public liability insurance coverage."

From the Viewpoint of the Architect/Engineer, via the Owner

"The law may hold me accountable for injury to your workers or to employees of your subcontractors because they may connect their injury with something they claim I did or should have done. You are being paid for your expertise in construction methods and your knowledge of safety rules. These are not activities in which I have been trained or in which I claim to have great skill or experience. For that reason if a claim is made against me for conduct that is your responsibility, I want you to hold me harmless and indemnify me."

SOURCE: Reprinted by permission from page 734 of *Legal Aspects of Architecture, Engineering and the Construction Process,* by Sweet; © 1989 by West Publishing Company. All rights reserved.

Construction, Supreme Court of Kentucky 694 S.W. 2d 702 (1984); "Liquidated Damages Must Be Reasonable," 1986].

10.6.6   Delay and No Damages for Delay

Delays in a construction project, while sometimes inevitable, are most always difficult for both parties and usually costly. Damages to the *owner* caused by delay of the *contractor* were covered briefly in Section 10.6.5. In the event of a nonexcusable delay attributable to the contractor, most contracts call for some sort of recovery by the owner, such as liquidated damages and/or contract termination.

There is another, equally important aspect to this question: What about damages to the contractor caused by *delays of the owner* or the owner's representatives?

Such delays are classified as follows: (1) excusable delays, for which the contract allows the contractor additional time equal to the time consumed by the delay, and (2) compensable delays, for which additional moneys are paid to the contractor equal to those lost as a result of the delay, both during and afterward. An excusable, compensable delay is thus one that provided for a time extension and additional payment ("Proving Entitlements for Damages Caused by Uncontrollable Delays," 1992). Such relief is not granted automatically. Suffice it to say that it is essential for the contractor to keep detailed records of finances, events, conversations, and any other information that will support a delay claim. Modern scheduling systems can be helpful to a contractor putting forth such a request. The amount, if any, that a contractor should be compensated to cover home office expenses during a delay period is a matter of long-standing controversy. The modern view seems to be that with proper documentation, unabsorbed office overhead is recoverable (Jervis and Levin, 1988).

Since damages for delay can be expensive to owners, a construction contract sometimes includes a *no-damages-for-delay* clause. As a rule, this provision states that a time extension shall be granted in such a case, but no compensation. A typical statement is the following:

> In the event the contractor is delayed by factors that are beyond the control of and without the fault of the contractor, the contractor may be entitled to an extension of the contractual performance period. This extension of time shall be the contractor's sole remedy in the event of a delay. The contractor shall not be entitled to an increase in the contract price as a result of any delay of any nature whatsoever, including delay caused by the acts or omissions of the owner. (Jervis and Leven, 1988)

Similar clauses are often inserted into contracts that involve construction management firms acting on behalf of owners in contracting with subcontractors, thus protecting the owner, the construction manager, and sometimes other subcontractors (Blinn, 1989).

Such provisions have been tested in court and found to be enforceable. In New York City, a declared moratorium on street closings delayed the performance of a sewer contractor under contract to the city. When the contractor brought suit for damages, the court found that a no-damages-for-delay clause was sufficient to bar recovery [Corrino Civetta Construction Corporation v. City of New York, 493 N.E. 2d 905 (N.Y. 1986)]. While such clauses can sometimes be waived, based on the conduct of the parties (Loulakis, 1986a), the prevailing view is that "absent a showing that a particular delay falls within a recognized exception, courts are generally reluctant to set aside the literal language of a no-damages-for-delay clause" ("No-Damages-for-Delay Clauses Can be Strictly Enforced," 1991).

10.6.7 Termination

Termination of a contract before the project is complete is an expensive, time-consuming, and difficult undertaking for all concerned. In many general conditions sections, the owner reserves the right to terminate the contract for two reasons, default and convenience. *Termination for default* is perhaps the more straightforward situation. If Contractor A breaches the contract, the owner is free to terminate that firm and have the work done by Contractor B, by its own forces, or by the bonding company. Often, in this situation, moneys due the contractor will not be paid at this time. Instead, upon completion of the work, some or all of these moneys will be used to offset any extra costs arising from the termination, and the terminated contractor will be paid whatever remains. Sometimes, the owner is permitted to take over equipment and material at the site to assist with project completion.

Sometimes an owner terminates a contract *for convenience*—because, for example, the completed project is no longer needed or economical, or project funds have been withdrawn unexpectedly. In such a case, the contractor is dealt with less harshly and typically is paid for work performed up to the date of termination, plus profit for that work. Sometimes the provision will also allow payment to the contractor for reasonable expenses incurred in the termination process (Schoumacher, 1986; Sweet, 1989).

10.6.8   Differing Site Conditions

One of the perplexing problems in the drafting and interpretation of construction contracts is the handling of unexpected conditions at the building site. For example, underground soil conditions may differ from those shown on the drawings or normally expected in the area; or, in a building renovation project, the existing building may contain hidden features that vary considerably from those contemplated in the contract documents. If a contract contains a clause to cover such an eventuality, it usually provides for an equitable price increase to cover the contractor's extra costs. However, the wise contractor will read the documents carefully, for sometimes they contain a disclaimer to the effect that the owner assumes no responsibility "for the accuracy or completeness of information on the drawings concerning existing conditions and the work required" ("Contractor Cannot Recover Costs for Changed Condition in the Face of Disclaimers," 1989).

The U.S. government has pioneered the use of differing site conditions clauses. Those contracts, and those of other owners who have followed this lead, divide such conditions into two categories.

- *Category I.* A condition encountered at the work site that is different from that represented in the contract documents. Simon (1989) calls this category "the administrative equivalent to a breach of contract," since, in essence, the owner failed to provide the consideration it promised when making the contract.
- *Category II.* A condition that is materially different from what a contractor would reasonably expect to encounter on the project. In this case, relief is predicated on having a clause in the contract that spells out how such relief is to be provided.

To succeed in being granted relief in either of these situations, the contractor must follow closely the contract requirements for notification of the owner and keeping proper documentation. A category II type claim may be more difficult to prove, since the trier of the facts will have to look at the contractor's interpretation of the contract documents, the contractor's experience, and the customs and practices of the industry in that place and on similar projects. As mentioned in Section 9.5.5, this author represented an owner in the defense of a claim of this type in which the plaintiff contractor cited "unexpected" permafrost (permanently frozen ground) underlying a rock and gravel material site as an excuse for claiming additional money and time. Unfortunately for the contractor, the defendant successfully argued that the presence of permafrost is common knowledge to anyone having reasonable familiarity with conditions in that area of interior Alaska.

An example of a successful category I claim was decided by the Armed Forces Board of Contract Appeals in 1990. The plans showed a buried, solid pipe that conveyed water to a trench. Upon commencing excavation for some structural foundations, the contractor encountered water flowing from that pipe. In fact, the pipe was perforated (a "french drain") to allow some water to be absorbed by the soil. The board granted an additional payment to the contractor because it "found an implicit representation of a changed pipe condition because the pipe was depicted as a solid-walled pipe rather than a perforated pipe,...a condition materially different from that which was indicated in the contract" (Loulakis and Cregger, 1991b).

10.7   DISCUSSION QUESTIONS

1. For each of the fee arrangements presented in Section 10.1.3, list the relative advantages and disadvantages to the owner and the designer.
2. Give some examples of situations in which the amount of a "fixed price" contract might appropriately be changed upward or downward.
3. Suppose a construction contract provided for the payment to the contractor of reimbursable costs plus 8% of those costs as a fee, with the following stipulations:

    The target price for reimbursable costs is $700,000.
    If the reimbursable costs are $550,000 or lower, the savings will be shared equally between owner and contractor.
    If the reimbursable costs exceed $800,000, the excess will be shared, with the owner covering 30% and the contractor absorbing 70% of the excess.
    In no case will the contractor be paid more than $900,000, including reimbursable costs and fee.

Find the total amount the contractor will be paid for each of the following verified reimbursable cost totals: (a) $700,000, (b) $620,000, (c) $510,000, (d) $750,000, (e) $820,000, (f) $850,000, (g) $930,000.

4. Santucci submitted a bid of $1.1 million for a drainage project, of which $775,000 was for the labor and materials for the drain pipe installation. The engineer's estimate was $1.4 million for the drain pipe and $1.9 million for the total project. Other total bids ranged between $1.7 and $1.8 million, with drain pipe estimates between $1.2 and $1.4 million. One day after the bid opening, Santucci notified the owner of a clerical error in the $1.1 million bid and sought to withdraw it. The owner refused and notified Santucci of its intent to contract, but Santucci refused. The owner retained Santucci's bid deposit, whereupon Santucci commenced a lawsuit to recover it.

   What did the court decide? [Santucci Construction Company v. County of Cook, 21 Ill. App. 3d 527, 315 N.E. 2d 565 (1974)]

5. An excavation contractor encountered soil conditions that were different from those shown in the contract documents. To cope with these conditions, the contractor, without notifying the owner, changed its construction method in a way that resulted in increased cost. Later, the contractor sought to recover its added costs from the owner, citing the contract's differing site conditions cause. What was the result? [Schnip Building Company v. United States, 645 F. 2d 950 (1981)]

6. The specifications described a dry excavation procedure, but the contractor encountered water. Is this a Category I or II condition according to the definitions of Section 10.6.8? Do you think the contractor obtained relief? Why? [Foster Const. C.A. and Williams Brothers Company v. United States, 193 Ct. Cl. 587,435 F. 2d 873, 887 (1970)]

7. Suppose the actual quantity of excavation for the project shown in Table 10.1 is 9800 cy instead of the 5800 cy shown. What is likely to happen to the $10 per cubic yard unit price for that item? Suggest an approach the parties might use to determine the revised unit price.

8. A condominium construction contract stipulated that written authorization from the owner was required prior to the performance of any extra work. The contractor had to change the roof structure because of errors in the owner's plans. The owner's representative orally directed the contractor to proceed with the changes, assured the contractor that the extra costs would be paid, and observed the contractor's activities while the changes were carried out. Later, when the contractor filed a claim for extra payment for this work, the owner refused to pay. What was the result? [Udevco, Inc. v. Wagner, 678 P. 2d 679 (1984)]

9. One item in a bridge construction project was a concrete barrier railing, which the low bidder priced at $42 per linear foot. The contractor used a slip-forming method for this railing, rather than a set of fixed forms as anticipated by the owner. Because the slip-forming method saved the contractor considerable money, the owner sought a price reduction. What

was the result? [Hardaway Constructors, Inc., v. North Carolina Department of Transportation, 342 N.E. 2d 52 (1986)]

10. A contractor failed to complete a building project by the specified date, and the owner rejected the last portion with respect to both materials and performance. Before the deficiencies had been remedied, the building was destroyed by fire. Progress payments had been made to the contractor each month. The contract made the contractor responsible for any damage during the project. The contractor had furnished a performance bond, as required. What, if anything, did the owner recover? From whom? Why? [United States v. United States Fidelity & Guaranty Company, 236 U.S. 512 (1915)]

11. A road-building contract stipulated that the builder would be paid $3000 per mile and that the owner would make payments every 2 weeks based on satisfactory performance of portions of the work. If the contractor completed only a part of the total project and then left the job, was he entitled to be paid for the value of his labor and materials? [Dillon & Harrison v. Suburban Land Company, 73 W. Va. 363, 80 S.E. 471 (1913)]

12. In connection with a condominium construction project, the owner was contractually obligated to furnish the contractor with certain patented molds that would be used in one phase of the construction work. Another provision obligated the contractor to give the owner written notice of any delay claim within a reasonable time after the occurrence of the delay. The contractor's work was delayed because the owner had not provided enough molds. The contractor notified the owner in writing of the delay and said it would furnish the dollar amount of its claim later. When the claim was submitted, the owner refused to pay because the earlier written notice had not included the alleged extra costs. What was the result? [Eagle's Nest Limited Partnership v. S.M. Brunzell, 669 P. 2d 714 (1983)]

13. A contract contained a no-damages-for-delay clause favoring the owner. The owner directed the contractor to proceed before some prerequisite work was complete. A delay of 175 days resulted. The owner granted a time extension but no extra payment for delay damages. On what basis might the court determine that the contractor *was* entitled to such payment? [United States Steel Corporation v. Missouri Pacific Railroad Company, 668 F. 2d 435 (8th Cir. 1982)]

14. A contract required a construction project located in South Carolina to be completed by September 15. The owner's engineer directed the contractor to suspend operations for the month of June, and the completion date was revised to October 15. Was that change in completion date appropriate? If the project had been located in the Northwest Territories instead of South Carolina, would the change in completion date have been appropriate?

15. C & H Sugar Company contracted with Sun Ship, Inc., to build a new shipping vessel for $25,405,000, to be delivered on June 30, 1981. A liquidated damages clause of $17,000 per day was a part of the contract. The vessel was completed on March 16, 1982. Actual damages suffered by

C & H due to the unavailability of the ship for the 1981 season came to about $370,000. Sun Ship refused to pay $4,413,000 in liquidated damages, C & H filed suit, and the district court found in favor of C & H. What was the result of the appeal? [California and Hawaiian Sugar Company v. Sun Ship, Inc., 794 FR. 2d 1433 (1986)]

16. A recreational waterslide constructed of concrete developed cracks in its concrete flumes soon after construction was completed. The cracks did not interfere with operations, although they had to be repaired at substantial cost. The contractor billed the owner $550,000, based on full performance of the contract, but when the owner paid only $150,000, the contractor filed suit to recover the full contract price. The court judgment to the contractor was based on a total award of $352,000 (less than the $550,000, to recognize the cost of repairs). The owner appealed, claiming the contractor had not substantially performed the contract. What should be the result of the appeal? [W. E. Erickson Construction, Inc. v. Congress-Kenilworth Corporation, 503 N.E. 2d 233 (1986)]

10.8   REFERENCES

American Institute of Architects. 1987. Standard Form of Agreement Between Owner and Architect, Document B141, 1987 edition. New York: AIA.

Blinn, K. W. 1989. *Legal and Ethical Concepts in Engineering.* Englewood Cliffs, NJ: Prentice-Hall.

Bockrath, J. T. 1986. *Dunham and Young's Contracts, Specifications for Law for Engineers* New York: McGraw-Hill.

"Contractor Cannot Recover Costs for Changed Condition in the Face of Disclaimers." 1989. *PEC Reporter,* Professional Engineers in Construction, National Society of Professional Engineers, Vol. 12, No. 2, November/December, 2.

"Contractor Must Outline All Charges in Contract." 1992. *Civil Engineering,* Vol. 62, No. 12, December, 28.

"Contractor Names Subs Late, Loses Bid." 1993. *Civil Engineering,* Vol. 63, No. 6, June, 32.

"Court Upholds Contract Price Despite Added Costs." 1993. *Civil Engineering,* Vol. 63, No. 2, February, 27.

"Courts: Follow Bid Instructions." 1990. *Civil Engineering,* Vol. 60, No. 8, August, 24.

Firmage, D. A. 1980. *Modern Engineering Practice.* New York: Garland STPM Press.

Grennberg, T. no date. "The Function Contract," Luleå University of Technology, Department of Civil Engineering, Luleå, Sweden.

Jervis, B. M., and P. Levin. 1988. *Construction Law: Principles and Practice.* New York: McGraw-Hill.

Jessup, W. E., Jr., and W. E. Jessup. 1963. *Law and Specifications for Engineers and Scientists.* Englewood Cliffs, NJ: Prentice-Hall.

"Liquidated Damages Must Be Reasonable." 1986. *Civil Engineering,* Vol. 56, No. 6, June, 40.

Loulakis, M. C. 1985. "Deletion of Work on Unit-Price Contracts." *Civil Engineering,* Vol. 55, No. 8, August, 34.

Loulakis, M. C. 1986a. "No Damages for Delay Clauses Can Be Waived." *Civil Engineering,* Vol. 56, No. 5, May, 51.

Loulakis, M. C. 1986b. "Overly Strict Inspections: A Contractor Can Get Even." *Civil Engineering,* Vol. 56, No. 8, August, 34.

Loulakis, M. C., and J. C. Thompson. 1990. "Design Professionals as Agents Under the AIA Documents." *Civil Engineering,* Vol. 60, No. 5, May, 45.

Loulakis, M. C., and M. A. Ingberg. 1988. "Indemnity Clauses: Broader Coverage than You Think." *Civil Engineering,* Vol. 58, No. 9, September, 42.

Loulakis, M. C., and W. L. Cregger. 1991a. "Contractor Beware: Public Bids Must Be Timely." *Civil Engineering,* Vol. 61, No. 1, January, 41.

Loulakis, M. C., and W. L. Cregger. 1991b. "Industry Standards of Practice May Affect Contractor Claims." *Civil Engineering,* Vol. 61, No. 5, May, 36.

Manzi, J. E.. 1984. "The A/E's Liability in the Design/Construction Process." *Construction Claims Monthly,* Vol. 6, No. 4, April, 1, 7.

"Monitored Construction Protects Contractor." 1992. *Civil Engineering,* Vol. 62, No. 2, February, 28.

"No-Damages-for-Delay Clauses Can Be Strictly Enforced." 1991. *Civil Engineering,* Vol. 61, No. 3, March, 46.

"Proving Entitlements for Damages Caused by Uncontrollable Delays." 1992. *PEC Reporter,* Professional Engineers in Construction, National Society of Professional Engineers, Vol. 14, No. 5, March, 5.

Schoumacher, B. 1986. *Engineers and the Law: An Overview.* New York: Van Nostrand Reinhold.

Simon, M. S. 1989. *Construction Claims and Liability.* New York: Wiley.

Stokes, M. 1977. *Construction Law in Contractors' Language.* New York: McGraw-Hill.

Sweet, J. 1989. *Legal Aspects of Architecture, Engineering and the Construction Process,* 4th ed. St. Paul, MN: West Publishing.

Tarricone, P. 1993a. "Deliverance." *Civil Engineering,* Vol. 63, No. 2, February, 36–39.

Tarricone, P. 1993b. "What Do You Mean by That?" *Civil Engineering,* Vol. 63, No. 4, April, 60–62.

"Total Price Error: Unit Price Upheld." 1986. *Civil Engineering,* Vol. 56, No. 10, October, 22.

Vaughn, R. C. 1983. *Legal Aspects of Engineering,* 4th ed. Dubuque, IA: Kendall/Hunt Publishing.

"West Ridge Natural Sciences Facility." 1992. University of Alaska Fairbanks, Project Manual, March 16.

"48538 Legal." 1993. Classified advertisement appearing in *Fairbanks Daily News-Miner,* Fairbanks, AK, June 27.

_____11
Legal Structures of Business Organizations

The choice of the organization's legal form may be an important decision for the engineering manager. When one goes into business, one must decide not only how the internal relationships will be arranged, as described in Chapter 2, but also what legal relationships will be established with owners, funding sources, creditors, and tax authorities.

In order of increasing complexity, the basic forms are the sole proprietorship, the partnership, and the corporation. We discuss each in turn, highlighting their respective advantages and limitations. In addition, we review the use of the joint venture, a temporary arrangement often used for large construction projects. The chapter closes with a brief description of some other organizations that may be useful to the engineering manager: including the not-for-profit corporation, the franchise, the limited liability company, the unincorporated association, and the share-office arrangement.

11.1 SOLE PROPRIETORSHIP

The sole proprietorship can be described rather quickly. It is a noncorporate business that is owned and operated by one person. Sole proprietorships are the most common form of business organization in the United States. Many small businesses, and some large ones, operate under this arrangement. Many private design professionals and construction companies use this form, which is a logical way to operate when you are small and just starting out.

Compared with other forms of business organization, the sole proprietorship is simple to create and run. To get started, you must obtain whatever licenses your state requires, generally beginning with a state business license. If you will be operating as an engineer, you probably will be required to show that you are a registered professional engineer. If sales tax laws require that you add sales taxes to your fees, you may be required to register your business with the taxing authority. Also, if you will use a business name other than your own, it may be necessary to file a request to "do business as" the name you select.

As sole proprietor, you will have sole responsibility for the business and sole enjoyment of any profits that result. You will be able to work with a minimum of regulations, and you will set your own operating rules. You answer only to yourself for all business decisions, employment conditions, and the like, provided they do not violate the law. Income and losses are reported on your individual income tax return. The situation is clearly a "good news–bad news" arrangement; you share the profits and other positive aspects with no one, but you also share none of the burdens, risks, and losses.

Disadvantages of the sole proprietorship include the limitation of available capital to personal funds and whatever can be obtained as loans, and the limitation of expertise to that of the sole owner. Also, the individual proprietor is responsible for the full amount of business debts, even if they exceed the proprietorship's assets. That is, there is *unlimited personal liability* for such debts, which means that unsatisfied creditors may recover against any personal assets such as your house, vehicles, and camping gear. Furthermore, the individual owner may be held personally liable for physical loss or personal injury (so-called tort liability, to be discussed in Chapter 12) resulting from the operation of the business. Table 11.1 summarizes the advantages and limitations of the sole proprietorship form of organization.

Cheeseman (1992) says that in a sole proprietorship, "the owner is actually the business." To illustrate, he cites the case of an engineering company and its owner. Each party had a savings account in each of two banks, and both banks failed. In one bank, the owner had $91,593 on deposit, and the company also had $91,593 on deposit; in the other, the owner had $30,605, and the company had $108,119. In deciding how much insurance benefit the owner was to receive, the Federal Savings and Loan Insurance Corporation (FSLIC) combined the owner's and the company's deposits in each bank. Since each combination ($183,186 and $138,724) was greater than the insurance limit of $100,000, the owner received only $200,000, or $100,000 for each bank. This payment was based on the rationale that the business was a sole proprietorship and thus should be viewed as the same as the owner. The owner sought recovery for the total of $91,593, $91,593, $30,605, and $100,000, but a U.S. Court of Appeals ultimately found against the owner, sustaining the FSLIC position that the business was a sole proprietorship [Lambert v. Federal Savings and Loan Insurance Corporation, 871 F. 2d 30 (1989), U.S. Court of Appeals, 5th Cir.)].

11.2 PARTNERSHIP

11.2.1 General Partnership

A *general partnership,* also known as an *ordinary partnership,* or just a *partnership,* is a (1) voluntary association (2) of two or more persons engaged in carrying on a business (3) as co-owners (4) for profit. All four of these elements must be present for a partnership to exist. For example, an association that simply holds property or is a nonprofit fraternal organization is not a partnership. If an

TABLE 11.1 Advantages and Disadvantages of a Sole Proprietorship

Advantages of a Sole Proprietorship

- *Ease of Formation.* There is less formality associated with establishing a sole proprietorship, and there are fewer legal restrictions. Little or no governmental approval is
necessary, and usually this form of organization is less expensive than a partnership
or corporation.
- *Sole Ownership of Profits.* The proprietor is not required to share profits with anyone.
- *Control and Decision Making Vested in One Owner.* There are no co-owners to consult. (Except possibly the sole proprietor's spouse.)
- *Flexibility.* Management is able to respond quickly to business needs in the form of
day-to-day decisions as governed by various laws and good sense.
- *Relative Freedom from Government Control and Special Taxation.*
- *Ease of Transfer.* The company can be transferred or sold if and when the owner
wishes to do so; there are no partners or stockholders whose approval must
be obtained.

Disadvantages of a Sole Proprietorship

- *Unlimited Liability.* The individual proprietor is responsible for the full amount of
debts incurred by himself or herself or the firm's employees in the course of employment, even if they exceed the proprietor's total investment. This liability extends to all
the proprietor's assets, such as house and car. Liability for physical loss or personal
injury is also unlimited but may be lessened by obtaining proper insurance coverage.
- *Unstable Business Life.* The enterprise may be crippled or terminated upon illness or
death of the owner.
- *Less Available Capital.* Ordinarily, since personal contributions are limited to those
of one person, there is not as much capital available as in business organizations of
other types.
- *Limitations on Other Contributions.* Contributions such as equipment are available
from one person alone.
- *Taxes.* Unlike the owners of a corporation, sole proprietors pay income tax on their
full profits, whether withdrawn or not. In addition, sole proprietors must pay self-
employment tax.
- *Relative Difficulty in Obtaining Long-Term Financing.* Unlike corporations, sole proprietorships cannot issue stock. Banks may be reluctant to provide loans to sole
owners.
- *Relatively Limited Viewpoint and Experience.* This is more often the case with one
owner than with several.

SOURCE: Based in part on Olmi (1982).

association meets the four criteria, however, it will be held to be a partnership even
if the participants did not intend it to be so.

Partnerships are more complicated than sole proprietorships simply because
more owners are involved and also because they are subject to more government

regulations. It will be necessary to obtain certain licenses and to file other paperwork, just as with a sole proprietorship. Although partnerships, like sole proprietorships, are not considered to be separate entities for the purpose of income taxation, the partnership is required to file informational returns with the Internal Revenue Service. Each partner then reports his or her portion of the partnership's profit or loss on his or her individual income tax return.

Partnerships have the advantage of being able to draw on a greater variety of expertise and background, compared to the sole proprietorship; generally, more funding, equipment, and other assets are available, as well.

Just as with the sole proprietorship, each partner has unlimited personal liability for partnership debts and for injuries and physical losses caused by the partnership or its members in conducting partnership business. Because you are not acting as a single owner any more, this potential liability is a primary disadvantage of the partnership form of business organization.

An extremely important feature of the partnership is that each partner is an agent of the partnership. Thus, much of basic agency law applies to such arrangements. In a sense, every partner is both principal and agent, since other partners act on one partner's behalf, and this one partner also acts as an agent on their behalf. "Mutual agency" is a term that is used to describe this feature of the typical partnership. Section 9(1) of the federal Uniform Partnership Act provides

> Every partner is an agent of the partnership for the purpose of its business, and the act of every partner, including the execution in the partnership name of any instrument, for apparently carrying on in the usual way the business of the partnership of which he is a member binds the partnership, unless the partner so acting has in fact no authority to act for the partnership in the particular matter, and the person with whom he is dealing has knowledge of the fact that he has no such authority. (quoted in Blinn, 1989)

Thus, an important conclusion is that the engineer, or anyone else who contemplates entering a business association with one or more others, must *choose partners wisely!*

The notion of *apparent authority* is suggested in the foregoing quotation from the Uniform Partnership Act. In our discussion of agency in Chapter 8, we noted that a principal may clothe an agent with apparent authority to act on behalf of the principal if the principal's conduct is such that a third party believes the agent to have that authority and relies on that belief. Certainly that idea applies in partnerships. Even if the partnership agreement limits the authority of a partner, if that partner overreaches his or her authority (say, in signing a supply contract) and the other partners make it look to the supplier as if the partner in question has that authority, a binding contract has probably been created.

The liability of members of a partnership for *contracts* made in the name of the partnership is said to be *joint* only, which means that a plaintiff must name all partners as defendants in a suit to recover on a contract. In the case of personal injuries and physical losses (*torts*, to be discussed in Chapter 12), the partners are jointly and severally liable; an injured party can bring suit against any or all of the partners, even if they did not commit the tort. In the case of both contract and tort liability, the successful plaintiff can recover the judgment from any or all of the named defendants.

Suppose you, a mechanical engineer, and your colleague, a structural engineer, form a two-person partnership, and he produces a faulty design for a structural frame. Suppose further that the frame collapses and it is determined that your partner had acted negligently in producing the design that caused personal injury. If you were named in the suit, you could become liable for those injuries. Furthermore, if both members of the partnership were found jointly and severally liable you might have to pay for any part of his share that he could not pay. Again, *choose your partners carefully* is sage advice. We discuss this matter further in Chapter 12.

Criminal acts committed by individual partners are the responsibility of the individual. Copartners are not liable unless they approved or participated in the act.

In general, according to the statute of frauds, partnership agreements do not have to be in writing to be enforceable unless they are to exist for more than one year or are organized to deal in real estate. However, a written partnership agreement is commonly executed. Such a document, called *articles of partnership,* is signed by the partners, and it sets forth the terms and conditions of the agreement. A typical set of articles might contain the following sections:

- Name of the firm, address of its principal office, and names and addresses of the partners
- Nature and scope of the business
- Date of beginning and intended duration
- Capital contributions of each partner
- Distribution of the respective shares of profits and losses
- Duties and restrictions of partners regarding management of the business
- Salaries, if any, to be paid to the partners
- Limitations, if any, on the authority of the partners to bind the partnership
- Provisions for settlement of disputes
- Provisions for adding or withdrawing partners, and for continuing or dissolving the partnership in the case of withdrawals or death
- Provisions for such financial matters as banking, accounting and inventories

Note that all of these articles are not required. (In fact, as we have noted, partnerships can exist without written agreements.) If a statement about continuity in case of the death of a partner is not included, for example, the partnership would be dissolved in such a case.

Some of the advantages and disadvantages of partnerships have been discussed. Table 11.2 gives a more complete listing.

Termination of a partnership involves two steps: dissolution of the firm and winding up of its activity. *Dissolution* commonly occurs upon the death, bankrupt-

TABLE 11.2 Advantages and Disadvantages of a Partnership

Advantages of a Partnership

- *Ease of Formation.* Legal informalities and expenses are few compared with the requirements for creating a corporation.
- *Direct Rewards.* Partners are motivated to apply their best abilities by the certainty of sharing directly in any profits.
- *Growth and Performance Facilitated.* In a partnership, it is often possible to obtain more capital and a better range of skills than can be accessed by a sole proprietorship.
- *Close Owner Control over Ownership and Management.* A partner cannot sell his or her portion of the company without the consent of all the other partners.
- *Flexibility.* A partnership may be relatively more flexible in the decision-making process than in a corporation. But it may be less so than in a sole proprietorship.
- *Relative Freedom from Government Control and Special Taxation.*

Disadvantages of a Partnership

- *Unlimited Liability of at Least One Partner.* The partners are responsible for the full amount of debts incurred by partners or employees in the course of employment, even if the amounts exceed the partners' total investment. This liability extends to all the partners' assets, such as houses and cars. Liability for physical loss or personal injury is also unlimited but may be lessened by obtaining proper insurance coverage.
- *Unstable Life.* Elimination of any partner constitutes automatic dissolution of the partnership (unless otherwise provided for in the partnership agreement), although operation of the business can continue based on the right of survivorship and possible creation of a new partnership.
- *Relative Difficulty in Obtaining Large Sums of Capital.* This is particularly true of long-term financing when compared to the advantages held by a corporation. Since, however, individual partners' assets may be used, opportunities are probably greater than in a proprietorship.
- *Mutual Agency.* The acts of any one partner as agent can bind the firm and all other partners.
- *Profit Sharing.* Each partner must share profits with the other partners.
- *Income Taxes.* Unlike the owners of a corporation, partners pay income tax on their allocated share of profits, regardless of whether withdrawn.
- *Difficulty of Disposing of Partnership Interest.* The buying out of a partner may be difficult unless this contingency is specifically arranged for in the written agreement.

SOURCE: Based in part on Olmi (1982).

cy, or withdrawal of one of the partners, with the mutual consent of the partners, or upon the expiration of the time set by the articles of partnership. Notice of the dissolution should be sent to those with whom the partnership has dealt. The law treats a dissolved partnership as still in existence until the completion of such winding-up activities as payment of accounts payable, collection of accounts receivable, and completion of obligations undertaken prior to dissolution. The reduction of partnership property to cash and the subsequent distribution of the proceeds takes place in a strict order. Bockrath (1986) lists the steps as follows:

1. In the vanguard come the creditors of the firm other than the partners.
2. Next are the claims of partners for loans and advances beyond the amount of their agreed capital contributions.
3. After the foregoing categories of claims have been satisfied, under the usual partnership agreement members are entitled to a return of their respective capital contributions. If there is insufficient property available for this purpose, the loss is normally to be shared by the solvent partners in the proportions in which they were supposed to share profits.
4. If the dissolved partnership has been a very successful enterprise and some undistributed assets remain after the various demands outlined above are taken care of, the balance will be disposed of among the partners in the form of profits.

To illustrate the winding-up process, and to emphasize the partners' unlimited personal liability, consider the land surveying partnership of Ace, Base, and Case. Suppose their initial contributions to the partnership were as follows: Ace, $60,000; Base, $40,000; Case, $5000. Suppose further that the partnership agreement provides for an equal sharing of any profits or losses. When the partnership is terminated, it has assets of $70,000, all in cash (for simplicity!), and liabilities of $25,000, all in accounts payable. Assume that none of the partners has withdrawn any due share of the profits and that none has made any additional contributions or loans to the partnership.

This example illustrates a case in which the partnership *lost money*. The total initial contributions were $105,000; at termination, the partnership had net assets of $70,000 - $25,000 = $45,000. This $60,000 loss will be distributed equally. The steps in the winding-up process are as follows:

1. The creditors are paid their $25,000 accounts payable, reducing the available cash to $45,000.
2. No loans or advances are due to the partners at this step.
3. Since the partnership lost a net of $60,000, each partner is responsible for $20,000 of the loss.
4. No undistributed assets remain.

The result of these steps leads to the following summary of distributions:

Party	Payment
Creditors	$25,000 paid *to* the creditors
Ace	$60,000 - $20,000 = $40,000 paid *to* Ace
Base	$40,000 - $20,000 = $20,000 paid *to* Base
Case	$5000 - $20,000 = -$15,000 paid *by* Case

So, Ace and Base will receive money as a result of the winding-up process, but Case will be required to contribute $15,000 from personal assets. To check our process, we can add the net amounts paid to or by the four parties to be certain that the total is the $70,000 available for distribution.

11.2.2   Limited Partnership and Subpartnership

A limited partnership has partners of two types: general and limited. General partners have the same relation to the firm as that described for the general partnership. They invest capital, manage the business, and are personally liable for partnership debts and obligations. *Limited partners* (sometimes called special partners), on the other hand, invest capital but do not participate in the management of the business. In exchange for this restriction on management activities, limited partners are liable for the debts and obligations of the partnership only up to their capital contributions.

The object of the limited partnership is to provide a means by which persons can invest in the business but remain exempt from general liability and place only their invested capital in jeopardy. The tradeoff is that limited partners may not participate directly in management decisions.

A *subpartnership* is a separate arrangement between one partner and an outside person called a subpartner. Under such an agreement, this outsider agrees to participate in the contracting partner's share of the profits and losses. The subpartner is in no sense a member of the firm; he or she performs no active function and has no control over partnership affairs.

Figure 11.1 is a diagram illustrating the relationships among general, limited, and subpartners and third parties in a typical limited partnership.

11.2.3   Partnering

This somewhat out-of-place section represents an attempt to alleviate the confusion often wrought by the term "partnering," which is not a form of business organization from the legal viewpoint. Instead, it is

> a long-term commitment between two or more organizations for the purpose of achieving specific business objectives by maximizing the effectiveness of each participant's resources. This requires changing traditional relationships to a shared culture without regard to organizational boundaries. The relationship is based upon trust, dedication to common goals, and an understanding of each other's individual expectations and values. Expected benefits include improved efficiency and cost effectiveness, increased opportunity for innovation, and continuous improvement of quality products and services. (Construction Division, 1991)

This rather lofty and idealistic arrangement is being incorporated into many government construction contracts in the United States, with considerable success. It has many of the marks of total quality management, which we considered in Chapters 3 and 4. The U.S. Army Corps of Engineers, the U.S. Bureau

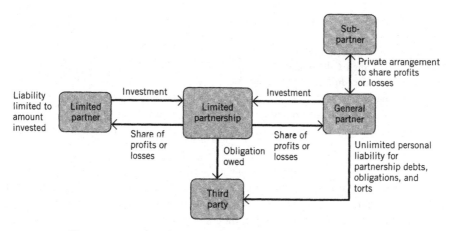

Figure 11.1. Limited partnership and subpartnership relationships.

of Reclamation, various transportation departments, and other agencies have urged such an approach for their contracts. Cynics might comment that if this arrangement can work successfully for a construction contract, where the parties' interests have traditionally been far from mutual, it ought to work most anywhere.

Basically, partnering recognizes the common goals of two or more parties and seeks to achieve them through cooperation and open communication. While the concept is hardly new, the intentional, formalized approach and commitment are features of such contracts that have appeared only since 1990. Partnering agreements seek to establish relationships in which the parties

- Seek win–win solutions
- Settle differences before they become disputes
- Place value on long-term relationships
- Consider trust and openness as normal
- Provide an environment for profit
- Are encouraged to address any problem openly
- Understand that neither will benefit from exploiting the other
- Encourage innovation
- Are aware of the other's needs, concerns, and objectives and are interested in helping the partners achieve each (Construction Division, 1991)

Cook and Hancher (1990) report the following examples of partnering relationships in the construction industry:

Fru-Con Construction Co.-U.S. Army Corps of Engineers	for construction
Bechtel, Inc. – Chevron Corporation Bechtel, Inc. – Shell Oil Company Fluor Daniel – E. I. du Pont de Nemours	for engineering, procurement, and construction management
RUST International Corporation- Great Northern Nekoosa Corporation	for engineering, procurement, and construction
M. W. Kellogg Company-Procter & Gamble	for engineering, purchasing, and construction

Partnering has found its way into a variety of relationships. An example is a partnering agreement signed by the U.S. Army Corps of Engineers and the National Society of Professional Engineers in 1993. The agreement states that the parties seek to promote their mutual engineering interests by working together toward such goals as promotion of ethical and competent engineering practice in the public interest, promoting quality through open communication, teamwork, and continuous improvement, and advancing the profession through support of education and continuing professional development (*Partnering Agreement,* 1993).

11.3   CORPORATION

Even though sole proprietorships may outnumber the other forms of business organizations in the United States, clearly the corporate form is most dominant, generating more than 85% of the country's gross business receipts (Cheeseman, 1992). The word "corporation" comes from the Latin *corpus,* or body. The primary distinguishing characteristic of the corporation is that it is a distinct legal entity— a legal person—having rights and obligations similar to those of an individual. In a famous early nineteenth century case [Dartmouth College v. Woodward, 4 Wheaton 518, 636 (1819)], Chief Justice John Marshall wrote:

> A corporation is an artificial being, invisible, intangible and existing only in the contemplation of law. Being the mere creature of the law, it possesses only those properties which the charter of its creation confers upon it, either expressly, or as incidental to its very existence. (quoted in Cheeseman, 1992)

Being distinct persons, corporations can buy and sell property, enter into contracts, have employees, borrow money, and sue and be sued. Furthermore, a corporation, as a separate entity, pays taxes, is subject to criminal prosecution and is entitled to the protection of the law (Lyden, Reitzel, and Roberts, 1985).

The most common type of corporation is the *private corporation,* which is owned by private individuals. A *public corporation* (e.g., a water district, a school district) is government-owned, formed to meet some specific government or political purpose. Our discussions are limited to the private corporation. We also distinguish between the publicly held and the closely held (or "close") corporation.

Publicly held corporations usually have large numbers of owners, and ownership shares are traded among the general public in organized securities markets. Shareholders rarely participate in the management of such corporations. *Closely held corporations* are owned by a few shareholders, who often are family members or friends. Frequently the owners are intensely involved in the management of the business. In most states, publicly held and closely held corporations are subject to the same general corporation statute.

11.3.1 Limited Liability of Owners

Because of the "legal fiction" (Blinn, 1989) that the corporation is an entity separate and distinct from its owners, the shareholders, the latter are protected from personal liability for the corporation's contract and tort debts. In the event that the corporation fails, the shareholders stand to lose only their investments. This limitation of liability to personal investments is a major difference between the corporation and both the sole proprietorship and the partnership.

Note that the limitation on liability is for the protection of the corporation's shareholders. Directors, officers, employees, and agents are still personally liable for negligent acts or crimes that they commit while acting on behalf of the corporation.

In some special cases, the corporate form can be disregarded and the owners held liable beyond the limit of their investments. If the company is really a shell with inadequate capitalization, intended as a conduit for the owners' personal affairs, third parties who have relied on what they believed to be a solvent corporation have standing to sue for damages. Disregarding the corporate form in such a case is referred to as *piercing the corporate veil*. When it happens at all, the organization at fault usually is a closely held corporation acting as the shareholders' alter ego. Examples of such cases are those involving commingling corporate and personal funds, converting corporate assets for personal use, and failing to follow required procedures for holding meetings, keeping minutes, and the like.

11.3.2 Formation of Corporations

Corporations are creatures of statute and thus can be formed only if certain procedures mandated by the laws of the state are followed. To protect the public from fraudulent sale of stock and abuse of power by corporate officers, a corporation cannot be formed without the approval of the securities and exchange commission (or equivalent) of the state and cannot be dissolved without approval of the same authority. The ordinary procedure begins with the taking of subscriptions for capital stock and the creation of a tentative organization. Then, the organization drafts and files *articles of incorporation* (sometimes called *certificate of incorporation* or *corporate charter*). When this document, stating the enterprise's powers and limitations, has been approved, and the proper fee paid, the corporation is ready to commence business (Blinn, 1989; Firmage, 1980).

Each state's corporation code sets forth the information that must be included in the articles of incorporation. Generally, the following information must be included:

- Corporate name
- Duration, which may be perpetual
- Purpose or purposes
- Aggregate number of shares of stock the corporation is authorized to issue, plus details about the various classes of stock
- Any preemption rights for shareholders (opportunity to purchase additional shares when offered)
- Address of initial registered office and name of initial registered agent
- Name and address of each incorporator; size of initial board of directors, and names and addresses of temporary board members who will serve until first annual meeting (Cheeseman, 1992)

Unlike a partnership, for which the agreement is a contract among the partners, the corporate charter is a public document recording the granting of certain powers by the state.

Recall our discussion in Section 9.2.1 regarding *ultra vires* contracts. We said that contracts made by a corporation that are beyond the scope of its charter will be held to be unenforceable. As suggested in the list above, the charter will state the corporation's purposes. Under earlier corporate law, such purposes were quite specific (e.g., shipbuilding, software design, crude oil transportation). Thus, a contract made by a shipbuilding company to manufacture steel will not be enforceable. The modern trend is toward broader corporate purposes; often the scope is stated as "for any lawful purpose," although some incorporators may certainly insist that the company concentrate its activities in a given area of expertise.

A corporation can be incorporated in only one state, even though it may do business in other states. As an artificial person with a "domicile" in its state of incorporation, it is called a *domestic corporation* in that state. In other states, it must qualify and be approved to operate as a *foreign corporation*. Generally, corporations incorporate in the state where they will do most of their business, although some corporations choose to incorporate in the state whose laws are most favorable to their internal operations; in the United States, Delaware is often chosen for this reason. An *alien corporation* is one that is incorporated in another country but is permitted to do business in a particular state, provided it is registered and has paid a fee.

After the articles of incorporation have been approved and a certificate has been issued, the directors elect a slate of officers who will be responsible for managing the company's affairs. Bylaws will be adopted, to serve as an internal charter for officers' authority, rules for director and shareholder meetings, and other corporate details.

11.3.3 Participation and Control

We shall discuss briefly three parties who have primary responsibility for the corporation's activities. The *shareholders* are the owners of the business. Their role in business decision making and management of business affairs, however, is limited to broad policy matters and the selection of directors. Most states require that the articles of incorporation provide for an annual shareholder meeting and a time for that meeting. Each shareholder has the right to participate in these meetings by offering resolutions, by taking part in debate, and by voting. Certain corporate transactions require affirmative vote of the shareholders; these include election and removal of directors, sale of all corporate assets, merger with another corporation, amendment of the articles of incorporation, and voluntary dissolution of the corporation.

Two methods are used to tally votes for directors, both of which are based on the number of shares of stock held by each shareholder. Under *straight,* or noncumulative, voting, each shareholder votes the number of shares he or she owns for each of the open positions. Some states require, and others permit, *cumulative* voting, the system under which a shareholder may accumulate his or her votes and cast all these votes (shares owned $\times$ number of open positions) for one (or more) candidates. Cumulative voting provides greater voice to minority shareholders. If you own 49% of the corporation's shares, and someone else owns the other 51%, under straight voting his candidates will all win. Under cumulative voting, if more than two directors are to be elected, you should be able to have at least one be "your" candidate. Suppose, for example, that there are a total of 10,000 shares, and you own 4900. Also suppose that three directors are to be elected from a slate of, say, seven candidates. You are entitled to $3 \times 4900 = 14,700$ votes, while the other shareholder is entitled to 15,300 votes. If you vote all your 14,700 for one candidate, that candidate will win one of the three positions because the other shareholder must spread his votes over two of the remaining candidates in order to maintain control. His votes might be 7,650 for each or 15,299 for one and one for the other, for example. In any case, with your 49%, you are assured that your candidate will emerge no worse than second place if you deploy all your votes for him, under cumulative voting.

The *board of directors* meets at regular intervals and at special meetings, as provided in the articles of incorporation. The board typically has power to enter into contracts, borrow money, select officers and determine their compensation, and declare corporate dividends. The directors cannot act as individuals to bind the corporation, since the board is expected to deliberate all matters as a body before reaching a decision.

The directors choose the *officers* of the corporation, including president, vice president, secretary, and treasurer, and sometimes general counsel, controller, and other officers. The officers are in charge of the day-to-day operation of the corporation. They are answerable to the directors but not directly to the shareholders. In large corporations, operational details will be delegated to other employees.

As might be expected, in a closely held corporation, which is what many engineering design firms are, the shareholders, directors, and officers are likely the same persons. Some states have enacted special close corporation statutes that permit the relaxation of some of the rules related to frequency of board meetings, required record keeping, and the like (Cheeseman, 1992).

In American law, there has been a marked trend toward requiring a fiduciary obligation of directors and officers. For example, it is not proper for a director or officer to use inside information to purchase shares of stock from an unknowing shareholder (Sweet, 1989).

11.3.4 Financing the Corporation

Corporations can raise funds to support operations by two means that are not available to sole proprietorships and partnerships, namely, equity financing and debt financing. When the corporation is first formed, the original incorporators purchase shares of stock, or *equity* securities, that provide the company's initial working capital. If more funds are needed, additional shares may be issued and sold, up to the limits imposed by the articles of incorporation. Furthermore, the articles may be amended with shareholder approval, to provide for the issuance of additional shares.

A share of stock issued by the corporation represents a proportionate share of interest, or equity, in the company. Shareholders do not have title to, or right to use, any specific corporation property. They are owners of the company, not its creditors. The extent of a shareholder's ownership interest is dependent on the number of shares he or she owns relative to the total number of shares that have been issued. Voting rights, as explained in Section 11.3.3, depend on the number of shares owned.

There are two classes of equity securities, common stock and preferred stock. If a corporation has only one class, it is *common stock*. Holders of common stock are permitted to vote at shareholder meetings and to receive dividends, based on the number of shares they own. These dividends may fluctuate from quarter to quarter, depending on the corporation's profits that quarter and the decision of the board of directors. Common shareholders are entitled to receive dividends only after preferred shareholders have been paid their dividends.

Preferred stock carries with it certain privileges. The articles of incorporation will typically provide that preferred shareholders will be paid dividends first, before common shareholders, and that they will be given liquidation preference, which means that if the corporation is dissolved, they will be entitled to some stated amount per share before common shareholders receive anything. Normally, preferred shareholders do not vote for corporate directors. Each issue of preferred stock comes with a stated dividend percentage. For example, a holder of 8.5% preferred stock with a value of $100 will receive $8.50 per share each year from the company's profits, before common shareholders receive any dividends. Some is designated *cumulative* preferred stock; if profits are not sufficient to pay preferred dividends in a certain year, the right to receive the fixed dividend amount remains active until a year when profits are sufficient (up to some set limit, such as 3 years). Noncumulative preferred stock carries no right of accumulation; in this case, the corporation does not have to pay any missed dividends.

Unlike equity securities, *debt securities* do not represent an ownership interest in the corporation. Instead, they establish a debtor–creditor relationship under which the investor loans funds to the corporation and the corporation issues the debt security, which is a promise to repay the funds with interest. The following types of debt security are used:

- *Debentures:* long-term unsecured debt instruments. That is, debentures may not be payable for 30 years or more, and they are backed up only by the firm's general credit standing rather than by any form of collateral such as real estate or equipment.
- *Bonds:* long-term instruments secured by collateral. If the corporation fails to repay the debt, the bondholders can foreclose by recovering the proceeds from the sale of the collateral.
- *Notes:* debt security with a maturity of 5 years or less. Notes can be secured or unsecured.

Both equity and debt securities of publicly held corporations are traded on the securities markets. In contrast, closely held corporations restrict the sale of their stock to certain persons, usually employees and their families. For a small, closely held engineering company that wishes to expand, "going public" represents a major and often traumatic decision.

Corporate finance is a topic to which millions of people around the world devote entire careers; we have dealt with it in less than two pages! The interested reader can find much additional information in such works as Altman (1986), Cheeseman (1992), Higgins (1989), and Morrow (1991). Lowenstein's *Sense and Nonsense in Corporate Finance* (1991) provides a mildly irreverent view of the topic.

11.3.5 Taxation: C Corporations and S Corporations

We have already learned that corporations are separate legal entities. As such, they are generally required to pay *corporate income taxes* to federal and state governments on any annual net income (generally, gross receipts minus expenses including depreciation). When a corporation distributes some or all of its aftertax profits to shareholders, the shareholders must pay *personal income taxes* on the dividends. This *double taxation* is a primary disadvantage of the corporate form of business organization.

Some closely held corporations are able to avoid this double taxation by filing to be taxed as *S corporations* (based on Subchapter S of the Internal Revenue Code). Under this status, shareholders pay personal income taxes on their proportionate shares of corporate income, regardless of whether the profits have been distributed as dividends, and the corporation pays no income taxes. Thus, the organization is treated like a partnership for income tax purposes. Incentives for filing for

S corporation status are twofold. First, if a profit is expected, and the personal income tax rate will be lower than the corporate income tax rate, there will be a savings. Second, if a loss is expected, those business losses can be offset against shareholders' other income.

Criteria for eligibility for S corporation status include

- Consent of all shareholders
- No more than 35 shareholders
- No nonresident alien shareholders
- Individuals, estates, or certain trusts as shareholders; not partnerships or corporations
- No more than one class of stock
- A domestic corporation
- Not a member of an affiliated group
- Passive investment income restricted to 20% of the total income

Once chosen, S corporation status can be rescinded by affirmative vote of the shareholders, but 5 years must pass before such status can be reinstituted. A *C corporation* is any other corporation; it is taxed as a corporation in the way we have described. Both C and S corporations enjoy the benefits of limited liability of shareholders described in Section 11.3.1 (Cheeseman, 1992).

11.3.6   Other Corporation Characteristics

Among the characteristics of corporations, as contrasted with sole proprietorships and partnerships, is the ease with which ownership interest can be transferred. Your 100 shares of Eastman Kodak can be easily sold with a single telephone call to your stock broker, and you can probably become a part owner of Duke Power with the same call. Also, corporations have the prospect of perpetual existence. The death of an owner does not end the company, as it does for a sole proprietorship and some partnerships. Instead, the owner's interest passes to the survivors as provided in the will or in statute, along with other assets in the estate, and the corporation continues to operate as before.

Furthermore, as alluded to throughout this discussion, an important characteristic of most corporations is the separation of ownership and control. Unless the corporation is closely held, shareholders have only indirect control over policy matters and very little participation in the day-to-day operations. In Table 11.3, we list several advantages and disadvantages of the corporate form, and in Table 11.4, we compare the primary characteristics of partnerships and corporations.

11.3.7   Professional Corporations

Until relatively recently, professionals were not permitted to form corporations, apparently on the grounds that the personal and confidential nature of the rela-

TABLE 11.3 Advantages and Disadvantages of a Corporation

Advantages of a Corporationzm
* *Limitations of the Stockholder's Liability to a Fixed Amount of Investment.*
* *Ownership is Readily Transferable.*
* *Separate Legal Existence.*
* *Perpetual Succession.* In the case of illness, death, or other cause for loss of an officer or owner, the corporation continues to exist and do business.
* *Relative Ease of Securing Capital in Large Amounts and from Many Investors.* Capital may be acquired through the issuance of various stocks and long-term bonds. Pledge of corporate assets as collateral makes long-term financing from lending institutions relatively easy.
* *Delegated Authority.* Centralized control is secured when owners delegate authority to hired managers, although owners and managers are often the same.
* *Ability of the Corporation to Draw on the Expertise and Skills of More than One Individual.*

Disadvantages of a Corporation
* *Activities Limited by the Charter and by Various Laws.* However, some states allow very broad charters.
* *Manipulation.* Minority stockholders are sometimes exploited.
* *Extensive Governmental Regulations and Required Local, State, and Federal Reports.*
* *Less Incentive If Manager Does Not Share in the Profits.*
* *Expense of Forming a Corporation.*
* *Double Tax.* Income tax is levied on corporate net income and on individual dividends. But S corporations allow corporate income to be declared as personal income for tax purposes.

SOURCE: Based in part on Olmi (1982).

tionship between professional and client "imposed a standard not consistent with the ordinary commercial relationship" (Bockrath, 1986). However, all U.S. states now permit the formation of *professional corporations,* although only certain professions are designated for eligibility in some states. It is common to see such corporate names as Jones Law Office, P.C.; Smith, Wiley, and Wiley, P.C., Certified Public Accountants; George A. Brown, MD, P.A.; and Professional Design Associates, S.C., where P.C. stands for professional corporation, P.A. is professional association and S.C. is service corporation. Such corporations are formed like ordinary corporations and have common corporate attributes.

The primary incentives for the formation of professional corporations are (1) personal limits on professional liability, (2) corporate income tax rates that may be lower than those for personal income, and (3) laws that allow fringe benefits to be paid to employees of corporations without including them as taxable income. The treatment of the liability matter varies among the states, with some requiring

TABLE 11.4 Comparison of Partnerships and Corporations

Aspect of Organization	Partnership	Corporation
Entity	Not a separate entity, although it has some characteristics of one, such as the ability to own property.	A legal entity or "person," taxable as such.
Creation	Created by formal or informal agreement of its members.	Created by the sate in which it is domiciled, upon application in required legal form.
Purposes and powers	Determined by members, with no state approval.	Limited to those stated in articles of incorporation and state corporation code.
Liability of owners	General partners' personal liability unlimited for partnership debts and obligations; limited partners' liability limited to their investment.	Shareholders' personal liability limited to their investment in shares.
Management	Management equally by general partners, who also act as agents of the partnership; no management by limited partners.	No direct management by shareholders, who elect directors, who, in turn, appoint corporate managers.
Duration	Terminable at will and upon death, incapacity, or withdrawal of a partner, unless otherwise provided.	Perpetual corporate life; shareholders may dissolve corporation.
Taxation	Not taxed as separate entities; profits and losses appear on partners' tax returns.	C corporations taxed as separate entities; shareholders taxed on dividends received. S corporations treated as partnerships for tax purposes.
Transferability of ownership	Consent required of all partners for transfer of ownership or admission of new partner.	Shares ordinarily transferable without consent of corporation or other shareholders.

SOURCE: Based on Henry R. Cheeseman, *Business Law: The Legal, Ethical and International Environment.* © 1992, page 780. Reprinted by permission of Prentice-Hall, Englewood Cliffs, New Jersey. And D.P. Lyden, J.D. Reitzel, and N.J. Roberts, *Business and the Law,* 1985, page 649. Reprinted by permission of McGraw-Hill.

that shareholders be personally liable for their own negligence and that of anyone they supervise. With respect to contract obligations, some states use the same rules that apply to partnerships, thus making shareholders' individual assets available to satisfy those obligations.

In a New York State case, the professional corporation of Cohen, Stracher & Bloom, P.C. entered into an agreement to lease space for their law office from We're Associates Company. When the landlord filed suit against the corporation and its three shareholders to recover $9000 that was allegedly due under the lease, the trial court dismissed the suit against the three individuals, saying that they were protected from liability because of the corporate status of their business. The New York Supreme Court, Appellate Division, affirmed the lower court's decision, citing the state statute that makes such shareholders responsible only for any "negligent or wrongful act or misconduct committed by him or by any person under his direct supervision and control." The court wrote [We're Associates Company v. Cohen, Stracher and Bloom, P.C. 478 N.Y.S. 2d 670 (1984), quoted in Cheeseman 1992]:

> It is well established that in the absence of some constitutional, statutory or charter provision, the shareholders of a corporation are not liable for its contractual obligations and that parties having business dealings with a corporation must look to the corporation itself and not the shareholders for payment of their claims.

11.3.8 Termination

Dissolution of a corporation occurs either voluntarily or involuntarily and either judicially or nonjudicially. A judicial dissolution, which involves court proceedings, can be instituted by shareholders, creditors or the state. A corporation might be dissolved nonjudicially because the term set forth in its charter expired, because of voluntary unanimous consent of the shareholders, through an affirmative vote of shareholders in response to a proposal by the board of directors, or because of an act by the state's legislature. Voluntary dissolution requires the filing of *articles of dissolution* and the issuance of a *certificate of dissolution*.

Once dissolved, the corporation ceases carrying on its business, but *termination* does not occur until the affairs have been wound up, the assets have been liquidated, and proceeds have been distributed to claimants. Claims are given priority in the following order: (1) expenses of liquidation and creditors, (2) preferred shareholders, and (3) common shareholders.

11.4 JOINT VENTURE

A *joint venture* is a voluntary association of two or more parties formed to conduct a single project with a limited duration. The parties, which may be individuals, partnerships, corporations, or other legal entities, are called *coventurers* or *joint adventurers*. The parties come together on a temporary basis to pool their talents and resources and thus make their services more attractive than they are separately.

As a typical example, consider two construction companies that form a joint venture to bid on a project too large or too complicated for either firm alone. Engineers sometimes join with other engineers or with architects for the same reason.

An agreement is made between the coventurers, and then, if they are successful in obtaining the sought-after project, the joint venture itself contracts with the client. Schoumacher (1986) provides a good discussion of the usual elements of a joint venture agreement, including scope, means of administration and decision-making, method of accounting, payment of expenses, distribution of profits, and mutual indemnity clauses.

The coventurers usually form a temporary partnership, which means that the principles of mutual agency, which we discussed in Section 11.2.1, apply here, and the acts of one of the parties can bind the other. (Choose your coventurer carefully!) In contracting with their client, the joint venture usually stipulates one of the coventurers as the *sponsoring coventurer*.

Although a joint venture partnership is generally governed by partnership law, some exceptions apply. They are summarized by Cheeseman (1992) as follows.

- Because of the limited nature of a joint venture, the parties have less implied and apparent authority than partners in a partnership.
- Although the death of a partner generally dissolves a partnership, the death of a joint adventurer does not usually terminate the joint venture.
- While a partner generally cannot sue copartners at law, joint adventurers are normally permitted to sue other joint adventurers at law.

To limit their liability, coventurers sometimes form corporations. Each party becomes a shareholder in the temporary corporation, and many of the regulations and procedures that apply to corporations in general must be followed by the joint venture.

11.5   OTHER ORGANIZATIONS

We note briefly five other organizational types that may be of interest to the engineering manager.

11.5.1   Not-for-Profit Corporation

As the name implies, these corporations are formed for purposes other than to earn a profit for their incorporators. They operate under most of the rules that apply to other corporations except that profits, if any, cannot be distributed to shareholders. Typical examples are hospitals, educational institutions, and other charitable institutions. Capital is raised from donations, grants, and, occasionally, sale of stock. If there are shareholders, they are exempt from personal liability for corporation debts. Not-for-profit corporations pay no corporate income taxes, but this tax-exempt status can be lost if the organization engages in certain political or profit-making activities not directly related to the scope of the charter.

11.5.2 Franchise

In a franchise arrangement, the owner of a trademark, product, patent, or trade secret (franchisor) licenses another party (the franchisee) to sell products or services under the franchisor's name. The franchisee pays a fee to the franchisor to cover such services as marketing and advertising. The two parties are separate legal entities. Fast-food restaurants, gasoline stations, and motels are often franchise operations. If your professional engineering firm contracts with McDonald's to design an addition and parking lot, it will be important for you to know with whom you are really contracting.

11.5.3 Limited Liability Company

When the liability of company owners, or "members," is limited, members are given "pass through" treatment for income tax purposes. Under a sort of hybrid structure that is neither wholly a corporation nor wholly a partnership, firm members are not personally liable for the company's general debts and obligations, but they maintain the control, flexibility, and tax advantages of a partnership. However, members are not shielded from their own professional liability, such as for damages caused by their professional errors and omissions ("Legal Corner," 1993).

11.5.4 Unincorporated Association

Examples of unincorporated associations are fraternal lodges, churches, and labor unions: groups of individuals who have banded together for a collective purpose without using any form of organization of the partnership or corporation type. These groups are considered not to be legal entities, which means that a member of the association who enters into a contract on behalf of the association will be personally liable on the contract. Also, a member whose actions are negligent while conducting association business will be personally liable. Despite lack of status as legal entities, however, such organizations are permitted by most states to hold property, make contracts, and sue or be sued in the name of the association. Design professionals who perform services for such associations will need to be very clear about the authority of the representatives with whom they contract.

11.5.5 Share-Office Arrangement

Often it is advantageous for sole proprietors to join together to share an office, equipment, and clerical assistance, without any sort of partnership or corporation arrangement. They may even do work for each other. The advantage should be quite apparent for two young registered engineers, for example, one working in hydrology and the other in bridge design (although complementary interests are certainly not a prerequisite). It is well to write an agreement about how contributions and expenses will be handled. If the participants work for each other, legal difficulties may arise if one performs improperly or if a client does not pay. Furthermore, if they create the impression to the outside world that they are a partnership, by the use of common stationery or a common listing in the telephone directory, they very likely will be treated as such.

11.6   DISCUSSION QUESTIONS

1. In the winding-up illustration in Section 11.2.1, suppose the cash assets of Ace, Base, and Case at termination are $160,000 instead of $70,000, and all other aspects of the problem remain unchanged. How much will be distributed to the creditors and to each partner?

2. Answer Question 1 if the cash assets at termination are $10,000 and all other aspects of the problem remain unchanged.

3. Maryanne Geneer, after receiving her P.E. registration, decides to open her own consulting office specializing in work for the forest products industry. She contributes $20,000 of her personal funds and obtains a $30,000 loan from a local bank to establish working capital for her sole proprietorship. An unexpected downturn in the forest products industry causes Maryanne to close her business after several months. At that time, the consultancy has $1500 in cash and no other assets; $30,000 is owed to the bank, as well as $7000 in rent and other unpaid bills. How will these debts be paid?

4. Sue, a relatively well-to-do registered mechanical engineer, and Fred, a very poor registered electrical engineer, share an office, a secretary, and a computer system. They maintain separate checking accounts, but the sign on their office door says "Sue and Fred, Engineers." Fred performs some design services that result in a negligence claim against him for $25,000. Since Fred is broke, the aggrieved party's lawyer includes Sue in the claim. Sometimes Sue had been present with Fred in meetings with this client. Will Sue be responsible for this judgment, if it is sustained?

5. Distinguish between partnering and partnership.

6. Three partners were in the business of manufacturing and selling timber trusses and laminated timber structural members. One of the partners made a contract with Far South Contractors to sell $60,000 worth of the firm's timber products for $48,000. What can the other partners do about this contract? Suppose the one partner makes an agreement to sell the business; can the other partners do anything about this arrangement?

7. Group ABC is a closely held corporation; there are 20,000 shares, owned by only five persons. At an election for seven directors, the owner of 6000 shares wants to be assured that her two favorite candidates will be elected. (a) Under straight voting, is she assured of success? (b) Under cumulative voting, is she assured of success?

8. At the time of dissolution, a corporation had 50,000 shares of preferred stock at $100 par value and 150,000 shares of common stock at $0 par value outstanding. The net assets, after satisfying all creditors, were worth $11,375,000. How much per share will each common stockholder receive as part of the termination process?

9. Suppose the net assets in Question 8 were $5 million. How much per share would the common stockholders receive?

10. Suppose the net assets in Question 8 were $2 million. How much per share would the common stockholders have to pay to the corporation?

11. According to its charter, a certain professional corporation had as its sole purpose "to design and manage the construction of residential and industrial piping systems." Now the corporation, without amending its charter, wants to expand its operations to include pressure vessels and machine parts. Can it do so?

12. A scheme was evolved under which the directors and certain high level employees could purchase stock in a corporation for $25 per share when the market value was $112 per share. The stockholders approved the plan because they had been convinced that it would benefit employees, and the wording of the motion made it unclear that directors were also eligible to purchase the bargain-priced stock. In fact, more than half the shares were purchased by directors. Rogers, who was a very minority stockholder, brought suit to require the directors to return the shares they received under this scheme. Was he successful? [Rogers v. Guaranty Trust Company of New York, 288 U.S. 123; 53 Sup. Ct. 295, 77 L. Ed. 652 (1933)]

13. Could a joint venture agreement be oral? If so, under what circumstances would it be enforceable?

11.7   REFERENCES

Altman, E. I. 1986. *Handbook of Corporate Finance.* New York: Wiley.

Blinn, K. W. 1989. *Legal and Ethical Concepts in Engineering.* Englewood Cliffs, NJ: Prentice-Hall.

Bockrath, J. T. 1986. *Dunham and Young's Contracts, Specifications and Law for Engineers.* New York: McGraw-Hill.

Cheeseman, H. R. 1992. *Business Law: The Legal, Ethical and International Environment.* Englewood Cliffs, NJ: Prentice-Hall.

Construction Division, U.S. Department of Interior Bureau of Reclamation. 1991. *Partnering: A New Concept for Attaining Construction Goals.* Final Draft, December 2.

Cook, E. L. and D. E. Hancher. 1990. "Partnering: Contracting for the Future." *Journal of Management in Engineering,* American Society of Civil Engineers, Vol. 6, No. 4, October, 431–446.

Firmage, D. A. 1980. *Modern Engineering Practice.* New York: Garland STPM Press.

Higgins, R. C. 1989. *Analysis for Financial Management.* Homewood, IL: Dow Jones-Irwin.

"Legal Corner." 1993. Engineering Times, Vol. 15, No. 6, June, 2.

Lowenstein, L. 1991. *Sense and Nonsense in Corporate Finance.* Reading, MA: Addison-Wesley.

Lyden, D.P., J.D. Reitzel, and N.J. Roberts. 1985. *Business and the Law.* New York: McGraw-Hill.

Morrow, V. 1991. *Handbook of Financial Analysis for Corporate Managers.* Englewood Cliffs, NJ: Prentice-Hall.

Olmi, A.M. 1982. *Selecting the Legal Structure for Your Firm.* U.S. Small Business Administration Management Aids Number 6.004. Washington, DC: SBA.

Partnering Agreement. 1993. U.S. Army Corps of Engineers and National Society of Professional Engineers. March 30.

Schoumacher, B. 1986. *Engineers and the Law: An Overview.* New York: Van Nostrand Reinhold.

Sweet, J. 1989. *Legal Aspects of Architecture, Engineering and the Construction Process,* 4th ed. St. Paul, MN: West Publishing.

───12
Professional Liability

In earlier chapters we have considered the liability of the engineer arising from contracts. A design engineer who breaches a contract, for example, may be found liable for the damages resulting from that breach. In this chapter, we introduce a different kind of liability, that associated with private injuries caused by the failure of the professional to perform in accordance with required standards of care.

To understand some ideas associated with this increasingly important topic, we must first introduce some general background on tort law and distinguish among intentional torts, negligence, and strict liability. We then cite several important trends in professional liability as related to engineering practice. This discussion will then lead to some suggestions for reducing professional liability exposure. Finally, we give some recent history on "tort reform," the umbrella name for the movement that attempts to control what many believe to be excesses in court decisions on liability cases.

12.1 INTRODUCTION TO TORT LAW

The strange term "tort" is based on the Latin word *tortum,* meaning twisted, as in twisting a body in a form of torture. (The English noun, as well, reflects the Latin root.) Thus, tortious conduct is twisted or crooked rather than straight, and it results in an injury that may be personal injury, property damage or, sometimes, economic (Sweet, 1989). A dictionary definition of a tort is "a wrongful act, damage or injury done willfully, negligently or in circumstances involving strict liability, but not involving breach of contract, for which a civil suit can be brought" (*American Heritage Dictionary,* 1982).

We distinguish torts, which are private injuries, from crimes, which are offenses against the public for which the state brings legal action through criminal prosecution. It is possible for a given action to constitute both a crime and a tort. For example, a driver who exceeds the speed limit and causes personal injury and property damage may be charged in a court with the crime of speeding and also may become the defendant in a negligence suit because of the injury and damage.

Torts against the *person* include libel, assault, and battery and can involve psychic or emotional interests, while those involving *property* include trespass and nuisance and can be both tangible and intangible (Blinn, 1989). *Economic*

349

losses are unconnected to personal harm or property damage; some courts have held that the design engineer who causes a contractor economic loss by failing to act promptly is liable in tort.

The balance of this section is devoted to discussing the three primary bases for tort liability: intentional torts, negligence, and strict liability. Table 12.1 pairs statements of the required duty and the degree of fault required to establish liability for each of these three types. Because intentional torts are seldom a part of engineering practice, our summary of this type is brief.

12.1.1   Intentional Torts

An *intentional tort,* sometimes referred to as *intentional interference,* is one in which the defendant possessed the intent to do the act that caused the plaintiff's injuries. They are of two main types, *wrongs against the person* and *wrongs against property.* The basis for intentional tort theory is that each member of society has a duty not to harm others' bodies, property, or character; the injured party can sue the perpetrator for any damages resulting from commission of the intentional tort.

Among those torts intended to *harm the person,* we list the following:

Assault: includes the threat by one person against another to cause immediate harm or offensive contact, or some action that causes reasonable apprehension of imminent harm.

Battery: unauthorized and harmful or offensive physical contact with another person. Physical contact does not have to be direct to qualify as battery: examples include rock throwing, poisoning a drink, and pulling a chair out from under someone, if injury results. Note that assault and battery often occur together, although they can happen separately.

False Imprisonment: the intentional confinement or restraint of another person, without justification and without that person's consent, through physical force, threats, or false assertion of legal authority. Some statutes provide that under certain conditions, shopkeepers who detain persons suspected of shoplifting are not committing false imprisonment.

Defamation of Character: making false statements about another person. Included are *slander,* which is an oral defamatory statement, and *libel,* which is a false statement that appears in written or other visual form. Usually the publication of opinions is not sufficient basis for action. The statement,

TABLE 12.1   Three Bases for Tort Liability

Theory of Fault	Duty	Degree of Fault for Liability
Intentional tort	Not to injure intentionally	Actions with intent to harm
Negligence	To use care	Failure to use care
Strict liability (liability without fault)	Not to injure	None

"That architect is a jerk," is an opinion, whereas "That architect undersized the columns," when she did not, is an untrue statement.

Infliction of Emotional Distress: "extreme and outrageous conduct [that] intentionally or recklessly causes severe emotional distress to another person" (Cheeseman, 1992). Note that the conduct must be extreme and the results must be severe. If such conduct is unintentional, the behavior is considered to be negligence and does not fall within this listing.

Invasion of Privacy: the violation of the right of a person to unwarranted and undesired publicity. Wiretapping and reading someone's mail without permission are examples. Unlike defamation, invasion of privacy involves distributing information that is or may be true. Public figures relinquish much of their rights to privacy.

Abuse of the Legal Process: behaviors including *malicious prosecution* (bringing a frivolous lawsuit) and *abuse of process* [using the legal system for an improper purpose (e.g., making unfounded complaints to authorities about a neighbor's behavior with the intent of pressuring the neighbor to move].

We list three types of tort that bring intentional harm to property, as follows:

Trespass to Real Property: the interference with an owner's right to exclusive possession of land or other real property. Trespass may occur even if there is no harm to the property and even if the owner is not using it.

Trespass to Personal Property: one person's action(s) that cause harm to another's personal property or temporarily interfere with that person's use of the property.

Conversion of Personal Property: similar to trespass but involves longer term use of another's property without the person's consent, by exercising ownership rights and thus depriving the true owner of its use and enjoyment.

12.1.2   Negligence

In contrast to an intentional tort, negligence could be called an *unintentional tort.* As Table 12.1 indicates, a person who has a duty to exercise some degree of care and unintentionally fails to do so, with resulting injury, can be found liable for negligence. Blinn (1989) says that negligence is "unintentional careless conduct representing a middle ground of blameworthiness, which translates legally into breach of duty by the defendant resulting in harm to another."

The law is quite settled as to the elements that the plaintiff must prove for the defendant to be found guilty of negligence. These four elements are as follows:

- A duty owed by the defendant to the plaintiff to conform to a certain standard of conduct that is supposed to protect the plaintiff against unreasonable risk of harm; a *"legal duty to care."*
- Failure by the defendant to exercise that required standard of conduct; a "breach of that legal duty."

- A close causal connection between the conduct of the defendant and the injury sustained by the plaintiff; the breach of duty must be the "proximate cause" (discussed below) of the plaintiff's harm.
- Actual damage must be suffered by the plaintiff; a "legally protected interest" must be invaded (some types of mental distress may not be considered valid).

In a construction-related case illustrating the first of these elements [Intamin, Inc. v. Figley-Wright Contractor, Inc. 608 F. Supp. 408 (1985)], design and construction defects were found in a roller coaster built for the Marriott Corporation. The design engineer settled with Marriott's construction manager and then filed a negligence suit against the contractor. The court dismissed the claim, because the engineer did not show that it was owed a duty to care by the contractor. Note that in this case, the "harm" sustained by the engineer was economic, not personal injury or property damage. Such harm would have been a valid basis for recovery if the duty to care had been established ("Design Error: Engineer v. Builder," 1985).

Generally, in a negligence suit, the plaintiff must prove that the defendant is responsible, using the four elements in the preceding bulleted list. However, sometimes it is impossible to show the defendant's specific acts or omissions even though it is likely the defendant is liable. In such a case, the doctrine of *res ipsa loquitur* may apply. Literally "the thing speaks for itself," this doctrine is often applied in such cases as tank explosions, airplane crashes, and falling elevators. This theory is appropriate when three conditions are present:

1. The accident is one that ordinarily does not occur in the absence of someone's negligence.
2. The thing causing the injury was within the exclusive control of the defendant.
3. The occurrence was not due to any voluntary action or contribution on the part of the plaintiff.

The result is not proof; rather, it is circumstantial evidence that tends to show that the occurrence was probably due to the defendant's negligence. This approach allows the plaintiff to get his or her claim to the jury even though direct evidence is not available, and the burden of proof shifts to the defendant to show absence of negligence. The Alaska Supreme Court allowed the *res ipsa* doctrine to be applied in the case of a water main break on a hillside in Anchorage. They decided that frost jacking of the pipe, which caused the break, can usually be avoided with proper safety precautions. Thus, the first element stated above was satisfied. The other two elements were satisfied as well, since the city's water utility had exclusive control of the system and the injured party did not contribute to the pipe breakage [State Farm Fire and Casualty Company v. Municipality of Anchorage 1990 788 P. 2d 726 (1990)].

Similar to *res ipsa loquitur* is violation of statutory duty. Examples include a law that requires protection of land adjacent to an excavation, a code requiring engineers to design all roofs for specified snow loads, and an ordinance stipulating that homeowners are responsible for upkeep of their sidewalks. The plaintiff need only show that the statute existed and was violated, that the plaintiff was injured as a result, and that the plaintiff is in a class of persons intended to be protected by the statute. This type of negligence is known as *negligence per se* (Streeter, 1988).

Usually it is clear what act caused injury to the plaintiff. Sometimes, however, intervening actions make it unclear who should be held responsible. The notion of *proximate cause* attempts to identify the negligent act that will be held responsible. Proximate means *near*, but in legal usage it means *responsible* or *substantial*. Many courts use the "natural and probable consequences" test, which defines proximate cause as any cause that "in natural and continuous sequence, unbroken by any efficient intervening cause, produces the injury complained of and without which the result would not have occurred." (Schoumacher, 1986) That is, no negligence would have resulted without the existence of the proximate cause. There must be a direct connection between cause and effect. If innocent acts intervene, the initial negligent act is responsible; if these intervening acts are wrongful, the last wrongful act is considered the proximate cause of the injury (Jessup and Jessup, 1963; Schoumacher, 1986; Vaughn, 1983).

An old English case [Scott v. Shepherd, 3 Wills. K.B. 403, 95 Eng. Rep. 1124 (1773)] is helpful in illustrating the concept of proximate cause. A lighted firecracker, or squib, was thrown into a crowd at a fair. Before the device exploded, it was tossed from one person to another in self-protection, until it finally burst and demolished Scott's eye. Who should be held responsible, the last person to toss the squib, or Shepherd, who initially threw it into the crowd? The court decided that the intervening acts were instinctive and innocent and that the initial throwing of the device was the proximate cause of the injury to Scott's eye.

Probably the most cited and controversial case regarding proximate cause is Palsgraf v. Long Island Railroad Co. [248 N.Y. 339, 162 N.E. 99 (1928)]. The plaintiff was standing on a railroad platform many feet from the eventual action. A train arrived at the station and two passengers ran to catch it. One got aboard safely but the other had to be assisted by a railroad employee. In the process, an innocent-looking package, which actually contained fireworks, was dislodged from the passenger's arms, fell to the rail, and exploded. Somehow, whether because of the force of the explosion or the stampede of the excited crowd, a large penny-in-the-slot weighing scale was overturned, striking and injuring the plaintiff. This was unfortunate for the plaintiff, but

> fortunate for legal scholars and generations of law students who dissected the subsequent appellate court decision. In a four-to-three opinion, the court held that the plaintiff was outside the zone of risk. As an unforeseeable plaintiff, the railroad company did not owe any duty to her despite its employees being careless toward someone else. (Sweet, 1989)

Contributory negligence is conduct on the part of the injured party that causes his injury. Under this theory, no recovery for the plaintiff is possible. Seemingly harsh outcomes from applying this theory have resulted; even slightly negligent plaintiffs may recover nothing to offset serious injuries received because of the defendant's actions. A construction injury case in which a worker foolishly walked underneath a heavy girder being lifted by a crane provides a harsh example. The sling broke, the girder fell, and the worker was killed instantly. Under the circumstances, the court concluded that the worker was guilty of contributory negligence, and his heirs were precluded from any recovery (Bockrath, 1986). Many courts have held that contributory negligence cannot be used as a defense if the defendant had the "last clear chance" to avoid harm and did not take it. If you spot someone lying in a roadway and can avoid hitting her with your car, you are obliged to do so; you have the last clear chance to prevent the accident even though she had contributed to the situation (by, e.g., becoming intoxicated).

Somewhat similar to the idea of contributory negligence is the *assumption of risk* defense, which states that people cannot voluntarily put themselves in positions of manifest danger and then expect to collect for injuries that result. If you have a heart attack after riding on a roller coaster, you probably cannot sue the operator successfully. But assumed risks are only those normally involved in the undertaking. If you are injured as a result of a roller coaster malfunction, recovery would likely be possible.

Modern courts have tended to set aside the "all or nothing" doctrine of contributory negligence in favor of *comparative negligence,* wherein recovery by a plaintiff who is partially responsible for his or her injuries is not barred but is reduced in proportion to the plaintiff's relative negligence. An almost unbelievable true case involved another construction crane. This time, the equipment was being used to lift roofing material to the top of a building. During a lull in the work, an employee grabbed the cable, began doing pull-ups and hung upside down on the cable. Unaware of the impromptu trapeze demonstration, the operator put the crane in motion and extended the boom to its full length, at which point the cable snapped and the ball fell on top of the subsequent plaintiff. Severely injured, the employee sued both the operator and the owner of the crane (who, incidentally, had lent it to the man's employer). The court found that the plaintiff was entitled to only partial damages because he was 60% responsible for his injuries. The owner of the crane had to pay the other 40% because he was held "vicariously liable" for the negligence of the crane operator [Mann v. Pensacola Concrete Construction Co. 527 So. 2d 279 (1988)].

Comparative negligence has been adopted in many states by legislation and in a few by court decision. In California, the adoption of comparative negligence did away with the notion of last clear chance and subsumed the assumption of risk doctrine by taking the approach that each case would be studied in total and the plaintiff's conduct would be treated as negligence in proportion to fault (Blinn, 1989). Some states have adopted a modified comparative negligence form under which the plaintiff cannot recover anything if his or her negligence is greater than or equal to that of the defendant (Schoumacher, 1986).

When an injurious act is done with reckless or wanton disregard for the consequences, it is no longer the common variety of negligence but is, instead, *gross negligence.* Chances for recovery by the victim are considerably improved if gross negligence can be shown. A defense of contributory negligence will not be allowed in such a case. In a case of extreme importance to the engineering profession, two engineers were found guilty of gross negligence for failing to review certain shop drawings, while one of them was also guilty of failing to review a certain design drawing prior to affixing his seal. This negligence and other misconduct led to the collapse of walkways at the Kansas City Hyatt Regency Hotel in 1981, killing 114 people and injuring many others. A commission of Missouri's Board for Architects, Professional Engineers and Land Surveyors defined gross negligence as "an act or course of conduct which demonstrates a conscious indifference to a professional duty." The Missouri Court of Appeals upheld this definition, stating that it "imposes discipline for more than mere inadvertence and requires a finding that the conduct is so egregious as to warrant an inference of a mental state unacceptable in a professional engineer" [Duncan v. Missouri Board for Architects, Professional Engineers and Land Surveyors, 744 S.W. 2d 524 (1988); Sweet, 1989].

If a visitor is injured while on your construction site or in your factory, to what degree are you responsible? The law establishes different degrees of responsibility for the owner or person in control of real property, depending on the status of the visitor. A *trespasser* is one who enters another's property neither with the occupier's consent nor as the bearer of legal privilege (e.g., a police official with a search warrant). The general rule is that the owner owes the trespasser no duty other than to refrain from inflicting intentional injury. You may not set traps or loaded guns that would inflict injury on trespassers. Beyond that, you have no duty to these unwanted visitors. An open question is whether it is proper for the contractor to keep savage dogs or barbed wire on a construction site to discourage trespassers. Many jurisdictions now recognize an *attractive nuisance* doctrine, under which the possessor of property that is appealing to children but also dangerous is under the duty to exercise reasonable care to protect juvenile trespassers.

A *licensee,* or "tolerated intruder," is an uninvited person who is permitted on the premises for that person's own benefit or convenience. In this case, the owner owes a duty to refrain from "active negligence" by exercising reasonable care to repair known defects and warn of dangerous conditions. Examples of licensees are social guests, door-to-door salespeople, and persons who take shortcuts across property with permission. Police or firefighters who enter a home are usually considered to be licensees, unless they have been called by the owner, in which case they have invitee status.

Invitees, or business visitors, are owed the greatest protection. Such persons have entered the premises for business of common interest or mutual advantage to them and the owner. A familiar example is the store customer, even one who buys nothing or has no intention of buying anything. The owner must protect the invitee not only against obvious dangers but also against those that could have been neutralized with reasonable care.

Often it is difficult to distinguish between a licensee and an invitee. Those who visit or perform services at your construction site may fall into either category, depending on the circumstance. In a case from 1951, an electrical subcontractor employee was given permission to use a scaffolding that had been installed by the steamfitting subcontractor primarily to do steamfitting work. The employee fell 15 feet and was seriously injured when a long supporting plank broke. Whether the steamfitting subcontractor was liable turned on whether the injured employee was a licensee or an invitee. The court wrote,

> It is well settled that an invitor owes the invitee the duty of furnishing him with reasonable safe premises or appliances and will be held liable for injuries caused by a negligent failure to perform the duty. It is equally well established that a licensor is not liable in damages for the injuries of a mere licensee, unless the injuries were the result of the licensor's active negligence. [Arthur v. Standard Engineering Co. 193 F. 2d 903 (D.C. 1951); quoted in Blinn, 1989]

In deciding in favor of the defendant steamfitting subcontractor, the court found that employee was a "mere licensee," since he was not using the scaffold for the mutual benefit of the two parties but only for his own company's benefit.

We should also note a recent trend away from the threefold classification described above. The test now being applied in some jurisdictions is whether the owner has acted as a "reasonable person" in view of the probability of injury to others without regard to their status as trespassers, licensees, or invitees. A California court said that such factors as "closeness of the connection between the injury and the defendant's conduct, the moral blame attached to the defendant's conduct, the policy of preventing future harm, and the prevalence and availability of insurance" can bear little relationship to the classifications cited above [Rowland v. Christian, 69 Cal. 2d 108, 443 P. 2d 561, 567–68, 70 Cal. Rptr. 97, 103–4 (1968); Blinn, 1989; Sweet, 1989]

12.1.3   Strict Liability

The third and final basis for tort liability is *strict liability,* or liability without fault. As indicated in Table 12.1, the duty here is not to injure; since negligence need not be shown, the party causing injury may be found liable even though he or she is not at fault. This extreme point on the tort liability spectrum has assumed increasing importance in engineering practice. We give an introduction to it here and then expand on certain points in Section 12.2.4. We describe three types of strict liability that may apply to engineering management: abnormally dangerous things and activities, employment accidents, and product liability.

Examples of *abnormally dangerous things and activities* are the keeping of animals (does this apply to the vicious dog kept at the construction site?), the practice of crop dusting near livestock, the storing of explosives in a populated area, and the execution of construction blasting activities. The idea is that despite all manner of reasonable caution and care exercised by the user or possessor, that person may be found liable under strict liability if an injury occurs. The New York State Court of Appeals dealt with such a case in Doundoulakis v. Town of Hempstead [42 N.Y. 2d 440, 368 N.E. 2d 24, 398 N.Y.S. 2d 401 (1977)], which involved a hydraulic landfill project that allegedly caused subsidence of a sand fill on which the plaintiff's homes were located. It was decided that although such a dredging operation is not "normal," there was insufficient evidence to hold that the activity was abnormally dangerous (Sweet, 1989).

Under workers' compensation laws, employers' liability for *employment accidents* is strict. Recovery against the employer does not require showing that the employer was negligent. The test is whether the injury was work related; if so, the employer pays even in the absence of no wrongful conduct. Employers can, and often are required to, insure their compensation liability burden.

Strict liability is having its greatest impact in the area of *product liability,* which has been greatly in the public eye as a result of large jury awards from cases involving injuries due to defective products. Under this theory, the injured party can recover from the manufacturer if he or she can prove that (1) the injury or damage resulted from a condition of the product, (2) the condition was an unreasonably dangerous one, and (3) the condition existed when the product left the control of the manufacturer (Bockrath, 1986). Once again, the idea is that recovery may be successfully sought even if the manufacturer exercised great care in the manufacturing process. Under strict liability, the duty is *not to injure.* For a time, one state sought, by statute, to make gun manufacturers liable for injuries caused by guns, even if the guns had performed properly.

For the moment, let us pose two questions that extend from the basic ideas about product liability. First, what about the seller of goods that injure and the middleman who sells to the seller? Are these parties liable under strict liability? (Answer: Maybe.) Second, how about the designers of water crafts, snow machines, bungee jumping apparatus, and so on? (Answer: No, not yet, at least.) We return to these ideas as related to both products and services after a review of some historical trends that lead to the current status of professional liability in engineering.

12.2   LIABILITY TRENDS OF IMPORTANCE TO THE PROFESSIONAL

12.2.1   Historical Overview

The idea of assessing blame for professional malpractice can be traced at least as far back as 1800 B.C., when Hammurabi was king of Babylon. Among other things, Hammurabi's famous code provided strict penalties for craftspersons who built faulty products. The injured party, however, received no compensation "other than the satisfaction derived from knowing of the severe punishment of the offender" (Thorpe and Middendorf, 1979). Early in this century, a stone pillar inscribed with the code's provisions was unearthed; the following is a small portion of a modern translation:

> If a builder has built a house for a man and has not made strong his work, and the house he built has fallen, and he has caused the death of the owner of the house, that builder shall be put to death. If he has caused the son of the owner to die, one shall put to death the son of the builder. (quoted in Streeter, 1988)

Down through the ages, the philosophy of *caveat emptor*—"let the buyer beware"—has persisted from Roman law. It was accepted by both buyer and seller as being a fair rule that prevented parties who sustained injuries from faulty products or services, either physically or economically, from recovering from the designer, producer, or seller. At the beginning of the twentieth century, with the development of more complicated products and greater production capacity, exceptions to the rule of caveat emptor began to be made. A quick review of some of those exceptions will be useful.

As we have learned, because of inconsistencies in rulings at all levels of the U.S. judiciary system, not all court decisions are representative of trends in case law. The selected cases listed next seem to reflect the evolution of the law related to professional liability over the years since the mid-nineteenth century (Blinn, 1989; Bockrath, 1986; Schoumacher, 1986; Sweet, 1989; Thorpe and Middendorf, 1979). Other cases are cited in the sections that follow this listing.

Winterbottom v. Wright [152 Eng. Rep. 402 (1842)]. In this English case, Wright had a contract with the postmaster general to maintain a mail coach "in a fit, safe and secure condition." Winterbottom was the driver of the coach and was injured as a result of Wright's negligent maintenance. The court dismissed Winterbottom's complaint because there was no contractual relationship between the parties to the dispute. The court said, "Unless we confine the operation of such contracts as this to the parties who entered into them, the most absurd and outrageous consequences, to which I see no limit, would ensue." This finding established the requirement for *privity,* or direct contractual relationship, in order for there to be liability.

Thomas v. Winchester [6 N.Y. 397 (1852)]. Thomas was injured when he took, as medicine, a small quantity of poison that was mislabeled as dandelion extract. Winchester was the wholesaler who had sold the wrongly labeled substance to the druggist from whom Thomas purchased it. Since the drug remained in its original condition all the way along the distribution chain from wholesaler to user, the court found Winchester liable. In a small way, the defense of privity of contract had thus begun to erode.

MacPherson v. Buick Motor Co. [217 N.Y. 382, 111 N.E. 1050 (1916)]. MacPherson was injured when a defective wheel fell off his car, which had been purchased from a dealer. The wheel, however, had been supplied to the manufacturer by an independent supplier. Buick claimed privity of contract, but the court, in a split decision, decided that the manufacturer's liability did extend to the user, even if the product was not "inherently dangerous." The court wrote, "If the nature of a thing is such that it is reasonably certain to place life and limb in peril when negligently made, it is a thing of danger; and if to the element of danger there is added knowledge that the thing will be used by persons other than the purchaser, then the manufacturer of the thing of danger is under a duty to make it carefully." Thus, the concept of third-party liability was established.

Wisdom v. Morris Hardware Co. [274 P. 1050 (1929)]. Wisdom purchased a fruit spray that caused injury to his crop. He sought damages from the manufacturer. The state of Washington court found that there was an agency relationship between the manufacturer and the retailer from whom Wisdom had bought the spray, and therefore the manufacturer, as principal, was responsible. Thus, the privity defense was denied.

Marsh Wood Products v. Babcock and Wilcox [240 N.W. 392 (Wisc. 1932)]. The plaintiff was injured by an exploding boiler that had been hydrostatically tested. The defendant attempted to show that by using this test, it had exercised reasonable care. However, the court found for the plaintiff on the grounds that a more sophisticated test was available and should have been used, even though this test represented a greater degree of care than required by government or industry guidelines.

Escola v. Coca Cola Bottling Company [24 Cal. 2d 497, 150 P. 2d 436 (Cal. 1944)]. Escola was a waitress in a restaurant. A bottle of Coca Cola exploded in her hand, causing severe injuries. In finding in favor of Escola, the court used the doctrine of *res ipsa loquitur,* stating that the evidence permitted a reasonable interference that the bottle "was not accessible to extraneous forces and that it was handled carefully by plaintiff or any third party who may have moved or touched it."

Biankanja v. Irving [320 P. 2d 16 (1958)]. Irving was a notary public who was negligent in failing to have a will properly witnessed when signed. As a result, Biankanja actually received only one-eighth of the amount provided for her in the faulty will. She sued Irving successfully, even though she was not his client and therefore was not in privity with him. This case has significance to engineers because it extended third-party liability to service contracts.

Pastorelli v. Associated Engineers, Inc. [176 F. Supp., 159 (D.C.R.I. 1959)]. In this Rhode Island federal court case, an architect was found liable for injuries to a racetrack clubhouse patron caused by a falling heating duct whose hangers had been designed improperly. Once again, the lack of privity was not a successful defense.

Henningson v. Bloomfield Motors [32 N.J. 358, 161 A. 2d 69 (1960)]. Mrs. Henningson received a new car from her husband as a gift. She was driving at 20 mph when the steering wheel broke, causing the car to crash into a wall. The damage was so severe that it was impossible to determine whether there had been negligence by the manufacturer. Thus despite an absence of privity of contract between Mrs. Henningson and the dealer, her suit was successful. The case is significant because it said that disclaimers regarding safety are invalid and that an implied warranty existed between the product user and the manufacturer. Again, the requirement of privity was struck down: "where the commodities sold are such that if defectively manufactured they will be dangerous to life or limb, then society's interests can only be protected by eliminating the requirement of privity between the maker and his dealers and the reasonably expected ultimate consumer."

Greenman v. Yuba Power Products, Inc. [59 Cal. 2d 57, 377 P. 2d 897, 27 Cal. Rptr. 697 (1963)]. Greenman was seriously injured when a piece of wood flew out of a combination lathe–saw–drill press and struck him on the forehead. His suit against the manufacturer charged both negligence and breach of warranty. The negligence portion was dismissed and the breach of warranty was sustained. Upon appeal, the California Supreme Court established clearly the strict-liability-in-tort theory when it wrote, "a manufacturer is strictly liable in tort when an article he places on the market, knowing that it is to be used without inspection for defects, proves to have a defect that causes injury to a human being.... [T]he purpose of such liability is to insure that the costs of injuries resulting from defective products are borne by the manufacturers that put such products on the market rather than by the injured persons who are powerless to protect themselves. Sales warranties serve this purpose fitfully at best."

Miller v. DeWitt [37 Ill. 2d 273, 226 N.E. 2d 630 (1967)]. Miller and two other construction workers were injured when temporary shoring collapsed as they were removing a portion of a gymnasium roof. They filed suit against the architect DeWitt, among others, on the basis that the "supervision of work" section of the contract between the architect and owner implied that the architect "had the right to interfere and even stop the work if the contractor began to shore in an unsafe and hazardous manner in violation of its contract with the owner." The architect did not design the temporary shoring, nor compute the loads that would be placed thereon, nor oversee and inspect the shoring; nevertheless, the Illinois appellate court found in favor of the injured workers. Note that the lack of privity of contract between the workers and the architect did not prevent their recovery.

Cooper v. Jevne [56 Cal. App. 3d 860, 128 Cal Rptr. 724 (1976)]. Architects who drew construction plans and specifications and acted as supervising architects were charged by third-party condominium purchases with professional malpractice. Architects answered that they were not liable for economic losses alone, relying on Seely v. White Motor Company [63 Cal. 2d 9, 403 P. 2d 145, 45 Cal. Rptr. 17 (1965)]. However, the court, in finding for the plaintiff, said that "unfortunately for the architects, the liability at issue in this case is the malpractice liability of a professional for negligence in the rendition of his services and not that of a manufacturer for defects in his product."

Caldwell v. Bechtel [631 F. 2d 989 (1980)]. Bechtel had a contract as a consultant on a subway construction project to provide safety engineering services, including overseeing enforcement of safety codes and inspecting job sites for violations. Caldwell contracted silicosis while working in the tunnel. The court found that Bechtel had a duty "to take steps reasonable under the circumstances to protect appellant from the foreseeable risk of harm posed by the unsafe level of silica dust." The court noted that Bechtel possessed special knowledge about safety engineering and thus had a special duty to care in the sense discussed in Section 12.1.2.

Ossining Union Free School District v. Anderson LaRocca Anderson [541 N.Y.S. 2d 335 (1989)]. In making its decision to close a school building and pay for space elsewhere, the school district relied on flawed reports by the defendants, who were structural consultants. Since the consultants had been hired not by the district but by its prime consultant, they defended the professional negligence suit against them by claiming lack of privity. Even through the losses to the school district were solely economic, the court found for the district, a ruling that is an exception to the general rule in New York State regarding economic losses to third parties.

12.2.2 Demise of the Privity Defense

After the overview in Section 12.2.1, we could almost close this chapter and go on to the next! But it is important to highlight the current status of some of the concepts that have emerged. That is, where are we *now*? It should be very clear that a lack of privity is no longer sufficient protection for a professional involved in a liability suit. Streeter (1988) defines privity as "mutual or successive relationship to the same rights of property" and privity of contract as "the relationship or connection between the contracting parties."

As the line of cases cited in Section 12.2.1 indicates, it is quite clear that third parties (those not "in privity") *do* have sufficient status to launch negligence claims against parties responsible for losses of some kinds. Manufacturers are now sued by people who have been injured as a result of defective product design or manufacture; plaintiffs' arguments are as old as MacPherson v. Buick. Architects and engineers, who typically contract with owners and have no contractual relationship with construction contractors or subcontractors, are often sued successfully by those contractors and subcontractors.

One aspect of the third-party liability question of great interest to engineering managers is confusing at best. The law applying to cases of economic loss is quite settled with respect to personal injury and property. Many jurisdictions, however, still uphold the principle that architects and engineers will not be liable for economic loss to parties with whom they have not contracted. The last several cases in Section 12.2.1 make this confusion evident.

Michael C. Loulakis writes on legal trends for *Civil Engineering* magazine. He has dealt with liability for economic loss to third parties in at least three columns between 1988 and 1991 (Loulakis 1988; 1989; Loulakis and Creger, 1991b). One report cites a Delaware decision that a contractor who lacked privity with a design engineer could assert negligence and negligent misrepresentation claims against the engineer for purely economic damages. Nine days after that decision, a Pennsylvania court, in a similar case involving a subcontractor and an architect/ engineer, reached the opposite conclusion. In 1991 Loulakis offered the following not very helpful status report:

Today it is difficult for A/E's to know exactly where they stand as to their potential exposure to suits by third parties for economic loss due to negligence or negligent misrepresentation. Courts in Maryland, Arizona, Florida and Tennessee have held that privity of contract is not required in order to sustain such a suit. Courts in Virginia, Illinois, Ohio and Rhode Island have held that privity of contract is required. And there are cases in some states where the courts have gone both ways. The design professional has reason to be confused.

12.2.3 Standards of Care

If a finding of negligence requires, in part, the showing that the defendant had a "legal duty to care," then it is essential that a required level of care be identified. Just how careful must a design professional be? Three points are important at the outset. First, infallibility is not required! Nobody is perfect. Second, persons with specialized knowledge are expected to practice a higher degree of care than those with less specialized knowledge. The designer of long span bridges will be held to a higher degree of care than the designer of concrete box culverts; the neurosurgeon is expected to have knowledge and skills not required of the physician in general practice. And third, standards of care evolve over time.

For example, during the early development of the airplane, there were many crashes but few successful negligence suits; in those years, there was little understanding of fatigue failure of metals, and thus the standard of care that would have been required to prevent those failures was beyond the reach of even the best airplane designers. Another vivid example is found in the failure of the Tacoma Narrows Bridge in 1940; the engineering profession at that time had almost no understanding that harmonic motion would be induced in a bridge of such dimensions and characteristics, leading to wild vibrations that destroyed the bridge in Washington. No negligence actions were filed in that case (Firmage, 1980).

The five standards of care described below have evolved over the years. To know which might be applied in a particular case is difficult, but some guidelines can be offered (Peck and Hoch, 1988; Sweet, 1989).

- *The Professional Practice Standard.* Under this standard, the engineer is expected to exercise skill and knowledge normally possessed by members of that profession in good standing in similar settings. As a starting point, the design engineer must abide by applicable building codes. But this approach may set a relatively low standard, since codes are often several years behind the publication of new knowledge. The Caldwell v. Bechtel case in Section 12.2.1 illustrates this standard of care; the consultant was held to a standard of care consistent with its special expertise as a safety engineer. A decision on another case [Forte Brothers v. National Amusements, Inc. 525 A. 2d 1301 (1987); "A/E Duty to Contractor Upheld," 1987] stated that "a supervising A/E must exercise the ability, skill and care customarily expected of A/E's in similar circumstances" in ruling that the architect/engineer had a duty to project contractors with whom they had an economic relationship and community of interest.

- *The Duty to Stay Informed.* This approach places a higher burden on the engineer than simply meeting a building code or following the current "normal" practice. It suggests a requirement to keep abreast of new developments by attending seminars, reading professional journals and the like, and applying that current knowledge to the work at hand. The term "state of the art" may apply here in referring to "scientific knowledge at the time of manufacture or construction that can be incorporated into the product in an economically and practically feasible manner" (Peck and Hoch, 1988). Blinn (1989) suggests that this is a "technology forcing" criterion similar to what is sometimes called "best available technology," or BAT.

- *Implied Warranty: The Outcome Standard.* With this approach, sometimes applied to cases in which the defendant both designed and built the final product, the court will inquire whether the product developed from the design meets the communicated or understood expectations of the client. Note the distinction from the first two entries in this list, where the comparison studies the process used rather than the product developed. In a 1960 Pennsylvania case involving an allegedly poorly designed built-up roof, the court stated, "While an architect is not an absolute insurer of perfect plans, he is called upon to prepare plans and specifications which will give the structure so designed reasonable fitness for its intended purpose, and he impliedly warrants their sufficiency for that purpose" [Bloomsburg Mills, Inc. v. Sordoni Construction Co., 401 Pa. 358, 164 A. 2d 201 (1960); Sweet, 1989].

- *Strict Liability.* As we have already noted, this standard imposes a duty not to injure, without regard to fault. In theory, even if the latest and best approaches to design were used (which is to say that the "duty to stay informed" had been met), a structure that failed because of a design error could result in a finding of liability against the designer. To date, strict liability has been applied only against "sellers" of "products." Courts have not applied it to structural engineers unless these professionals have participated in the assembly or manufacture of a structure or building components. Analogous applications have been made in other areas of the profession, but, for the most part, strict liability has not been applied to engineering work.

- *Regulatory Standard.* This is a lower standard that has been used on a trial basis, for example, by the U.S. Environmental Protection Agency as a way of encouraging innovation. In return for trying new methods, engineers are given contractual immunity from liability for innovative designs in the absence of gross negligence. For projects involving large safety risks to the public, the regulatory standard is unlikely to be used.

Recent case law seems to place the usual required standard of care at the professional practice standard or the duty to stay informed, or somewhere in between. A few courts have used implied warranty, but strict liability has yet to be applied to engineering design. The regulatory standard seems limited to relatively few practical applications.

12.2.4 Product Liability

In 1976 a Pennsylvania farm manager named John Newlin ordered an International Harvester Front End Skid Loader. The machine was equipped with a roll bar, but Newlin had it removed because his barn door was too low to accommodate this safety feature. For several months, Jim Hammond, a farm employee, operated the machine; then one day, in a freak accident, the machine overturned and Hammond was killed. Hammond's widow sued International Harvester and won, on the grounds that the loader was defective by not having a roll bar, even though the roll bar had been removed at the request of the purchaser (Neely, 1991). This is an example, albeit rather extreme, of a product liability case. One is inclined to speculate as to who "really" was at fault. Perhaps Farmer Newlin had some responsibility because he allowed his employee to use equipment that had been modified in a way that made it unsafe. Perhaps the local dealer could be criticized for not clearing the modification with the manufacturer before removing the roll bar. Among other things, the case illustrates the tendency of plaintiffs to "go for the deep pockets" by suing those parties whose financial resources make them most likely to be able to pay large damage awards even though their connection to the case is tenuous at best.

We have already presented many ideas and examples with respect to liability for injuries caused by defective products. This section summarizes some of those topics. At the outset, we note that the focus is on manufacturers, but it also includes the sellers and designers of defective products (Kemper, 1990). We consider three possible bases for product liability action: negligence, breach of warranty, and strict liability (Brown, 1991).

In a *negligence* action, the plaintiff must prove negligence, based on the four elements described in Section 12.1.2. In such an action, the injured party must show that the manufacturer made the product and did so without meeting the required standard of care. A mere defect is not sufficient to prove the manufacturer's negligence; the plaintiff must prove that negligence resulted in the defect. Historically, negligence actions related to product liability have involved the inspection and testing of the product and its components, the product's design, or the manufacturer's failure to warn potential users of a possible hazard (Schoumacher, 1986).

Warranties are of two types, express and implied. *Express warranties* are specific representations about the characteristics of the product. An actual statement is required, although it need not be in writing and need not include the word "warranty" or "guarantee." To prove that a defendant is liable for breach of express warranty, the plaintiff purchaser must show that he or she relied on the promise and that the promise was not fulfilled. *Implied warranties* are inferred by law. They are not based on writing or other forms of representation. A product is expected to have an implied warranty of merchantability, which guarantees it for ordinary uses for which such goods are normally used, and, if the seller knows how the product will be used, an implied warranty of fitness, which guarantees it for that specific use. The Henningson v. Bloomfield Motors case in Section 12.2.1 turned, in part, on a finding of implied warranty between the purchaser and the manufacturer. In some sense a warranty is a contract, under which the seller promises a certain condition of the product; breach of warranty is therefore related to breach of contract (Settle and Spigelmyer, 1984; Streeter, 1988).

The rationale for *strict liability* in product injury cases has been stated as follows:

1. People injured by an unreasonably dangerous product should be compensated in today's society of mass production and automation, despite the fact that manufacturers have used due care;
2. The cost of accidents should be spread among society and to all users of the products; and
3. Strict liability will serve as impetus to business to produce safer products (Settle and Spigelmyer, 1984).

In Greenman v. Yuba Power Products, the landmark 1963 decision cited in Section 12.2.1, the court found that a manufacturer is strictly liable if its product "proves to have a defect that causes injury to a human being." The American Law Institute provides the following rule regarding strict tort liability:

One who sells any product in a defective condition unreasonably dangerous to the user or consumer or to his property is subject to liability for physical harm thereby caused to the ultimate user or consumer, or to his property, if
a) the seller is engaged in the business of selling such a product, and
b) it is expected to and does reach the user or consumer without substantial change in the condition in which it is sold. (Kemper, 1990)

The foregoing rule applies even if the seller has provided all possible care in the manufacture and sale of a product and even if there is no privity between seller and injured party. Note that the word "seller" is used; manufacturers are necessarily sellers, and their design engineers probably are also. Note also that words like "defective" and "unreasonably dangerous" are (necessarily) vague enough to provide interesting tasks for juries for a long time to come.

To recover in a strict liability case, the plaintiff must prove five things, as follows (Schoumacher, 1986; Smith, 1981):

1. The defendant had a relationship to the product by being the manufacturer or a member of the distribution chain.
2. The product was defective and unreasonably dangerous.
3. The defective condition existed at the time it left the defendant's control.
4. The plaintiff sustained injury.
5. The defect was the proximate cause of the injury.

In such cases, the defect may be in the manufacturing process (an improper weld), in the design (failure to provide proper safety guards), or in the lack of adequate warning about the product's potential danger. Personal injury and property damages can be the basis for recovery, but some courts have not allowed recovery for purely economic loss in strict liability cases.

To summarize, if the designer and the manufacturer of a defective product are charged with negligence, the court will determine whether they acted reasonably in the design and manufacturing process. If the charge is breach of warranty, the court will attempt to ascertain what promise was given or implied to the purchaser. And if strict liability is the basis of the suit, the court will look at the product itself to determine whether it is defective.

12.2.5   Services Liability

Although it is not always possible to separate the responsibility for the design of a product from that for its manufacture, engineers' activities are often confined to providing *services* rather than producing products. These services range from planning and design to inspection of the fabrication or construction of a design. In this section we summarize the engineer's liability for such services, where the "product" may be a set of design drawings or a report. Some of these concepts have been mentioned, and, indeed, much of the material related to product liability applies here as well. But it is important to clarify similarities and differences between product and services liability.

The simplest part has to do with negligence. A number of the cases cited make it clear that if negligence in the preparation of plans and specifications can be proved, the designer will be found liable, provided the defect is not patent or discoverable (Gamble, 1987). (We also admitted that the question of economic loss, rather than physical injury or property damage, is still unsettled.) Liability as a result of design is usually based on one or more of the following (Thorpe and Middendorf, 1979):

1. The design contains a concealed danger.
2. The design omits needed safety devices.
3. The design included materials of insufficient strength, size, or other characteristics or failed to comply with accepted standards.
4. The designer failed to consider possibly unsafe conditions caused by reasonably foreseeable product use or misuse.

With some exceptions, attempts to hold those who perform services, such as engineers and architects, strictly liable or liable based on breach of warranty have not been successful. More than 20 years ago, a California court said, "Those who hire such services are not justified in expecting infallibility, but can expect only reasonable care and confidence. They purchase service, not insurance" (Bockrath, 1986). More recently, a Minnesota court wrote, "we decline to extend the implied warranty/strict liability doctrine to cover vendors of professional services" [City of Mounds View v. Walijarvi, 263 N.W. 2d 420 (1978)].

A few paragraphs from a New Mexico decision will be helpful. The defendant architects had designed renovations to a penitentiary that included a bay window with large sheets of bullet-proof glass. During a riot, prisoners gained access to the central control area by breaking the glass; death and injury to prisoners, injury to officers, and extensive damage to property resulted. The state filed suit against the architect, using as one of its arguments breach of implied warranty of "sufficiency of the recommendations, designs, plans and specifications to provide a control center adequate to serve as a central stronghold in the event of an inmate uprising." In dismissing that part of the suit (and even ordering the state to pay costs of the appeal!), the court wrote, in part:

> In the present case, the state parties wish to utilize an implied warranty of the sufficiency of plans and specifications against the architects. This would not entail a showing that the architects were negligent in performing their work, but only that the design or specifications for the central control area were not fit for the intended purpose. . . .
>
> The minority of cases make the view that the architect or engineer impliedly warrants the fitness of his design or plans for their intended purposes. . . .
>
> However, the majority view is that an implied warranty of sufficiency of plans and specifications will not be imposed against services performed by "professionals," including architects and engineers. In refusing to recognize the implied warranty, some courts have emphasized their reluctance to apply the concept of implied warranties to contracts for services generally, while other courts specify that contracts only for "professional" services do not carry the implied warranty. . . .
>
> Traditionally, architects, along with other professionals such as doctors and lawyers, do not promise a certain result. The professional is usually employed to exercise the customary or reasonable skills of his profession for a particular job. He "warranties" his work only to the extent that he will use the skill customarily demanded of his profession.
>
> . . . the Florida court of Appeals . . . stated:
>
>> With respect to the alleged "implied warranty of fitness," we see no reason for application of this theory in circumstances involving professional liability. . . . An engineer, or any other so-called professional, does not "warrant" his service or the tangible evidence of his skill to be "merchantable" or "fit for an intended use." . . . Rather, in the preparation of design and specifications as the basis for construction, the engineer or architect "warrants" that he will or has exercised his skill according to a certain standard of care, that he acted reasonably and without neglect. Breach of this "warranty" occurs if he was negligent. . . .
>
> We do not extend the concept of implied warranty, which was developed in reference to the sale of goods, to professional services [State v. Gathman-Matotan, 98 N.M. 740, 653 P. 2d 166 (1982); quoted in Streeter, 1988]

We should point out the minority view, however. That position was stated in Section 12.2.3 in connection with Bloomsburg Mills v. Sordoni, where the Pennsylvania court held that the architect impliedly warrants the sufficiency of the plans and specifications for their intended purpose.

So, the general rule is that professionals who provide services are not subject to strict liability or breach of warranty claims. Exceptions have been made in the case of design–build projects, where the end result is more a "product" than a "service" (Streeter, 1988). Also those who mass-produce homes or building lots have been held strictly liable in some states (Sweet, 1989).

12.2.6   Statutes of Limitation and Statutes of Repose

In 1988, a bridge over Connecticut's Mianus River collapsed. Although the design for that project dated from more than 25 years before the collapse, the design engineer was named as a defendant in a professional liability suit. After extended legal proceedings, however, he was relieved of any liability, based on statutes that limit the period within which such suits may be brought (Van Hemert, 1991).

Statutes of limitations and repose have been enacted for very practical reasons. A long period of time after project completion or the discovery of a defect, the evidence can be both stale and difficult to acquire. Furthermore, there is a belief that possible defendants are entitled to know that eventually they need not be concerned over a claim (Firmage, 1980). Reasonable time limits strike a balance between the interests of potential plaintiffs and those of potential defendants.

If the statute provides that plaintiffs cannot bring a liability action after, say, 5 years "from the time the cause of action accrued," it is unclear whether the action accrued (1) when the wrongful act took place (negligent design or construction), (2) when the injured occurred, or (3) when the plaintiff discovered the defect. For this reason two types of statutes have been adopted.

A *statute of limitations* limits the period of time within which the plaintiff must file an action after the time of injury or damage was first discovered or should have been discovered. A *statute of repose,* on the other hand, stipulates a time limit following substantial project completion, after which claims will be dismissed. Statutes of limitations are typically in the range of 2–3 years, while statutes of repose vary more widely but average between 6 and 8 years (Lunch, 1990; Van Hemert, 1991).

In the area of product liability, the Georgia statute provides that no product liability suit may be filed more than 10 years after the first sale or use of the product. In Oregon, strict liability suits must be filed within 2 years after the date of injury but no more than 8 years following the first sale of the product (Schoumacher, 1986).

Statutes of repose have been challenged in many states, with the result that about one-quarter of those tested in court have been declared unconstitutional. The basis seems to be the U.S. Constitution's equal protection clause, since to exempt some but not all parties from a liability finding through the common law rule of joint and several liability would result in failure to treat all parties equally.

Time limits in the statutes may differ depending on the type of injury and whether it is personal injury or property damage (or economic loss, if that is allowed). Special time limits are placed on asbestos claims, for example (Schoumacher, 1986).

12.2.7 Review of Shop Drawings and Other Submittals

The construction contractor is usually required to submit information to the design professional with certain details about how the contractor intends to do the work, as noted in Section 10.4.5. Submittals consist of shop drawings, brochures, material samples, and the like. As pointed out in Section 10.4.5, the law is quite settled that review of this information by the design professional does not constitute approval, and thus if the review is conducted according to the usual standards of care, the professional should not be subject to a liability claim.

In Waggoner v. W&W Steel Co. [657 P. 2d 147 (1982)], two workers were killed and another was injured when a steel frame collapsed during construction. It was determined that if temporary connections had been provided on some expansion joints, the accident most likely would not have occurred. The plaintiffs contended that the architects' approval of the shop drawings that did not specify such temporary connections constituted negligence. The court said that the central question was the following. "Is the architect who designed this building responsible for ensuring that the contractor employ safe methods and procedures in performing the work?" The court answered, "No," referring to a provision of the construction contract stating that the shop drawings were to be submitted to the architect for approval "only for conformance with the design concept of the project and with the information given in the Contract Documents." It was the duty of the contractor, not the architect, to see that the shop drawings included provisions for the temporary connections (Sweet, 1989).

But the design professional is not always immune in matters involving shop drawings. We mentioned in Section 12.1.2 the tragic collapse of walkways at the Hyatt Regency Hotel in Kansas City. Two engineers were found guilty of gross negligence because, in their shop drawing review, they failed to notice that six key connections had been inadvertently redesigned by the shop drawing drafter. Not until 114 persons had been killed and 185 injured in the collapse did the error become noticed; the result of the redesign had been to double the load on each connection. Figure 12.1 is a reproduction of the original detail and the connection as it was shown on the "approved" shop drawing and subsequently installed.

Shop drawings and other submittals do require care, as do contract provisions stipulating how they will be handled. One writer (Boehmig, 1990) has said, "Shop drawings are frequently ignored—until they land in court."

12.2.8 Job Site Safety

One particularly sticky issue for design professionals involved in construction projects is their potential exposure to responsibility for injuries to workers on those projects. Like the shop drawing issue, it seems that the matter turns on the extent to which the design professional is responsible for managing the construction operations. The general principle is that "a design professional firm performing routine and customary professional services at a construction site is not engaged in 'construction work'" (Schwartz 1993b) and is thus not responsible for job site safety.

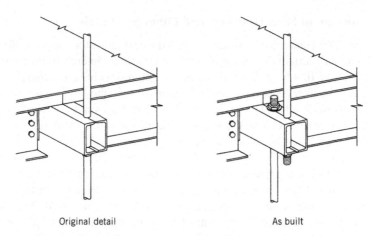

Original detail   As built

Figure 12.1. Connection details: the Hyatt Regency disaster. (Reprinted by permission from page 170 of *Legal Aspects of Architecture, Engineering and the Construction Process,* by J. Sweet; © 1989 by West Publishing Company. All rights reserved.)

According to Reynolds (1983), responsibility for worker safety on the construction site on the part of the design professional can be created in three ways:

1. *Out of Contract.* The contract between the architect/engineer and the owner may establish that the A/E is explicitly responsible for a safe work site. (Why the A/E would sign such an agreement is unclear!)
2. *Common Law Negligence* (the real topic of this section). It might be decided that the "legal duty to care" that is imposed on a construction site inspector includes this responsibility or that the architect/engineer, as an agent of the owner, has such a responsibility because the owner does.
3. *Violation of Statutes or Regulations.* Some jurisdictions have laws that at least imply responsibility by the design professional for job site safety, or they make the owner responsible, which could involve the design professional under the agency relationship.

In 1987 legislation was introduced to Congress that would have required the involvement of a state-registered professional engineer or architect in the oversight of construction projects by assigning to the A/E the job of issuing a permit before the start of specified construction activities (Dinges, 1988). As might be expected, a lively debate ensued in which is was shown that such an approach was unrealistic and unworkable, since most design engineers are neither construction experts nor safety experts. They would be less than enthusiastic about acquiring additional liability exposure. The suggested legislation was abandoned in favor of a requirement that the contractor employ a construction safety specialist to be in charge of safety precautions and procedures (Lunch, 1990).

The confused nature of this issue is illustrated in two 1990s cases. In the first, a Florida court ruled that the construction manager was liable for injuries to two employees of a steel erection subcontractor who were knocked from a work platform when a bundle of steel grating shifted. Although the construction manager is not the same as the designer, the facts are of interest. The construction manager was engaged to oversee construction of the project. Among other duties, the CM was to develop a safety manual, review contractors' safety programs, provide daily surveillance for compliance with the safety program, develop methods for notifying contractors of violations, and assure that such violations were taken care of.

The Florida contract also provided that "the construction manager shall have no responsibility or right to exercise any actual or direct control over employees of the contractors.... The furnishing of such services shall not make the construction manager responsible for construction means, methods techniques, work sequences or procedures." The construction manager notified the general contractor that the steel erection subcontractor's employees were using neither safety nets nor tied-off safety belts, at least one of which was required by the safety rules; this warning was repeated 2 months later. Eight months after the first notice, the two workers fell, one landing on a beam after falling 18 feet and the other severing an arm after a fall of 37 feet. The court ruling made the CM responsible [Wenzel v. Boyles Galvanizing Co. 920 F. 2d 778 (1991); Loulakis and Cregger, 1991a].

The second case took place in Indiana, where a suit against a design engineer was ruled in his favor. Two employees were injured when the boom of a crane struck a 34,500-volt overhead power line. In suing the engineer, among others, the injured employees alleged that he was the "prime contractor" and thus had responsibility for the entire project. The court said, "The two employees' argument fails to recognize the distinction between 'the obligation to assure that the construction conforms to the authorized plans' and 'the obligation to assume responsibility for the safety of persons lawfully on the construction site'" (Schwartz, 1992).

Table 12.2 relates the experience of engineer of record for the 16-story L'Ambiance Plaza project in Connecticut, which collapsed while under construction in 1987.

12.2.9 Computer-Aided Design

Increasingly, engineering designers are relying on computer software developed by someone else. If that software has a bug, who should be responsible? Regardless of whether it seems fair, generally the software developer is responsible only up to the cost of the single copy of the software that you bought and used. Most mass-produced software is considered to be a good and is therefore subject to the rule that remedies for defects that cause economic loss are limited to the contract, which, in the case of software, is probably printed on the diskette envelope. Look closely and you will probably find a *limit of responsibility equal to the software price*. It may be possible to recover under an implied warranty theory, but the prevailing view gives that approach little chance.

If a faulty design resulted from defective software, courts seem to place the burden on the engineer to show that the software developer, not the user, was

TABLE 12.2 "Every Engineer's Nightmare," as described by the Engineer of Record for the L'Ambiance Plaza

The sixteen story L'Ambiance Plaza residential complex in Bridgeport, Connecticut, was under construction in 1987 when it collapsed, killing twenty-eight workers. A National Bureau of Standards report identified the lift-slab method used by a subcontractor to position concrete floor slabs as the cause of the collapse. James O'Kon of O'Kon and Company, Atlanta Georgia, was the structural engineer of record for the project. Excerpts of the first public account of his experience and his survival strategy follow:

I was present at the site for three days, observing the carnage, in constant admiration for the courage . . . of the rescue workers searching for signs of life. However, I concentrated on . . . taking photos, making notes on the collapse mode and assessing minute details.

Overall, I kept a low profile. I could not imagine what would happen if I met a family member of one of the dead or missing workers. I felt they would assume the death and destruction were my doing and act accordingly.

I relied on my background as a forensic engineer and my knowledge of the techniques that would be used by other investigators. I knew which areas might be vulnerable to attack. My strategy was to deflect those attacks and be proactive.

My project manager for L'Ambiance Plaza and I began a thorough post-mortem of design, calculations, drawings, shop drawings, correspondence and testing reports, searching for potential weaknesses in the design and construction process. Our hypothesis of a construction failure was divulged only to our attorney and our records were boxed, sealed and shipped to him.

Survival tactic no. 1 — place your records in a protected place, safe from review by investigators or the media.

Next, I personally contacted all our clients, told them of our involvement in L'Ambiance Plaza, and our conclusion that the cause was a construction error. I told each of them that I would personally keep them abreast of the investigation and do everything in my power to avoid embarrassing them with this incident. When those clients were approached by others, they told them that they knew all about the failure and were sure I would be found innocent. This was survival tactic No. 2.

We were concerned that if we gave interviews, our comments would be taken out of context, and we would be placed in a position that could jeopardize our defense and our reputation. So survival tactic No. 3 was not to talk to the press during the investigation.

Connecticut's attorney general undertook a criminal investigation of the failure. When we were subpoenaed to appear in Hartford for questioning by the state police, I must tell you I was scared. No one was indicted on criminal charges in this case, but the possibility caused many sleepless nights. Survival tactic No 4: If you find yourself in a similar position, try to arrange for the questioning to be on your turf, tell the truth, but don't volunteer anything.

Tactic No. 5 — don't let your license lapse. Mine expired during the investigation. When I went to pay the fees, the clerk refused to accept and said I had to apply to the state board for reinstatement. This was contrary to state law which . . . provided for reinstatement up to five years with late fees.

When the National Bureau of Standards report was released, we were elated. The report said construction error was the failure cause and OSHA levied large penalties on the contractors. O'Kon was not charged with errors, criticized by NBS, or penalized by OSHA.

TABLE 12.2   continued

A two-judge panel was appointed to develop a settlement, and announced a goal of $40 million. I was put in the group with the contractors found guilty. I realized later that this was because I had $1 million in Errors & Omissions insurance. The judge assigned to me asked for an additional $1 million from me personally.

I was thunderstruck, and said, "We were found innocent. How can you penalize me?" The judge held up a piece of paper—"You checked the shop drawings, your staff visited the site. I could go on, but we have enough basis to bring you deeply into litigation. If you don't agree to our recommended contribution, you can be cast outside the protected group who will contribute. You can be sued by dozens of people, brought into hundreds of depositions, spend hundreds of days in court.... You will be unable to carry on your business. The choice is yours."

I told the judge, "I have been found innocent by my peers. This disaster was caused by construction workers whose families will become wealthy as a result of this disaster. This year I have experienced a personal disaster. My daughter is now paralyzed by a horseback riding accident. All my funds have been and will continue to be spent on her medical care and rehabilitation."

The judge backed off. However, he made history as he raised a $41 million settlement fund through this type of mediation. The hot dog man who provided coffee and sandwiches to the site had to pay $75,000. The dry wall installers paid $150,000, and they would not have started their work for six months.

The supervising architect paid only $7,500, because he did not have insurance. The judge felt sorry for him.

No suits or claims were filed against my firm. The contractors did not survive; the developer did not survive. O'Kon and Company survived. I was declared guilty by the press, investigated for murder, almost lost my professional license, and threatened with business failure by a federal judge.

My own instincts and my attorneys' advice led to survival. I have a larger client base now, and we continue to work on state-of-the-art buildings. But the collapse has left its mark on the profession and on me. My pain will not be forgotten.

SOURCE: "L'Ambiance Plaza: What Have We Learned?" V. Fairweather, *Civil Engineering Magazine,* Vol. 62, No. 2, © 1992. Reprinted by permission of American Society of Civil Engineers.

responsible. A contractor underbid a project by $245,000 and sued Lotus Development Corporation, claiming that a bug in its Symphony spreadsheet caused the problem. The claim was dismissed, with the court finding the cause to be user error. Similar results are likely for the foreseeable future (Backman, 1993).

12.3   STRATEGIES FOR PROTECTION OF THE PROFESSIONAL

The preceding sections make it obvious that the nature of the profession places the engineer in a vulnerable position with respect to liability. The suggestions that follow are ways to minimize one's chance of being sued or, if a suit is filed, to minimize the prospect of a guilty verdict. This material is intended to apply to the work of the design professional who works on manufactured products or construction projects.

Somewhat arbitrarily, we have divided these strategies into three categories: technical, administrative, and legal. It becomes quite clear that a key to minimizing professional liability problems is found in competent, thorough, well-documented professional design, manufacturing, and construction practices. The quality management principles covered in Chapters 3 and 4, and good engineering management practice in general, are apparent in many of the suggestions that follow. We have drawn from a large number of sources, which expand on the items listed below (Firmage, 1980; Forster and Schreyer, 1989; Gamble, 1987; Lepatner and Banner, 1993; Lurie, 1983; Morton, 1983; Peck and Hock, 1988; Schwartz, 1993a; Settle and Spigelmyer, 1984; Smith, 1981; Streeter, 1988; Thorpe and Middendorf, 1979).

Technical Strategies

Design Standards. As a minimum, the designer should be aware of the company's design standards and those of the industry, such as Underwriters' Laboratories standards and the ASME pressure vessel codes.

Design Review. The more formal the better. This in-house process occurs several times during the project's development; the process should emphasize safety as well as performance.

Quality-Built Products. Build quality into the product, and follow with testing and inspection conducted by qualified personnel and documented in writing. Inspect machinery, processes, and procedures regularly.

Peer Review. A complete review of your design, by a professional individual or organization outside the firm, is recommended for all projects above a certain size and complexity.

Design Assurance Audit Checklist. A simple Yes/No/NA checklist can be used to remind the designer about such considerations as identifying the ultimate user, codes and standards, design reviews, and warning labels. Smith (1981) suggests a 60-item checklist.

Testing. Scale model or full-scale strength testing of critical elements or innovative designs might be considered. Accelerated aging tests of new products can help uncover defects.

Proper Packaging. Consider the effects that handling and shipping will have on the integrity of the product and its components.

Professional Competency. Read professional journals and attend seminars to acquire knowledge of advances prior to their inclusion in design manuals. In firms, an individual or group should have most of this responsibility.

Administrative Strategies

Refusing Certain Projects. Be selective about the contracts you sign. If a potential client has a history of adversarial dealings, is known to be quick to file lawsuits with little justification, or has a shaky financial condition, deciding not to enter into an agreement may save many headaches later. At the same time, the design professional must evaluate whether he or she has sufficient background, experience, auxiliary personnel, time, resources, and equipment to do the job.

A Clear, Complete Contract. The contract must be clear about design services, construction-related services, and ownership of plans and specifications after completion.

Proper Project Management. Necessary components include:

Project team members with sufficient skill and experience

A project manager with sufficient authority and responsibility to get the job done properly

A written program plan, prepared before the work begins

A time schedule and budget allocation for each project phase

Methods for monitoring progress and cost status

Procedures for verifying the accuracy and completeness of calculations, drawings, and specifications

Postproject evaluation to identify problem areas and plan for improvements on future projects.

Documentation of Effort. Written evidence is essential in supporting your case. Documents include correspondence; memoranda of meetings, conversations, and telephone calls; journals; photographs; and logs of daily activities. All should be dated.

Strict Lines of Communication. Write correspondence only to those with whom you have a contract. Otherwise, a court may imply a contract that was not intended.

Construction Schedule. Avoid directing that a construction schedule be changed or that the job be stopped. To do so with the purpose of review, analysis, and the like often leads to major claims by the contractor.

Effective Shop Drawing Review. The design professional should limit required submittals to those he or she intends to review. All submittals should be logged and time stamped. A suggested "approval" stamp is the following (Lurie, 1983):

_____ Conforms with design concept

_____ Rejected

_____ Revise and resubmit as noted

This review is made for the limited purposes set forth
in the contract documents. This stamp does not constitute
a change order nor does it relieve the
contractor of his obligations under the
contract documents.

Review of Existing Products. A product liability claim might be prevented with a feedback system on product performance and problems from distributors and users.

Accurate Product Advertising. The marketing department must avoid overstating product performance, which can lead to warranty difficulties.

Legal Strategies

Contract Clauses. Some contracts allow indemnity, or hold harmless, clauses, which release the engineer from liability. However, some states prohibit such clauses, and there is confusion about distinctions between indemnity and limitation of liability clauses.

Incorporation of the Business. Corporate assets will be at risk, but personal assets are generally safe, if the service-oriented design professional does all work through a corporation.

Professional Liability Insurance. Also known as "errors and omissions insurance," this is often a major expense for the design professional. A single liability policy covers all projects the design professional is developing, while a project liability policy covers a single project. Exclusions will typically stipulate that the insurance will not cover fraudulent or willful design defects, only "honest" errors; also, uninsurable risks such as excavating next to an existing structure are likely to be excluded.

Unified Risk Insurance. An alternate to professional liability insurance, this type puts all parties on a single construction project in a single package, with the costs passed directly to the owner.

Self Insurance. A single large firm, or groups of firms that are part of an association, may establish a fund for their own protection.

Going Bare. This strange term means simply not carrying insurance. The "little guy" who has very few assets, or any engineer who has assigned most personal assets to a spouse or a corporation, is unlikely to be the object of liability suits. If the engineer is uninsured, the prospective plaintiff has little chance to recover very much and tends to be discouraged from filing suit.

Labeling, Warning, and Instructions. Review government standards, consider all potential risks and all potential users, and warn against as many hazards as feasible. Helpful guidelines are available (Settle and Spigelmyer, 1984).

Avoiding Inadvertent Warranties. Make sure that correspondence and other written materials do not imply warranties of the design product if none are intended.

Working for a Product-Oriented Corporation. If this kind of employment is appealing and available, it can provide protection for the individual. Individual engineers in such positions are seldom listed as codefendants; the company is usually the primary defendant. If an engineer is responsible for harm, however, sometimes the company will attempt to recover its loss from that individual.

12.4   TORT REFORM

During the 1980s, the number of lawsuits filed against design professionals rose significantly, and average jury awards for damages in such cases and others also grew considerably larger. In 1993 it was estimated that of every 100 architectural or engineering firms, 25 to 30 would have claims filed against them each year in the remainder of the decade, compared with an annual rate of 40 per hundred firms in the 1980s and 12.5 in 1960 (Schwartz, 1993c). Often it takes 3–5 years to settle a liability case, and only 30–40% of the cost of reaching a settlement goes to the victim. These large jury awards have resulted in increases in the cost of liability insurance or the outright cancellation of policies in some cases.

We close this chapter with a brief look at some of the proposals that have been considered in the tort reform movement, whose adherents are attempting to change the laws governing tort liability. The design professions are not alone in this effort. Persons and organizations with a variety of interests—doctors, dentists, physical therapists, midwives, manufacturers, air carriers, truckers, daycare operators, hotel and restaurant owners, and bartenders, all of whom serve the public in one way or another—have often joined together because of their common interest in minimizing their liability exposure and keeping liability insurance available and reasonably priced. Insurance companies have joined with these groups to attempt to influence legislation. Primary opposition has come from trial lawyers, for obvious reasons, and consumer interest groups, who perceive such changes as placing increased risk on members of the public (Citizens Coalition for Tort Reform, no date; Lunch, 1990; Streeter, 1988).

Legislative changes have been made in the following areas.

Joint and Several Liability. Under this concept if two or more parties are found to be at fault and the judgment is apportioned, any one of the parties is liable for the full amount if the other parties are unable to pay their proportionate amount. Legislation has been passed in various states that limits the individual responsibility to 100, 150, or 200% of the amount of the more affluent party's share.

Awards for Noneconomic Damages. These awards compensate the victim for intangible losses for which no economic value exists: pain and suffering, disfigurement, loss of consortium, traumatic experiences, and so on. Attempts have been made to place limits on the amounts juries can award for such damages, such as $100,000 or $250,000. Some state courts have upheld such statutes, and others have rejected them.

Awards for Punitive Damages. These awards are assessed against the guilty party as punishment for particularly offensive behavior such as gross negligence. There have been some efforts to limit the amounts of these awards and also to have them paid to the state rather than the victim, since the victim is already being compensated with other awards.

Structured Settlements. These arrangements are schedules of payments to the victim over a period of years. The total value of the future losses is computed at its present value, and the recipient has the choice of accepting this lump sum amount or the periodic payments.

Collateral Income Sources. In some states, legislation now provides that juries must be told about insurance or workers' compensation payments that have already been paid to the plaintiff. Without such a statute, this information is withheld, and insurance companies that have fulfilled their obligations must sue the victim to recover their payments from the total award.

Contingent Fees. With such an arrangement, the plaintiff's attorney receives compensation based on the outcome of the case. The fee can be as much as 40% of the award (which, of course, equals zero if there is no award). A sliding scale has been implemented in some states, whereby this percentage decreases as the amount of the award increases.

Statute of Limitations. These statutes limit the amount of time within which a suit can be brought, as discussed in Section 12.2.6. Several states have recently revised and clarified their statutes, as part of tort reform.

Frivolous Suits. Lawsuits that are based on far-fetched theories of liability or are brought despite the absence of factual proof must still be defended. At the federal and state level, rules have been adopted to require the plaintiff's attorney to certify that there is a reasonable and meritorious cause for filing the action.

12.5   DISCUSSION QUESTIONS

1. An uninhabited house containing valuable antiques had been burglarized several times. The owner, having appealed unsuccessfully to law enforcement officials for protection, installed a spring gun that would discharge and injure the legs of any prowler who entered a particular room. Indeed, a would-be burglar was injured severely. The "visitor" was found guilty of a misdemeanor crime. He then brought a personal injury suit against the owner. What did the jury decide? [Katko v. Briney, 183 N.W. 2d 657 (1971)]

2. A child found an unexploded cartridge at a construction site, took it home, and was injured when it subsequently exploded. Was the construction company liable to this trespassing child? [Concrete Construction, Inc., of Lake Worth v. Petterson, 216 So. 2d 221 (1968)]

3. California has a law that encourages public use of privately owned recreational property by limiting owners' tort liability. A person on the way to a recreational beach was injured while exploring two nearby houses that were under construction. Can the owner invoke the statute successfully? [Potts v. Halsted Financial Corporation, 142 Cal. App. 3d 727, 191 Cal. Rptr. 160 (1983)]

4. How might the doctrine of *res ipsa loquitur* apply to objects that fall from buildings under construction, causing injury to passersby? How might the contractor defend against such a suit?

5. Ed and Ned are engaged in the repair of some overhead machinery, with Ed working overhead and Ned on the floor. Before ascending to his workplace, Ed turns off the power to the machinery at the main panel. A third worker, Jed, requires power for a separate task, so he throws a switch on the panel to the "on" position. Unfortunately, Jed selects the wrong switch, power is restored to the overhead machinery, Ed receives a shock that causes him to drop a heavy steel rod, and Ned sustains severe damage to his skull and left shoulder. If Ned takes his case to court, what is the likely result? Suggest several parties whom Ned might name as defendants.

6. Suppose you are driving along an icy roadway when another driver, acting negligently, causes an accident with your automobile. Your vehicle comes to rest, you step out, uninjured, to inspect the damage and consult with the other driver. Then you slip on the ice, cracking two vertebrae and one wrist. Can the other driver be held liable for your injuries? State some reasoning *for,* and some reasoning *against,* a finding of negligence for your injuries.

7. A design defect caused improper drainage of sewage water from a construction site and damage to an adjoining property. The architect said the design was intended to provide proper drainage after project completion, not during construction. Explain how the architect might have used privity of contract as a defense. What do you think the result was? [McCarthy v. J. P. Cullens & Son, Corporation, 199 N.W. 2d 362 (1972)]

8. A visitor to your construction site is injured in a fall through an open hole and sues for $250,000. The jury finds your company 65% responsible for the accident, with the other 35% being the responsibility of the injured visitor.
   (a) Under the doctrine of contributory negligence, how much will the visitor recover in damages?
   (b) Under the doctrine of comparative negligence, how much will the visitor recover in damages?

9. Pyrodyne Corporation contracted to display fireworks at the Western Washington State Fairgrounds. During the display, one of the rockets traveled improperly in a horizontal trajectory and exploded near Danny Klein, setting his clothes on fire and causing facial burns and serious eye injuries. In a lawsuit filed by Klein, Pyrodyne argued that negligence by the *manufacturer,* not the display contractor, was the cause of the injuries. What was the result? [Klein v. Pyrodyne Corporation, 810 P. 2d 917 (1991)]

10. The plaintiff was injured while operating a forklift truck in a high stack storage area. The truck had no protective overhead guard. Should the truck manufacturer be held liable for the operator's injury? [Posey v. Clark Equipment Company, 400 F. 2d 560 (1969)]

11. Dunham lost the sight of his right eye when a piece of the metal head of a hammer broke off while he was using the implement to fit a pin into a clevis to connect his tractor to his manure spreader. The hammer had been used for 11 months, and the head met hardness requirements. No design or manufacturing defect was found, nor was there a showing of a defect when

it left the manufacturer's control. Did Dunham recover in a lawsuit against the manufacturer? If so, on what basis? [Dunham v. Vaughan & Bushnell Manufacturing Company, 247 N.E. 2d 401 (1969)]

12. A 6-year-old child drowned in a 12-foot-wide, 6-foot-deep ditch that had been dug to bury tree stumps at a construction site in a residential area. Overnight rainfall had caused a pool to form in an area where children had customarily played. In such a case, what standards of care are the contractor and landowner required to meet? Were these parties found to be negligent in this instance? [Minneti v. Evans Construction Company, 160 Fed. Supp. 372]

13. Suppose a construction worker is injured while operating a pipe-bending machine. Suppose further that the manufacturer shows successfully that all industry standards were followed in the design and manufacture of the machine and that the equipment is "state of the art." Under what sort of argument might the manufacturer lose a negligence suit brought by the injured worker?

14. Suppose the original brake linings (or other component) failed and caused a plaintiff's injuries. Suppose further that these linings or other components had been supplied to the automobile producer by another manufacturer. Shouldn't the component supplier be held responsible for the injury rather than the producer who assembled the components into a finished product?

15. While incarcerated in a penal institution, a young man hanged himself by attaching a rope to an uncovered ceiling duct opening, which served as a "yardarm". His heirs brought suit against the designer of the facility on theories of negligence and strict liability, arguing that the design should have provided for a guard grill over the air conditioning duct. Was the designer found to be negligent? [Easterday v. Masiello, 518 So. 2d 260 (Fla. Sup. Ct., 1988)]

16. Gierach was injured when a ratchet wrench he was using to repair an automobile slipped and struck his face, mouth, and teeth. Evidence showed that he was using the tool properly and that a gear tooth had sheared during its use. Gierach brought a suit against the manufacturer, charging that the ratchet wrench was improperly designed or manufactured. The auto shop service manager testified that the teeth were not properly hardened. The manufacturer countered that the hardening was proper and that the problem occurred because of improper cleaning and maintenance (grease and debris buildup had prevented the gear pawl from seating properly to engage the gear teeth). The court, in its instructions to jury, said that Gierach had a duty to use ordinary care for his safety and that the manufacturer had a duty to exercise ordinary care in design and manufacture to make the product safe for its intended use. It also instructed the jury to consider *res ipsa loquitur*. [Gierach v. Snap-On Tools Corporation, 255 N.W. 2d 465 (1977)]

   (a) What would be the essence of the *res ipsa* instruction?

   (b) Would comparative negligence be a possible finding? On what basis?

   (c) What do you think was the result of the appeal?

17. A state has a statute of limitations of 3 years for property damage caused by design defects and a statute of repose of 7 years for design work. Suppose a building was substantially completed on July 1, 1985, and, on April 1, 1993, it collapsed as a result of a design defect. If a liability suit was filed on August 1, 1993, against the structural design engineer, what is the likely result?

18. An employer asks a carpenter to assist in the lifting of a heavy wall section with a crane. In accordance with the employer's instruction, no tag lines are used. Summarize the arguments *for* and *against* finding the architect negligent for the injuries sustained by the carpenter when the wall section falls on her.

12.6 REFERENCES

"A/E Duty to Contractor Upheld." *Civil Engineering*, Vol. 57, No. 11, November, 32.

American Heritage Dictionary, 2nd college ed. 1982. Boston: Houghton Mifflin.

Backman, L. 1993. "Computer-Aided Liability." *Civil Engineering,* Vol. 63, No. 6, June, 41–43.

Blinn, K. W. 1989. *Legal and Ethical Concepts in Engineering.* Englewood Cliffs, NJ: Prentice-Hall.

Bockrath, J. T. 1986. *Dunham and Young's Contracts, Specifications and Law for Engineers.* New York: McGraw-Hill.

Boehmig, R. L. 1990. "Shop Drawings: In Need of Respect." *Civil Engineering,* Vol. 60, No. 3, March, 80–82.

Brown, S. Ed. 1991. *The Product Liability Handbook: Prevention, Risk, Consequences, and Forensics of Product Failure.* New York: Van Nostrand Reinhold.

Cheeseman, H. R. 1992. *Business Law: The Legal, Ethical and International Environment.* Englewood Cliffs, NJ: Prentice-Hall.

Citizens Coalition for Tort Reform. No date. "Tort Reform: A Comprehensive Solution to the Crisis in Civil Justice and Insurance." Anchorage, AK.

"Design Error: Engineer v. Builder." 1985. *Civil Engineering,* Vol. 55, No. 12, December, 22.

Dinges, C. 1988. "New Role for CEs in Construction Safety?" *Civil Engineering,* Vol. 58, No. 9, September, 128.

Fairweather, V. 1992. "L'Ambiance Plaza: What Have We Learned?" *Civil Engineering,* Vol. 62, No. 2, February, 38–41.

Firmage, D. A. 1980. *Modern Engineering Practice.* New York: Garland STPM Press.

Forster, J. F., and P. R. Schreyer. 1989. "The Long Arm of Liability." *Civil Engineering,* Vol. 59, No. 9, September, 80–81.

Gamble, R. O., II. 1987. *How to Reduce Professional Liability for Engineers and Architects.* Park Ridge, NJ: Noyes Data Corporation.

Jessup, W. E., Jr., and W. E. Jessup. 1963. *Law and Specifications for Engineers and Scientists.* Englewood Cliffs, NJ: Prentice-Hall.

Kemper, J. D. 1990. *Engineers and Their Profession,* 4th ed. Philadelphia: Saunders College Publishing.

Lepatner, B. B., and R. A. Banner. 1993. "8 Tips for Avoiding Liability Claims." *American Consulting Engineer,* Vol. 4, No. 2, Second Quarter, 8.

Loulakis, M. C. 1988. "Privity as a Defense for Economic Loss." *Civil Engineering,* Vol. 58, No. 11, November, 40.

Loulakis, M. C. 1989. "Limiting the Economic Loss Defense." *Civil Engineering,* Vol. 59, No. 7, July, 33.

Loulakis, M. C., and W. L. Cregger. 1991a. "Confusion Reigns on Liability for Economic Loss." *Civil Engineering,* Vol. 61, No. 9, September, 41.

Loulakis, M. C., and W. L. Cregger. 1991b. "Liability for On-the-Job Safety." *Civil Engineering,* Vol. 61, No. 7, July, 38.

Lunch, M. F. 1990. "The Liability Crisis—Revisited." *Journal of Management in Engineering,* American Society of Civil Engineers, Vol. 6, No. 2, April, 197–202.

Lurie, P. M. 1983. "How Can I Manage the Legal Risks to Which I Am Exposed?" in Cushman, R. F., Ed., *Avoiding Liability in Architecture, Design and Construction.* New York: Wiley, pp. 233–249.

Morton, R. J. 1983. *Engineering Law, Design Liability, and Professional Ethics: An Introduction for Engineers.* Belmont, CA: Professional Publications.

Neely, R. 1991. "The Peculiar Problem of Product Liability." *National Forum: the Phi Kappa Phi Journal,* Vol. 71, No. 4, Fall, 30–32.

Peck, J. C., and W. A. Hoch. 1988. "Liability and the Standards of Care." *Civil Engineering,* Vol. 58, No. 11, November, 70–72.

Reynolds, H. E. 1983. "Do I Have Any Responsibility for Job-Site Safety?" In Cushman, R. F., Ed., *Avoiding Liability in Architecture, Design and Construction.* New York: Wiley, pp. 53–62.

Schoumacher, B. 1986. *Engineers and the Law: An Overview.* New York: Van Nostrand Reinhold.

Schwartz, A. 1992. "Engineer Not Liable for Safety, Says Indiana Court." *Engineering Times,* Vol. 14, No. 7, July, 12.

Schwartz, A. 1993a. "NSPE Goes to Bat for Limitation of A/E Liability." *Engineering Times,* Vol. 15, No. 2, February, 16.

Schwartz, A. 1993b. "Court Considers Case on A/E Job Site Liability" *Engineering Times,* Vol. 15, No. 5, May, 16.

Schwartz, A. 1993c. Personal conversation.

Settle, S. M., and S. Spigelmyer. 1984. *Product Liability: A Multibillion-Dollar Dilemma.* New York: American Management Associations.

Smith, C. O. 1981. *Products Liability: Are You Vulnerable?* Englewood Cliffs, NJ: Prentice-Hall.

Streeter, H. 1988. *Professional Liability of Architects and Engineers.* New York: Wiley.

Sweet, J. 1989. *Legal Aspects of Architecture, Engineering and the Construction Process,* 4th ed. St. Paul, MN: West Publishing.

Thorpe, J. F., and W. H. Middendorf. 1979. *What Every Engineer Should Know About Product Liability.* New York: Dekker.

Van Hermert, W. 1991. "Impacts of Statutes of Limitations and Repose." *Alaska Designs,* Alaska Professional Design Council, Vol. 14, No. 1, January, 1, 7, 11.

Vaughn, R. C. 1983. *Legal Aspects of Engineering,* 4th ed. Dubuque, IA: Kendall/Hunt Publishing.

_____13
Professional Ethics

In the summer of 1993, while the manuscript for this book was being prepared, the author was invited by a colleague to participate in a building condition survey. The building owner was preparing to sell the property, and some structural problems had been noted in an earlier inspection. We prepared our draft report to the owner, with proposed engineering solutions, and invited corrections of any issues of fact (dates, sequence of improvements, etc.). Within a short time, the realtor for the owner faxed a request to my colleague asking that the report be modified to "guarantee" that the solutions would be satisfactory. Such modifications of course would have increased the value of the property. My engineer friend replied that proposals in the report would stand as written and that to make the requested changes would constitute collusion to influence and perhaps collusion to defraud. He explained the differences between providing professional services and marketing services and declared the professional's responsibility not to mislead the layperson with such words as "insure."

Twenty years earlier, in October 1973, Spiro T. Agnew resigned as vice president of the United States in the midst of charges of bribery and tax evasion in connection with his previous employment as county executive of Baltimore County, Maryland. A civil engineer and lawyer who had assumed increasingly responsible positions in local government, Agnew had the authority, as county executive, to award contracts to engineering firms for public works projects. In an elaborate kickback scheme orchestrated by Agnew, certain consulting firms were given special consideration for contracts in exchange for payments to Agnew of 5% of its fees for these projects. Several of the many participants were tried, convicted, sentenced to fines and imprisonment, and expelled from the American Society of Civil Engineers and the Maryland Society of Professional Engineers (Firmage, 1980; Lewis, 1977; Martin and Schinzinger, 1989).

In small and large ways, the engineer is often confronted with situations in which ethics plays a role. How can we know how to act "correctly" in these situations?

In contrast to the subject matter of Chapter 12, which dealt with the legal liability an engineer may incur as a result of intentional or unintentional acts that cause injuries, property damage, or economic loss, this chapter examines the basic rightness or wrongness of actions from an ethical or moral standpoint. At the outset, we note that it is sometimes difficult to partition the two issues in a particular situation. Certainly an action can be both legally and morally wrong, and we sometimes tend to think of the two matters as one. But, it is also possible for an action to be legally right but morally wrong, and we might even think of a situation that we consider to be morally right but the law determines to carry some legal liability. Despite the tendency to commingle professional liability and professional ethics, the reader is urged to consider them separately when confronting a particular professional decision.

Does this topic have a place in a book on the management of engineering? Engineers are generally regarded as being quite ethical. A 1985 survey of 1000 Americans ranked engineers third behind clergy and medical doctors in terms of the percentage of respondents who ranked the professions as "high" or "very high" in honesty and ethical standards; 61% ranked clergy in this category, with 48% for medical doctors, 45% for engineers, and so forth, down to 16% for state political officeholders and 6% for car salespeople in this listing of 17 occupations (Kemper, 1990). Better than that, 1000 corporation executives, when asked in a forced-choice survey which types of professionals were most ethical, replied as follows (Zinckgraf, 1992):

Engineers	34%
CPAs	24%
Doctors	17%
Dentists	7%
Lawyers	8%
Others or don't know	10%

Despite these laudatory statistics, we must not become complacent, especially since we are far from perfect, as attested to by the fiascoes that occur from time to time, and also since most all engineers report that ethical considerations are at least occasionally a part of their decision making. In 1989 the National Society of Professional Engineers surveyed its members to ask the frequency with which they encountered decisions that required ethical judgment. The responses were as follows ("Survey Finds Ethics a Concern," 1990):

Occasionally	45%
Frequently	25%
Rarely	13%
Seldom	13%
No answer	4%

Thus, we believe it is altogether fitting to devote a chapter exclusively to ethical considerations in engineering management. We begin with some distinctions and definitions about morality and ethics and a discussion of personal morality as related to social justice and workplace issues. Then we review some information about codes of ethics and the three primary, sometimes conflicting, interests to which such codes usually speak. A major section deals with a number of ethical issues, including conflict of interest, contributions, whistleblowing, and confidentiality, as well as some global issues including environmental ethics and national defense. At the end, we consider the engineering manager's responsibility for ethics and offer some words of hope (!) for the engineer about to be faced with ethical considerations in the real world.

The chapter proceeds from the author's deeply held belief that there *are* absolutes when it comes to personal morality and the application of personal standards to the professional workplace. In an era when everything seems relative, it may be a source of either strength or frustration to approach the workplace with the view that some things are wrong, no matter what. The material presented in this chapter is intended, in part, to assure the reader that others have faced ethical dilemmas in the past and that we can learn from their experiences, especially if, in hindsight, it appears that the decisions they made were wrong.

13.1 MORALITY AND ETHICS

The terms "morals" and "ethics" are often used interchangeably, both denoting something about "good" or "right" as opposed to "bad" or "wrong." We commonly speak of an ethical or a moral person or action without distinguishing between the adjectives. One basis for ethical or moral behavior is the Golden Rule: "Do unto others as you would have them do unto you." In a similar vein, Confucius is reported to have replied, when asked for a single word to guide one's life, "Is not reciprocity such a word?" (Blinn, 1989).

It is possible that ethical and moral principles may not have corresponding legal obligations; there may be no legal requirement for you to report a cheating fellow employee or an environmental violation, for example. But those traditional values of honesty, integrity, and concern for the rights and needs of others are still excellent guidelines for ethical decisions in professional engineering situations, as they are in one's personal life.

Concern for ethical behavior is not new. One example is the statement by a bishop of the Methodist Church: Ivan Lee Holt wrote, in 1934:

> We take our ethics as we take our athletics—watching [others] play the game, to be commended now, and to be criticized later where there is a failure to give the best or abide by the rules. A sense of obligation has slipped from our shoulders. We have wanted the fruits of civilization, but none of its burdens. Downright honesty we must have, or else our whole social structure collapses. (quoted in Blinn, 1989)

Even though the two terms are commonly used interchangeably, it is helpful to distinguish between them. Gluck (1986) says, *"Morality* is concerned with *conduct* and *motives,* right and wrong, and good and bad *character. Ethics* is the philosophical study of morality...."

One helpful way to define *ethics* is to say that it is "the study of systematic methods which, when guided by individual moral values, can be useful in making value-laden decisions" (Vesilund, 1988). Thus, ethics is a framework, or an approach, that helps us to study moral dilemmas and arrive at an acceptable course of action.

Moral values, on the other hand, are "those standards or patterns of choice that guide us toward satisfaction, fulfillment or meaning" (Vesilund, 1988).

If ethics has something to do with a system for evaluating choices that involve moral values, then *engineering ethics* can be defined as "1) the study of the moral issues and decisions confronting individuals and organizations involved in engineering and 2) the study of related questions about moral conduct, character, ideals and relationships of people and organizations involved in technological development" (Martin and Schinzinger, 1989). We need to look further into the various bases, or frameworks, upon which such decisions might be made.

What is the proper basis for making ethical choices? In other words, what makes some actions right and others wrong? Ever since the time of Socrates, philosophers have argued this vital but elusive question. Five primary theories seem to have emerged, as follows (Babcock, 1991; Martin and Schinzinger, 1989; Vesilund, 1988):

- *Utilitarianism:* Holds that actions that produce the greatest utility ought to be chosen. Good and bad consequences are the only relevant moral considerations. Our actions should result in the greatest good for the greatest number of people.
- *Duty Ethics:* A system under which there are certain duties to be performed, without consideration of the consequences. Such universal moral imperatives as "Do not lie," "Do not steal," and "Be fair" are always "right," even though they may not produce the greatest good.
- *Rights Ethics:* Views actions as wrong if they violate certain fundamental moral rights. The framers of the Declaration of Independence took this view when they wrote about the self-evident truths that everyone is endowed with the unalienable rights of life, liberty, and the pursuit of happiness.
- *Virtue Ethics:* Distinguishes between right and wrong actions depending on whether they support good or bad character traits (virtues or vices). The emphasis is on the morally good person, rather than right actions. Those actions are right because they build good character.
- *Environmental Ethics:* A modern concept, developed in the mid-twentieth century. Whereas the other approaches include only people in the "moral community," this concept broadens the outlook to involve plants, animals, and even inanimate objects. Environmental ethics can be utilitarian, wherein a right action is that which promotes the greatest overall good, or duty-based, in which case there are some absolute wrongs with respect to the overall community including the environment.

Is it possible to teach morality and ethics? Educators have considered this question for a long while, and engineering educators, especially, have debated whether, and if so in what form, to include such topics in the curriculum. One reasonable position is suggested by Vesilund (1988); it is a helpful means of distinguishing between morality and ethics. If ethics is a framework for making value-laden choices, then ethics can be taught. The various approaches described above can be explained and contrasted, and students can be challenged to apply their own moral values to cases and to learn from the experiences of others. Engineers *can* be taught to "think ethically" just as they can be taught to "think scientifically."

On the other hand, Vesilund suggests that morals cannot be taught. It is simply impossible to explain, for example, why it is wrong to manipulate water quality data to achieve a passing result. "Teaching morals is simply not possible, since moral values 'are the product of strong parental, cultural and psychological factors and forces. They are not items of knowledge to be learned simply by "taking thought" ' " (Morrill, 1980, quoted in Vesilund, 1988). The best that educators can do is to influence by example. "Our students' antennas are up," Morrill continued, "and they are searching for the values they will carry with them for the remainder of their careers. We are their role models."

Is our obligation only to the social order and the planet? In other words, should we be emphasizing social policy questions to the exclusion of matters of personal morality? We answer with a strong "No!" and argue that both personal morals and social concerns are required by the professional engineer and all other members of society. Sommers (1991) provides a fitting illustration of this point. While many college ethics courses emphasize such current topics as abortion, capital punishment, euthanasia, transplant surgery, and DNA research, Sommers notes that there is little emphasis on private decency, honesty, personal responsibility, or honor. She argues that this is not proper and that both private and social morality should be attended to. A colleague of hers had the opposite opinion and continued to awaken social conscience by teaching about the oppression of women, corruption in business, transgressions of multinational corporations in the Third World, and the like. The other half was a waste of time, said this colleague, who said further, "You are not going to have moral people until you have moral institutions." At the end of the semester, however, she came to Sommers's office carrying a stack of exams and looking very upset: she had just found that more than half her students had cheated on their social justice take-home final exams by plagiarizing!

We may not be able to "teach moral values," but we can certainly make the case that personal morality is an essential ingredient in any balanced, meaningful life.

13.2   CODES OF ETHICS

One of the characteristics of the professions that distinguishes them from other human endeavors is the habit of developing codes of ethics to guide the actions of their members. In this section, we list the various roles played by such codes, describe how they are structured, discuss modes of enforcement, and report recent changes in some codes of ethics for engineers.

The book entitled *Codes of Professional Responsibility* (Gorlin, 1990) contains 43 codes, ranging from the Code of Professional Conduct of the American Institute of Certified Public Accountants to the Code of Ethics and Professional Responsibility of the National Association of Legal Assistants. The various codes are organized into three categories— business, health, and law—with engineering and architecture included in the business section along with accounting, banking, and financial planning. It is of interest that only two organizations are represented in the engineering section, the American Association of Engineering Societies and the National Society of Professional Engineers. Since it is known that all the discipline-related engineering associations, such as the American Society of Civil Engineers and the Institute of Electrical and Electronics Engineers, have their own codes, a reasonable conclusion is that the professions, in total, have several hundred such codes.

The codes of ethics for engineers are intended as guidelines to protect the public, to build and preserve the integrity and reputation of the profession, and to describe proper relations between engineers and their employers and clients (Wright, 1994). As just indicated, there is no single code for all of engineering, even though there is considerable agreement as to what constitutes ethical behavior and thus much similarity among the several codes.

What roles are these codes of ethics intended to play? Martin and Schinzinger (1989) suggest the seven prominent roles listed below, three of which may have negative as well as positive consequences.

- *Inspiration and Guidance.* They provide a positive stimulus for proper conduct and, to some degree, guidance concerning obligations in particular situations.
- *Support.* If a code has been proclaimed publicly, an engineer is able to say, "I am bound by the code of ethics of my profession, which, for this situation, says...."
- *Deterrence and Discipline.* Codes are used as the formal basis for investigating allegations of unethical conduct by professional societies and registration boards.
- *Education and Mutual Understanding.* Codes are used in the classroom, in professional meetings, and elsewhere as opportunities for professionals to "gather around" the principles and remind the participants of their obligations.
- *Contribution to the Profession's Public Image.* Codes present a positive public image, depicting a profession committed to high ethical standards. A result can be more opportunity for self-governance and less governmental regulation.
- *Protecting the Status Quo.* The ethical conventions become minimum standards of conduct, which may be difficult to change.
- *Promoting Business Interests.* Codes may unduly protect the profession at the expense of the public and thus become self-serving.

A typical code of ethics begins with general introductory section, followed by a series of fundamental statements or "canons." Then, the canons are expanded and explained as a means of providing guidance in particular situations. The Code of Ethics of Engineers, published by the Accreditation Board for Engineering and Technology (1985), begins with the following statement of fundamental principles:

> Engineers uphold and advance the integrity, honor and dignity of the engineering profession by:
>
> I. using their knowledge and skill for the enhancement of human welfare;
> II. being honest and impartial, and serving with fidelity the public, their employers and clients;
> III. striving to increase the competence and prestige of the engineering profession; and
> IV. supporting the professional and technical societies of their disciplines.

The code ends by providing seven fundamental canons. Separate from the official code of ethics, but related to it, is an expanded statement called "Suggested Guidelines for Use with the Fundamental Canons of Ethics." For example, for the first canon ("Engineers shall hold paramount the safety, health and welfare of the public in the performance of their professional duties."), there are six guidelines. We quote the first two:

> a. Engineers shall recognize that the lives, safety, health and welfare of the general public are dependent upon engineering judgments, decisions and practices incorporated into structures, machines, products, processes and devices.
> b. Engineers shall not approve nor seal plans and/or specifications that are not of a design safe to the public health and welfare and in conformity with accepted engineering standards.

We reproduce, as Table 13.1, the entire text of the Code of Ethics for Engineers of the National Society of Professional Engineers. Note that it consists of a preamble, five fundamental canons, rules of practice that support and explain the canons, a set of eleven professional obligations, and some concluding remarks. The practicing engineer does great honor to the public and the profession by displaying the code in a prominent place and heeding its provisions.

In an interesting analysis of several engineering codes of ethics, Oldenquist and Slowter (1979) were able to categorize the provisions of such codes into three key concepts: (1) the public interest, (2) qualities of truth, honesty, and fairness, and (3) professional performance. Table 13.2 shows the elements of these three core concepts.

Some have objected to engineering codes of ethics because they tend to cover situations that are most likely to arise only in design firms and other enterprises in private practice. And indeed, large numbers of engineers are employed in manufacturing, construction, and governmental organizations. In response to such charges of insufficient breadth of focus, 25 engineering and engineering-related societies joined together to produce and adopt a set of Guidelines to Professional

TABLE 13.1   Code of Ethics for Engineers

Code of Ethics For Engineers

PREAMBLE

Engineering is an important and learned profession. The members of the profession recognize that their work has a direct and vital impact on the quality of life for all people. Accordingly, the services provided by engineers require honesty, impartiality, fairness and equality, and must be dedicated to the protection of the public health, safety and welfare. In the practice of their profession, engineers must perform under a standard of professional behavior which requires adherence to the highest principles of ethical conduct on behalf of the public, clients, employers, and the profession.

I. FUNDAMENTAL CANONS

Engineers, in the fulfillment of their professional duties, shall:

1. Hold paramount the safety, health and welfare of the public in the performance of their professional duties.
2. Perform services only in areas of their competence.
3. Issue public statements only in an objective and truthful manner.
4. Act in professional matters for each employer or client as faithful agents or trustees.
5. Avoid deceptive acts in the solicitation of professional employment.

II. RULES OF PRACTICE

1. Engineers shall hold paramount the safety, health and welfare of the public in the performance of their professional duties.

   a. Engineers shall at all times recognize that their primary obligation is to protect the safety, health, property and welfare of the public. If their professional judgment is overruled under circumstances where the safety, health, property or welfare of the public are endangered, they shall notify their employer or client and such other authority as may be appropriate.

   b. Engineers shall approve only those engineering documents which are safe for public health, property and welfare in conformity with accepted standards.

   c. Engineers shall not reveal facts, data, or information obtained in a professional capacity without the prior consent of the client or employer except as authorized or required by law or this Code.

   d. Engineers shall not permit the use of their name or firm name nor associate in business ventures with any person or firm which they have reason to believe is engaging in fraudulent or dishonest business or professional practices.

   e. Engineers having knowledge of any alleged violation of this Code shall cooperate with the proper authorities in furnishing such information or assistance as may be required.

2. Engineers shall perform services only in the areas of their competence.

   a. Engineers shall undertake assignments only when qualified by education or experience in the specific technical fields involved.

   b. Engineers shall not affix their signatures to any plans or documents dealing with subject matter in which they lack competence, nor to any plan or document not prepared under their direction and control.

TABLE 13.1 continued

---

    c. Engineers may accept assignments and assume responsibility for coordination of an entire project and sign and seal the engineering documents for the entire project, provided that each technical segment is signed and sealed only by the qualified engineers who prepared the segment.

3. Engineers shall issue public statements only in an objective and truthful manner.

    a. Engineers shall be objective and truthful in professional reports, statements or testimony. They shall include all relevant and pertinent information in such reports, statements or testimony.

    b. Engineers may express publicly a professional opinion on technical subjects only when that opinion is founded upon adequate knowledge of the facts and competence in the subject matter.

    c. Engineers shall issue no statements, criticisms or arguments on technical matters which are inspired or paid for by interested parties, unless they have prefaced their comments by explicitly identifying the interested parties on whose behalf they are speaking, and by revealing the existence of any interest the engineers may have in the matters.

4. Engineers shall act in professional matters for each employer or client as faithful agents or trustees.

    a. Engineers shall disclose all known or potential conflicts of interest to their employers or clients by promptly informing them of any business association, interest, or other circumstances which could influence or appear to influence their judgment or the quality of their services.

    b. Engineers shall not accept compensation, financial or otherwise, from more than one party for services on the same project, or for services pertaining to the same project, unless the circumstances are fully disclosed to, and agreed to by, all interested parties.

    c. Engineers shall not solicit or accept financial or other valuable consideration directly or indirectly, from contractors, their agents, or other parties in connection with work for employers or clients for which they are responsible.

    d. Engineers in public service as members, advisors or employees of a governmental or quasi-governmental body or department shall not participate in decisions with respect to professional services solicited or provided by them or their organizations in private or public engineering practice.

    e. Engineers shall not solicit or accept a professional contract from a governmental body on which a principal or officer of their organization serves as a member.

5. Engineers shall avoid deceptive acts in the solicitation of professional employment.

    a. Engineers shall not falsify or permit misrepresentation of their, or their associates', academic or professional qualifications. They shall not misrepresent or exaggerate their degree of responsibility in or for the subject matter of prior assignments. Brochures or other presentations incident to the solicitation of employment shall not misrepresent pertinent facts concerning employers, employees, associates, joint venturers or past accomplishments with the intent and purpose of enhancing their qualifications and their work.

    b. Engineers shall not offer, give, solicit or receive, either directly or indirectly, any political contribution in an amount intended to influence the award of a contract by public authority, or which may be reasonably construed by the public of

TABLE 13.1   continued

having the effect of intent to influence the award of a contract. They shall not offer any gift, or other valuable consideration in order to secure work. They shall not pay a commission, percentage or brokerage fee in order to secure work except to a bona fide employee or bona fide established commercial or marketing agencies retained by them.

III. PROFESSIONAL OBLIGATIONS

1. Engineers shall be guided in all their professional relations by the highest standards of integrity.

   a. Engineers shall admit and accept their own errors when proven wrong and refrain from distorting or altering the facts in an attempt to justify their decisions.

   b. Engineers shall advise their clients or employers when they believe a project will not be successful.

   c. Engineers shall not accept outside employment to the detriment of their regular work or interest. Before accepting any outside employment they will notify their employers.

   d. Engineers shall not attempt to attract an engineer from another employer by false or misleading pretenses.

   e. Engineers shall not actively participate in strikes, picket lines, or other collective coercive action.

   f. Engineers shall avoid any act tending to promote their own interest at the expense of the dignity and integrity of the profession.

2. Engineers shall at all times strive to serve the public interest.

   a. Engineers shall seek opportunities to be of constructive service in civic affairs and work for the advancement of the safety, health and well-being of their community.

   b. Engineers shall not complete, sign or seal plans and/or specifications that are not of a design safe to the public health and welfare and in conformity with accepted engineering standards. If the client or employer insists on such unprofessional conduct, they shall notify the proper authorities and withdraw from further service on the project.

   c. Engineers shall endeavor to extend public knowledge and appreciation of engineering and its achievements and to protect the engineering profession from misrepresentation and misunderstanding.

3. Engineers shall avoid all conduct or practice which is likely to discredit the profession or deceive the public.

   a. Engineers shall avoid the use of statements containing a material misrepresentation of fact or omitting a material fact necessary to keep statements from being misleading or intended or likely to create an unjustified expectation; statements containing prediction of future success.

   b. Consistent with the foregoing, Engineers may advertise for recruitment of personnel.

   c. Consistent with the foregoing, Engineers may prepare articles for the lay or technical press, but such articles shall not imply credit to the author for work performed by others.

4. Engineers shall not disclose confidential information concerning the business affairs or technical processes of any present or former client or employer without his consent.

TABLE 13.1 continued

a. Engineers in the employ of others shall not without the consent of all interested parties enter promotional efforts or negotiations for work or make arrangements for other employment as a principal or to practice in connection with a specific profect for which the engineer has gained particular and specialized knowledge.

b. Engineers shall not, without the consent of all interested parties, participate in or represent an adversary interest in connection with a specific project or proceeding in which the engineer has gained particular specialized knowledge on behalf of a former client or employer.

5. Engineers shall not be influenced in their professional duties by conflicting interests.

a. Engineers shall not accept financial or other considerations, including free engineering designs, from material or equipment suppliers for specifying their product.

b. Engineers shall not accept commissions or allowances, directly or indirectly, from contractors or other parties dealing with clients or employers of the Engineer in connection with work for which the engineer is responsible.

6. Engineers shall uphold the principle of appropriate and adequate compensation for those engaged in engineering work.

a. Engineers shall not accept remuneration from either an employee or employment agency for giving employment.

b. Engineers, when employing other engineers, shall offer a salary according to professional qualifications.

7. Engineers shall not attempt to obtain employment or advancement or professional engagements by untruthfully criticizing other engineers, or by other improper or questionable methods.

a. Engineers shall not request, propose, or accept a professional commission on a contingent basis under circumstances in which their professional judgment may be compromised.

b. Engineers in salaried positions shall accept part-time engineering work only to the extent consistent with policies of the employer and in accordance with ethical considerations.

c. Engineers shall not use equipment, supplies, laboratory, or office facilities of an employer to carry on outside private practice without consent.

8. Engineers shall not attempt to injure, maliciously or falsely, directly or indirectly, the professional reputation, prospects, practice or employment of other engineers, nor untruthfully criticize other engineers' work. Engineers who believe others are guilty of unethical or illegal practice shall present such information to the proper authority for action.

a. Engineers in private practice shall not review the work of another engineer for the same client, except with the knowledge of such engineer, or unless the connection of such engineer with the work has been terminated.

b. Engineers in governmental, industrial or educational employ are entitled to review and evaluate the work of other engineers when so required by their employment duties.

TABLE 13.1   continued

   c. Engineers in sales or industrial employ are entitled to make engineering comparisons of represented products with products of other suppliers.

  9. Engineers shall accept personal responsibility for their professional activities; provided, however, that engineers may seek indemnification for professional services arising out of their practice for other than gross negligence, where the engineer's interests cannot otherwise be protected.

   a. Engineers shall conform with state registration laws in the practice of engineering.

   b. Engineers shall not use association with a nonengineer, a corporation, or partnership as a "cloak" for unethical acts, but must accept personal responsibility for all professional acts.

  10. Engineers shall give credit for engineering work to those to whom credit is due, and will recognize the proprietary interests of others.

   a. Engineers shall, whenever possible, name the person or persons who may be individually responsible for designs, inventions, writings, or other accomplishments.

   b. Engineers using designs supplied by a client recognize that the designs remain the property of the client and may not be duplicated by the engineer for others without express permission.

   c. Engineers, before undertaking work for others in connection with which the engineer may make improvements, plans, designs, inventions, or other records which may justify copyrights or patents, should enter into a positive agreement regarding ownership.

   d. Engineers' designs, data, records, and notes referring exclusively to an employer's work are the employer's property.

  11. Engineers shall cooperate in extending the effectiveness of the profession by interchanging information and experience with other engineers and students, and will endeavor to provide opportunity for the professional development and advancement of engineers under their supervision.

   a. Engineers shall encourage engineering employees' efforts to improve their education.

   b. Engineers shall encourage engineering employees to attend and present papers at professional and technical society meetings.

   c. Engineers shall urge engineering employees to become registered at the earliest possible date.

   d. Engineers shall assign a professional engineer duties of a nature to utilize full training and experience, insofar as possible, and delegate lesser functions to subprofessionals or technicians.

   e. Engineers shall provide a prospective engineering employee with complete information on working conditions and proposed status of employment, and after employment will keep employees informed of any changes.

"By order of the United States District Court for the District of Columbia, former Section 11(c) of the NSPE Code of Ethics prohibiting competitive bidding, and all policy statements, opinions, rulings, or other guidelines interpreting its scope, have been rescinded as

TABLE 13.1 continued

unlawfully interfering with the legal right of engineers, protected under the antitrust laws, to provide price information to prospective clients; accordingly, nothing contained in the NSPE Code of Ethics, policy statements, opinions, rulings or other guidelines prohibits the submission of price quotations or competitive bids for engineering services at any time or in any amount."

STATEMENT BY NSPE EXECUTIVE COMMITTEE

In order to correct misunderstandings which have been indicated in some instances since the issuance of the Supreme Court decision an the entry of the Final Judgment, it is noted that in its decision of April 25, 1978, the Supreme Court of the United States declared: "The Sherman Act does not require competitive bidding."

It is further noted that as made clear in the Supreme Court decision:

1. Engineers and firms may individually refuse to bid for engineering services.
2. Clients are not required to seek bids for engineering services.
3. Federal, state, and local laws governing procedures to procure engineering services are not affected, and remain in full force and effect.
4. State societies and local chapters are free to actively and aggressively seek legislation for professional selection and negotiation procedures by public agencies.
5. State registration board rules of professional conduct, including prohibiting competitive bidding for engineering services, are not affected and remain in full force and effect. State registration boards with authority to adopt rules of professional conduct may adopt rules governing procedures to obtain engineering services.
6. As noted by the Supreme Court, "nothing in the judgment prevents NSPE and its members from attempting to influence governmental action. . . ."

Note: In regard to the question of application of the Code to corporations vis-a-vis real persons, business form or type should not negate nor influence conformance of individuals to the Code. The Code deals with professional services, which services must be performed by real persons. Real persons in turn establish and implement policies within business structures. The Code is clearly written to apply to the Engineer and it is incumbent on a member of NSPE to endeavor to live up to its provisions. This applies to all pertinent sections of the Code.

National Society of Professional Engineers
1420 King Street
Alexandria, Virginia 22314-2794
703/684-2800 FAX: 703/836-4875
Publication date as revised: July 1993 • Publication #1102

SOURCE: Reprinted by permission of the National Society of Professional Engineers.

Employment for Engineers and Scientists. The objectives include "recognition of the responsibility to safeguard the public health, safety, and welfare," but the primary emphasis is on employment conditions, professional growth, employee loyalty and creativity, and equal opportunity. Sections are included on

TABLE 13.2   Core Concepts in Engineering Ethics

I. THE PUBLIC INTEREST

    A. Paramount responsibility to the public health, safety, and welfare, including that of future generations.

    B. Call attention to threats to the public health, safety, and welfare, and act to eliminate them.

    C. Work through professional societies to encourage and support engineers who follow these concepts.

    D. Apply knowledge, skill, and imagination to enhance human welfare and the quality of life for all.

    E. Work only with those who follow these concepts.

II. QUALITIES OF TRUTH, HONESTY, AND FAIRNESS

    A. Be honest and impartial.

    B. Advise employer, client, or public of all consequences of work.

    C. Maintain confidences; act as faithful agent or trustee.

    D. Avoid conflicts of interest.

    E. Give fair and equitable treatment to all others.

    F. Base decisions and actions on merit, competence, and knowledge, and without bias because of race, religion, sex, age, or national origin.

    G. Neither pay nor accept bribes, gifts, or gratuities.

    H. Be objective and truthful in discussions, reports, and actions.

III. PROFESSIONAL PERFORMANCE

    A. Competence for work undertaken.

    B. Strive to improve competence and assist others in so doing.

    C. Extend public and professional knowledge of technical projects and their results.

    D. Accept responsibility for actions and give appropriate credit to others.

SOURCE: A. G. Oldenquist and E. E. Slowter (1979); reprinted by permission of National Society of Professional Engineers.

recruitment, employment, professional development, and termination and transfer (Kemper, 1990).

Do codes of ethics have any "teeth?" Can they be enforced? The answer is "Yes!" Professional societies that promulgate codes of ethics have an interest in having their members abide by the written provisions and a responsibility for seeing that this is done. Likewise, state boards of registration often are vested by statute with authority to investigate allegations of violations of codes of ethics. The American Society of Civil Engineers offers an example of professional society activity in this regard. Alleged infractions of its code (by members) can be brought to the attention of the ASCE Executive Committee, which refers the case to the Committee on Professional Conduct. In a very structured, thorough, and confidential manner, the committee conducts its investigation and reports its findings to the society's board of direction. Hearings are held and penalties imposed, if appropriate. Only the board makes these decisions. Penalties can include a letter of reprimand with no disclosure of names, a letter of reprimand with disclosure to the ASCE membership of the nature of the infraction and the names of the guilty parties, dismissal from the society for a given number of years, or expulsion from the society with no possibility of later reinstatement (American Society of Civil Engineers, 1991; Firmage, 1980).

Like most other human endeavors, codes of ethics tend to evolve and change over time. An example is the elimination of sections that prohibited the submittal of priced proposals for engineering planning and design work. We outlined in Section 10.1.3 some reasons for the difficulty of estimating in advance the cost of such efforts. On the basis of such reasoning, ASCE, NSPE, and the American Institute of Architects (AIA) had for many years deemed it unethical to submit priced proposals under conditions that constitute price competition for professional services (and therefore providing penalties for violation). However, in 1972, the U.S. Department of Justice filed suit against these three societies, alleging that their codes of ethics violated the Sherman Antitrust Act by being in restraint of trade. The end of a rather long and complicated story is that ASCE and AIA decided to accept a consent decree to remove such stipulations from their codes and supporting documents, while NSPE opted to contest the action. At each stage, all the way through the U.S. Supreme Court, NSPE lost; the result was a modified NSPE code. The statement at the end of the code reproduced in Table 13.1— a paragraph required by the final court judgment and a statement by the NSPE Executive Committee—are the result of that long and bitter struggle.

A concise and interesting review of the events summarized in the preceding paragraph is provided by Firmage (1980), based on his active involvement with ASCE during those years.

We should note that just as professional societies have developed codes of ethics to guide the actions of their members, so too have private organizations. The Maguire Group, Architects/Engineers/Planners, headquartered in Foxboro, Massachusetts, established an Ethics and Business Conduct Task Force of 27 employees to draft and implement a comprehensive ethics program (*Maguire Group Ink*, 1992, 1993). The result was Maguire's *Code of Ethics and Business Conduct Guidelines* (1993). Table 13.3 reproduces the introduction to the program. The detailed code of ethics includes sections on business courtesies, gifts and gratuities, improper payments, commissions/contingency fees, political contributions, conflict of interest, confidentiality, competitor resources and accounting, and software use.

Codes of ethics are an excellent source of guidance for the engineer when confronting questions of right and wrong. As pointed out early in this section, they are intended to provide guidance in relations with the public, the profession, and the employer and client. The next section asks us to think briefly about situations in which ethical obligations to more than one of these groups may be in conflict.

13.3   OBLIGATIONS IN THREE SOMETIMES CONFLICTING DIRECTIONS

Keeping one's performance closely in line with one's professional code of ethics is not always easy and is sometimes impossible, as we shall see in some of the cases in Section 13.4. Even though the various elements shown in Table 13.2 all seem very laudatory, closer examination reveals that they may be in conflict. If

TABLE 13.3   Code of Ethics and Business Conduct Guidelines

OUR COMMITMENT TO EXCELLENCE

Conducting business in an ethical manner is the most important commitment a business can make to its clients and employees. For more than 55 years, Maguire has been providing high quality, personalized services to a variety of clients and we are proud of our heritage which is based upon integrity.

To reaffirm our commitment to setting and meeting high standards of ethical conduct in the workplace and beyond, Maguire Group Inc. has developed a formal code of ethics, detailing the standards of professional behavior to which all employees are expected to adhere. Maguire acknowledges that achieving and maintaining a good business reputation requires a commitment to excellence from *all* employees.

All members of Maguire are responsible for upholding and conforming to the high standards of ethical behavior and business conduct outlined in *Maguire Code of Ethics and Business Conduct Guidelines* and must strive to fulfill the following obligations:

- To be dedicated to the principle of integrity.
- To accept only the highest of standards of ethical behavior in all aspects of business practice.
- To provide expert, professional and quality services to clients and the public.
- To accurately document and report all business transactions.
- To be law abiding in all their activities.

The Maguire of today promises to be an even stronger company in the future.

SOURCE: Maguire Code of Ethics and Business Conduct Guidelines. Reprinted by permission of Maguire Group, Inc. © 1993

we identify the engineer's primary duties as being threefold, to the *public,* to the *profession,* and to the *employer or client,* fulfilling a duty to one may mean being unfaithful to another.

Consider an engineer whose employer requires that all correspondence be routed strictly "within channels." If the employee discovers a practice that is unsafe to the public and reports it to the employer, who chooses to ignore the report, what action is proper? Should the engineer go outside the regular communication channels to call attention to the unsafe condition? Loyalty to the employer and concern for public safety seem to be in conflict.

Concern for the profession may militate against genuine concern for the needs of the public. In their paper about ethical decision making for engineers, Nelson and Peterson (1979) include a section entitled "The Professional as an Honest Crook." In this context they warn that "it is possible for a professional to totally ignore the larger, moral dimensions of his or her professional scope of activities and yet be a paragon of virtue within the profession." Using some interesting word choices, the authors suggest that "professional" conduct of integrity may simply be a code of honor among thieves:

> being loyal to your accomplices, not betraying your partners in crime, keeping promises made to one's collaborators,... being reliable (i.e., don't be late with the getaway car), improving one's professional specialty and skills (safecracking, burglary, extortion techniques, etc.), not charging less than competitor colleagues.... "Professionalism" must mean more than this internal governing of accepted practice and rules or moral heroism will amount to no more than achieving technical perfection at any cost. (Nelson and Peterson, 1979)

One approach to resolving the potential dilemma inherent in trying to satisfy the three elements of the engineer's duty is to prioritize them. Blinn (1989) suggests the following order: "1) the effect on the welfare of the public, 2) the effect on the other members of the profession and the profession as a whole, and 3) giving adequate consideration to the matter of loyalty to the employer."

Next we describe some situations in which these conflicting duties may be major issues.

13.4 SOME ETHICAL ISSUES IN ENGINEERING

This section is far from exhaustive. We note some types of situations that can give rise to ethical dilemmas. By citing "major" cases from engineering practice, we do not wish to leave the impression that the engineer typically encounters such dilemmas in day-to-day practice. Rather, it is more often the seemingly minor issue that is encountered; furthermore, probably every world-scale fiasco began in some small way. Note also some overlap in these somewhat arbitrary categories. For example, whistleblowing may be a means of exposing improper contributions.

13.4.1 Conflict of Interest

Conflicts of interest can arise in a number of situations. In the broadest sense, conflict arises whenever taking an action for a given party prevents one from meeting at least one other obligation. Examples of pairs whose interests may not be congruent at all times include your employer and a supplier, a public body and your private practice, and your employer and a competitor. If you serve on a hospital board of directors, should you seek and be granted a contract to do design services for that hospital?

The engineer is often in the position of receiving gifts from suppliers or others. While there probably cannot be absolute rules for deciding what size gift is too large, the following case is illustrative. [Like many of the cases in this chapter, it is taken from the files of NSPE's Board of Ethical Review (BER).]

A piping supply company was interested in gaining acceptance in a certain area and having its products specified by design engineers. The company invited engineers to attend a free one-day seminar, including refreshments, buffet luncheon, and cocktail reception, at which advantages and disadvantages of various types of pipe were to be discussed. In deciding whether attendance by an engineer would constitute unethical practice, the NBER recognized that the seminar was primarily an educational event and was not a consideration in exchange for which the engineer would agree to specify the company's products. The food and social events were minor ("de minimis") activities, and thus the engineers' participation was considered to be ethical. The board noted that the acceptance of items of "substantial value," such as travel expenses and a multiday event at a resort location, would have led to a different conclusion (BER Case 87-5, 1987).

Sometimes a designer who is also the overseer of construction can be placed in a situation that borders on conflict of interest. A firm designed a major highway project in Peru. The Peruvian government hired the same firm to represent the country's interest in the construction work, which was being undertaken by another firm. When it became apparent that the design was inadequate and unsafe, resulting in major landslides and other problems, the oversight firm's representative refused to carry out orders to (1) allow padding of the payroll to cover slide cleanup costs and (2) assure the government that the work was proceeding properly. The representative was fired, but he later testified before officials of the General Accounting Office, revealing the misconduct of his former employer. While this case raises many ethical issues, a major one is certainly the conflict between representing a client government's interests and those of one's employer (Martin and Schinzinger, 1989).

In another case relating to specifying a product, a manufacturer offered an engineer indemnification if the client would bring suit against the engineer in the event the manufacturer's product did not perform according to the client's expectations. This promise was a sort of consideration in exchange for specifying the use of this particular product. NSPE's Board of Ethical Review looked into several sections of the society's Code of Ethics related to compensation and conflict of interest and advised that the engineer had an ethical obligation not to accept the indemnification for his personal benefit. The board wrote:

> We can foresee circumstances where such an arrangement would create a conflict of interest between the specifying engineer's obligation to specify products consistent with the best interests of the client and the specifying engineer's

self interest in achieving the maximum protection from potential liability. For example, the specifying engineer's self interest could lead to a decision to specify the product and not some other more appropriate or less expensive product. (BER Case 91-7, 1991)

A final conflict-of-interest case illustrates the thorny nature of many such situations. A consulting engineer contracted with a federal environmental agency to develop a hazardous waste remediation strategy. The contract provided for basic consulting services plus possible additional services at a later date, at the agency's request. The contract was silent as to the consultant's working for other clients. After 2 years, the engineer was retained by an industrial client in connection with a hazardous waste site for which the federal agency claimed that the industrial corporation was responsible. After the contract with the industrial client had been executed, the federal agency again contacted the engineer and asked that he provide additional services, as an extension of the original agreement. The services contemplated would have involved the hazardous waste site that is central to this case. The engineer responded that to accept the proposed contract extension would be a conflict of interest, and he declined to perform the work.

The especially thorny issue in the hazardous waste site case arises at step 2, not step 3. Was it unethical for the engineer to agree to perform services for the *industrial client* without prior consent from the federal agency? NSPE's Board of Ethical Review, by a vote of 4 to 3, decided that this action *was* unethical, with the minority holding that sufficient time had passed and that it was unreasonable to preclude the engineer who had developed the strategy the government had desired to have applied from ever working for a client over whom the agency had oversight responsibility (BER Case 91-6, 1991).

13.4.2 Confidentiality and Employee Loyalty

The apparent need to keep certain information confidential, and the obligation to remain loyal to one's employer, are issues that at times are one and the same. We deal with them as one issue here, even though an employee's loyalty may not involve confidential information, and confidentiality can include issues other than relations with an employer. An engineer who shares secret laboratory results with his or her employer's competitor has violated both confidentiality and company loyalty. An engineer who informs a regulatory agency that his or her employer has made improper changes in test procedures for the sake of more favorable engine emission test results is certainly discounting loyalty to the employer in exchange, presumably, for a more important concern about public health (Martin and Schinzinger, 1989). The ethical obligation not to share a former employer's trade secrets with a new employer is a confidentiality issue that protects the former, but not the current, employer.

Donald Wohlgemuth was a chemical engineer who managed B. F. Goodrich's space suit division in the 1960s. He left that company and joined

International Latex Corporation as manager of engineering for industrial products; he managed, among other assignments, a large government contract for developing space suits for the Apollo program. He had signed a confidentiality agreement with Goodrich under which he promised not to reveal the company's trade secrets to any new employer. While it may be possible for a person in such a position not to convey specific information, it seems virtually impossible not to share general knowledge, hence to "leak" information of advantage to the new employer.

Goodrich charged Wohlgemuth with unethical behavior for taking the job with International Latex and filed suit to prevent him from working for that or any other company engaged in developing space suits. The Ohio Court of Appeals was not willing to grant such an order, but it did issue an injunction that prohibited Wohlgemuth from revealing any Goodrich trade secrets. The proper balance, the court said, was to keep trade secrets from the new employer but still to allow the individual to advance his career (Martin and Schinzinger, 1989).

The tenants of an apartment building sued the owner to require that defects affecting their quality of life be repaired. The owner's attorney engaged an engineer to inspect the building and give expert testimony that would support the owner's position. During the inspection, the engineer discovered serious structural defects that he believed constituted an immediate threat to the safety of the tenants. Upon learning of these findings, the attorney directed the engineer to keep them confidential, as the information was part of a separate lawsuit. The tenants' suit was related only to "quality of life" and did not mention the safety-related defects. NSPE's Board of Ethical Review dealt with the question of whether it was ethical for the engineer to conceal his knowledge of the safety-related defects on the grounds that an attorney had told him that he was legally bound to maintain confidentiality. We quote at some length from the board's findings:

> The obligation of the engineer to protect the public health and safety has long been acknowledged by the Code of Ethics and by the Board of Ethical Review. This responsibility rests with the recognition that engineers with their education, training and experience possess a level of knowledge and understanding concerning technical matters which is superior to that of the lay public. It also is rooted in the implicit fact that as individuals who are granted a license by the state to practice, engineers have a duty to engage in practice which is consistent with the interests of the state and its citizenry.... [I]t is clear that there may be facts and circumstances in which the ethical obligation of engineers in protecting the public health and safety conflict with the ethical obligation of engineers to maintain the right of confidentiality in data and other information obtained on behalf of the client. While we recognize that this conflict is a natural tension which exists within the Code, we think that under the facts of this case, there were reasonable alternatives available to [the engineer] which could assist him in averting an ethical conflict.

> It appears that [the engineer], having become aware of the imminent danger to the structure, had an obligation to make absolutely certain that the tenants and

public authorities were made immediately aware of the dangers that existed. [The engineer's] client was the attorney and technically [the engineer] had an obligation not to reveal facts, data and other information in a professional capacity without the prior consent of the attorney. However, there were valid reasons why [the engineer] should have revealed the information directly to the tenants and public authorities....

Although the attorney retained [the engineer] directly, he did so on behalf and for the benefit of the owner....

[The Code of Ethics] makes a clear exception concerning the obligation of engineers not to reveal facts obtained in a professional capacity without the client's consent. That exception allows the disclosure of such information in cases authorized by the Code or required by law. We believe that in cases where the public health and safety is endangered, engineers not only have the right but also the ethical responsibility to reveal such facts to the proper persons. We also believe that state board rules of professional conduct might require such action by professional engineers.

CONCLUSION: It was unethical for [the engineer] not to report the information directly to the tenants and public authorities. (BER Case 90-5, 1990)

A significant case involving employee loyalty versus potential public safety problems arose in 1977. Virginia Edgerton, a member of the Institute of Electrical and Electronics Engineers and an engineer with 13 years experience in the data processing field, was hired as a senior information scientist to work on a project to install a criminal justice information system on a computer operated by the New York City police department. The same computer already operated SPRINT, an on-line police dispatching system that assists in finding the exact locations of police cars and directing the various units efficiently to emergencies. In the course of her assignment, Edgerton became convinced of the possibility that the addition of the criminal justice information system would degrade the performance of the SPRINT system to the extent that police responses would be slower and the public safety would be adversely affected. Her immediate supervisor disagreed with this assessment and refused to have the potential problem studied. Edgerton sought the advice of IEEE, which referred her to the manager of systems programming at the Columbia University Computer Center, who responded that she had raised a legitimate issue. She then wrote a memorandum to her supervisor, who was the project director, outlining the danger as she understood it; the director rejected her concerns.

Two weeks later, Edgerton wrote a memorandum expressing her continuing concerns to the committee responsible for the project (and thus her supervisor's "boss"). One week later, her employment was terminated because, as her supervisor declared, distribution of the last memorandum to the committee was in direct violation of policy established by her supervisor and against orders that all communications sent to committee members must be approved by her supervisor.

IEEE conducted its own investigation of the incident and concluded that Edgerton's action was fully in accord with the spirit and letter of the IEEE

Code of Ethics, that her discharge constituted seriously improper treatment of a professional, and that her action on behalf of public safety, at considerable personal sacrifice, was in the highest tradition of professionalism in engineering ("MCC Report in the Matter of Virginia Edgerton," 1983; "Professional Responsibility and the Dispatching of Police Cars—A Case Study, 1983).

At Ford Motor Company in the early 1970s, Frank Camps was a senior principal design engineer involved in the development of the new Pinto. Crash tests of early models showed that the cars had a high potential for ruptured fuel tanks in crashes at speeds exceeding 25 mph. An in-house study determined that the design could be improved at a cost of $11 per car, but when the benefits to society were quantified, this amount per car came out to be about $4. It was then decided not to make the improvements. When Camps became concerned about the prospect of large numbers of accidents, he was met by resistance from supervisors and lack of support from colleagues. To protect himself against liability, he sued Ford. The sad end of the story is that at least 53 people have died as a result of gas tank explosions in rear-end Pinto collisions. More than 50 lawsuits were filed against the company (Callahan, 1988; Glazer, 1988).

13.4.3 Contributions and Kickbacks

This chapter began with the story of a kickback scheme involving a vice president of the United States who had to answer for actions taken when he was a county executive responsible for public works design services. It seems impossible to give absolute guidance on the proper limits to engaging in social relationships with those responsible for contracting with you. But certainly the payment of bribes or kickbacks in exchange for public work is absolutely unethical. It is truly sad when the engineer participant in such a scheme justifies the action by claiming a higher loyalty to employees (to provide them with jobs) or to family (to provide them an income) than to the public whose resources are being diverted for personal gain. Perhaps this topic is one of the easiest for which to stipulate some absolutes!

Allan Kammerer was the 32-year-old principal of a small Virginia consulting firm. He was approached by the county engineer about doing highway design work for the county. A condition was that 25% of the payments to Kammerer's firm would be paid back covertly to the county engineer. Kammerer decided to go along, justifying his actions at the time by the need to keep his staff employed, and he kicked back a total of $100,000 over 4 years. When the U.S. attorney general began an investigation of the county's contracting practices, Kammerer had the chance to destroy his records. However, by then he had seen the error of his ways, and he became a witness against the county engineer, who was subsequently convicted of extortion and sentenced to a large fine and a jail term. After that, Kammerer endured "seven years of hell," as he was investigated by his state's department of transportation, suspended by ASCE, considered for a civil lawsuit (which never came to trial), and subjected to public exposure and embarrassment in front of family, friends, and colleagues (Kemper, 1990).

For an entertaining, almost unbelievable true account of an engineer's encounters with a kickback system, read *I Gave Up Ethics—To Eat!* (1975). An engineer describes his experience in setting up a new consulting firm and finding no work until a colleague explained the need to know how to practice the fine art of "political engineering." A key person to the neophyte's scheme was "the Reverend," who "had his own Code of Ethics, and...abided by it. He was so crooked he could have worn a corkscrew for a tie-clasp, but when he worked for me, I received his undivided loyalty." The Reverend had the right kind of political connections, which he willingly exercised in return for a 10% commission on work obtained.

Are most engineers engaged in such schemes? Certainly not! But, just as we can learn much about structural mechanics by studying examples of the small minority of structures that collapse, so too can examples of the "bad eggs" in professional practice help to remind us of the kind of activities we wish to avoid.

13.4.1 Whistleblowing

A dilemma often faced by the professional is whether to "blow the whistle" on an employer for acts or situations the employee believes to be ethically improper. Professionals, because of their superior knowledge and background in their respective areas of specialty, can find themselves suspecting or knowing about highly sensitive situations even though a lay-person, including the supervisor, does not. On the other hand, the employer may understand the situation fully and still choose not to act ethically. Several of the cases described above, including the Peruvian highway project, the Pinto case, and the New York City police department computer development, included elements of whistleblowing.

Author Sissela Bok begins a paper on "Whistleblowing and Professional Responsibilities" (1988) with the following:

> "Whistleblowing" is a new word in the glossary of labels generated by our increasing awareness of the ethical conflicts encountered at work. Whistleblowers sound an alarm from within the very organization in which they work, aiming to spotlight neglect or abuses that threaten the public interest.

> The stakes in whistleblowing are high. Take the nurse who alleges that physicians enrich themselves in her hospital through unnecessary surgery; the engineers who disclose safety defects in the braking systems of a fleet of new rapid-transit vehicles; the Defense Department official who alerts Congress to military graft and overspending: All know that they pose a threat to those whom they denounce and that their careers may be at risk....

> The whistleblower hopes to stop the game, but since he is neither referee nor coach, and since he blows the whistle on his own team, his act is seen as a violation of loyalty. In holding his position, he has assumed certain obligations to his colleagues and clients: stepping out of channels to level accusations is regarded as a violation of these obligations. Loyalty to colleagues and to clients comes to be pitted against loyalty to the public interest, to those who may be injured unless the revelation is made.

A survey revealed that 65% of managers had firsthand knowledge of fraud, waste, or mismanagement in a one-year period, but only 12% of their companies provided effective methods for revealing and reporting such conditions. Forty-two percent of the managers suspected that their companies punished those who reported problems internally. Thus, ethically minded employees tend to respond to such lack of support by taking the information outside formal channels; for their trouble, they are often transferred, demoted, suspended, or forced to resign ("The Whistleblowing Dilemma," 1992).

Kermit Vandiver was an instrumentation engineer, data analyst, and technical writer for the B. F. Goodrich Company. In 1968 he became involved in a celebrated whistleblowing case related to Goodrich's design and production of a wheel brake for the A7D light attack aircraft. His side of the story is told in highly readable form in a popular book entitled *In the Name of Profit* (Heilbroner et al., 1972). The contract for this relatively small, 106-pound, four-disk brake was awarded by the Ling-Temco-Vought Company (LTV), partly because the proposed brake was lighter than the usual five-disk type. In numerous qualification tests, the brake failed, despite some irregularities in the test procedures apparently tolerated in the hope of achieving passing results. Moreover, the report stating that the brake had finally passed contained numerous falsifications that Vandiver and a colleague had been directed to insert.

When the brake entered the flight test stage, difficulties continued; on one occasion, high heat caused the brake to weld together, and the plane skidded nearly 1500 feet. At this point, Vandiver approached his attorney, who advised him to go to the FBI. Soon after the FBI heard the story, the Air Force revoked its earlier approval of the brake test report. Vandiver and his engineer colleague resigned their positions at Goodrich, and, within 2 days, the company recalled its qualification report and announced that it would substitute a five-disk brake at no extra cost to LTV. Vandiver testified at hearings of Senator William Proxmire's Economy in Government Subcommittee, as did representatives from his former employer. Although the subcommittee reached no firm conclusions, the Department of Defense instituted major changes in its inspection, testing, and reporting procedures as a result of the Goodrich episode (Vandiver, 1972).

Are there two sides to this story? Of course! Goodrich was trying to push the edge of technology to save weight and money on an innovative design. Did its engineering managers exercise "engineering judgment" in interpreting test results? Yes! Did they deviate from specifications? Maybe! Did the company employ some test procedures generally accepted by the Air Force although not precisely in accord with requirements? Yes! Martin and Schinzinger (1989) conclude their description of this case thus: "It is clear that the lodging of a charge of wrong-doing against other persons carries with it an enormous responsibility, because those other persons have rights too."

The design of the Bay Area Rapid Transit (BART) system in northern California involved the innovative use of computers for system control. Three engineers identified dangers in the design of the automatic train control system. These problems, together with insufficient system testing and inadequate, operator and construction monitoring, resulted in several early accidents. During the development phase, the engineers wrote memos and voiced their concerns to their employers and colleagues, to no avail. They contacted members of BART's board of directors and a private engineering consultant. They were invited to address the board with their concerns but were unable to convince its members of the seriousness of the problems. A week later, they were invited to resign or be fired, on the basis of "insubordination, incompetence, lying to supervisors, causing staff disruptions and failing to follow understood organizational procedures." Their suit against BART was eventually settled out of court for $75,000 minus 40% for lawyers' fees (Martin and Schinzinger, 1989).

On January 27, 1986, 73 seconds after liftoff, the space shuttle *Challenger* exploded in flight, killing all seven astronauts, destroying a $1.2 billion shuttle, throwing 25,000 employees out of work, and setting the nation's space program back several years. The incident is an important case for the engineering community, because it represents responsible whistleblowing that went unheeded.

The primary cause of the explosion was a seal failure in the solid rocket booster joint. The joint is equipped with a primary and a secondary O-ring, which flex and maintain the seal when pressure from hot gas causes the joint to rotate (see Figure 13.1). Tests had shown that at low ambient temperatures, the rubber O-rings were less flexible and some gas blowby was possible. Roger M. Boisjoly was an engineer at Morton Thiokol, the company that supplied the O-ring assemblies to NASA. Boisjoly, who was considered to be the country's leading expert on O-rings and rocket seal joints, advised that the low temperature problem needed further work, and he predicted major difficulties if it was not addressed. On the day before the launch, the overnight temperature was predicted at 18°F; since 53° was the recommended minimum seal temperature, Boisjoly and colleagues urged that the launch be postponed. A critical meeting that day included a statement that has become a classic in misguided engineering management and ethics. One manager turned to another, who was opposing the launch, and asked him to take off his "engineering hat" and put on his "management hat."

Since the disaster, Boisjoly, who left his company one year afterward with posttraumatic stress disorder, has devoted his energies to explaining his unsuccessful attempt to have the launch postponed. Before his resignation, he testified to the commission that was investigating the incident and made documents available, thus violating the "company line." On one occasion, he said,

I have been asked by some if I would testify again if I knew in advance of the potential consequences to me and my career. My answer is always an immediate "yes." I couldn't live with any self respect if I tailored my actions based upon the

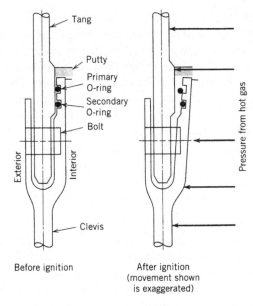

Cross-section of field joint

Figure 13.1. *Challenger* rocket seal. (From M.W. Martin and R. Schinzinger, *Ethics in Engineering*, 2nd ed. © 1989. Reprinted by permission of McGraw-Hill.)

personal consequences as a result of my honorable actions. (Boisjoly and Curtis, 1987; McCarthy, 1988)

A case in which the whistle was *not* blown involved the McDonnell Douglas DC-10 airplane. In 1974, nine minutes after takeoff from Paris, a DC-10's cargo door blew open, and the cabin floor collapsed as a result of decompression of the cargo compartment. Since in this aircraft, the control systems are routed through the floor, the collapse destroyed aileron and rudder control, the plane crashed, and all 346 passengers and crew died. A partial collapse of the cabin floor had also occurred over Windsor, Ontario, in 1972, but the pilot had been able to land safely. The key to these problems was the design of the cargo door latches. Convair Division of General Dynamics was the subcontractor to McDonnell Douglas responsible for fuselage design, including the cargo doors. Partway through the design process, Convair was directed to save weight by changing from hydraulic to electric actuators for the latches. There is an important distinction between these two modes of actuations: if the latches are not seated properly, only a small amount of pressure differential will cause a hydraulic latch to open, whereas an electric latch under the same conditions is likely to fail under higher pressure differentials. Since small pressure differentials occur on or near the ground and higher differentials occur at higher elevations, the potential for catastrophic failure was much higher with the electric actuators. After the Windsor incident, the Federal Aviation Administration prepared a draft of an airworthiness directive that would have required certain actions before the DC-10 could resume flying. That directive was never issued; instead, a "gentlemen's agreement" was arranged under which McDonnell Douglas would make certain modifications, including the installation of inspection windows and the use of deeper latch engagement. An inspector had certified that these modifications had been made to the plane that crashed near Paris, but in fact they had not. After the Windsor blowout, Convair's director of production engineering wrote a memo to his superiors expressing concern over the safety of the latch system. He wrote, in part, "It seems to me that, in the twenty years ahead of us, DC-10 cargo doors will come open and I would expect this usually to result in the loss of the airplane." Convair's management responded to the engineering manager about his memo but made no formal communication to McDonnell Douglas, and the matter was apparently dropped (French, 1982; Kemper, 1990).

What happens to whistleblowers? We have already mentioned the post-traumatic stress disorder suffered by Roger Boisjoly.

Bertrand G. Berube was a civil engineer and regional administrator for the U.S. General Services Administration, with responsibility for 7000 employees and a $1.2 billion annual budget. He blew the whistle on what he perceived to be neglected maintenance on major government buildings that resulted in unsafe workplaces; in 1983 he was fired. He wrote about his experience and identified four common methods of retaliation against whistleblowers: (1) make the dissenters, not the message, the issue; (2) isolate the dissenters in bureaucratic Siberia, both to make an example of them and to block their access to information; (3) put a dissenter on a pedestal of cards: appoint him or her to solve the problem, and then make the job impossible; and (4) when the problem is not solved, fire the employee for incompetence. He also talked about monetary and personal costs, as follows (Berube, 1988):

> If you blow the whistle on a boss, you are likely to be without a job for three or four months, and legal fees will be in the range of $30,000 to $40,000.
>
> For blowing the whistle on an agency, you may expect to be out of work for one to two years, and your legal fees may run from $125,000 to $150,000.
>
> If you blow the whistle on the political administration in power, you may be off the job four to seven years, and legal fees may be in the $400,000 to $550,000 range.
>
> Another aspect of cost is personal. Friends, for example, turned on me because they thought I had no right to imperil my family's security by telling the truth. I found this attitude in most people I knew.

This author was in the audience when Berube made a presentation to a professional meeting. During the discussion session, someone commented that Berube must have seen many "gray areas" during his experiences. His response went something like this: "I have seen so much that was starkly black and white that I haven't even had time to worry about the gray areas."

Some small consolation was afforded to Berube in 1988, when he was paid $560,000 to settle his claim to have been fired unjustly ("Ex-GSA Employee Gets $560,000 for Discharge," 1988).

Federal employee whistleblowers now come under the Whistleblower Protector Act of 1989, which revised earlier statutes by giving whistleblowers control of their cases, eliminating unrealistic burdens of proof, providing financial relief during the hearing and appeals process, and protecting disclosure channels. At least 20 states have similar statutes (Committees on Standards of Practice and Employment Conditions, 1990).

13.4.5 Professional Conduct

This section treats both assurances by the engineer that his or her products and services are competent and the topic of relationships among professionals. Such issues are covered to some degree in the typical code of ethics in paragraphs dealing with the signing of drawings, the role of the engineer as "faithful agent or trustee," and the standard injunctions against falsely injuring another's reputation or untruthfully criticizing another's work. We cite three cases reviewed by NSPE's Board of Ethical Review.

In the first, an engineer working in the computer manufacturing industry designed some equipment that although proper, resulted in a process that was deemed too costly. The manufacturing division suggested changes to her design to reduce the costs, but the engineer's analysis of these changes indicated that reliability would decrease and downstream costs would increase. The supervisor employed by the firm, who was not an engineer, requested that the designer of the equipment sign off on the changes. No risk to public safety and health was involved. After raising her concerns to the supervisor, the engineer agreed to sign off on the changes without further protest.

While noting that no health or safety issue was involved, the BER relied heavily on the engineer's ethical obligation to act as a faithful agent or trustee: "We cannot see how an engineer could be said to be acting as a "faithful agent or trustee' by silently assenting to a course of action which will have serious long-term ramifications for an employer." In deciding that the engineer had not fulfilled her ethical obligations, the board also suggested that since her immediate supervisor had not been receptive to her concerns, she had an obligation to advise the supervisor that she would bring the matter to the attention of those higher in management (BER Case 88-5, 1988).

How closely must the engineer be involved in the preparation of a drawing in order for that engineer to be ethical in signing them? Here is the BER's opinion as of 1990:

[W]e are not of the view that an engineer must personally prepare the drawings, plans and other documents involved. The key requirement is that an engineer possesses sufficient competence, assumes full responsibility for the work product and carefully directs, controls and reviews the material prepared under the engineer's responsible charge.

The case involved two questions related to the use of computer-aided design and drafting (CADD) systems. First, is it ethical for an engineer to sign and seal documents that he prepared using such a system? And, second, is it ethical for an engineer to sign and seal documents prepared by others under his direction and control using a CADD system? In both cases the board said "yes," provided the engineer checked and reviewed the work in some detail (BER Case 90-6, 1990).

To what degree is it proper to criticize the work of another engineer? The key here seems to be whether the intent is malicious and whether the criticism is untrue or simply a matter of opinion. In a 1990 case, an engineer who was an expert in safety and forensic engineering testified that a manufacturing company's safety practices were inadequate and might have contributed to certain employee injuries. Another engineer who worked for the company later commented to engineers and others, with no substantive support, that the safety expert had acted unprofessionally because he had ignored certain information and focused on circumstances helpful to his client. The board, in looking at a long history of similar cases, said that there may be honest differences of opinion among equally qualified engineers in interpreting known physical facts and stated that it is not unethical for engineers to offer conflicting opinions in such cases. In the case at hand, however, the second engineer's accusations were found to be unethical: "Disagreement on a technical or professional question... does not necessarily mean that the one advocating the position is acting unethically or unprofessionally." The board wrote that although the second engineer was presumed to be sincere in his view that the first engineer had acted improperly, that sincerity could not justify the injury that might have been caused to the first engineer's reputation (BER Case 90-2, 1990).

13.4.6   Global Issues

As the world becomes more complex, and linkages among peoples and systems become more pronounced, the impact on society of the engineer's decisions assumes increasing importance. In doing their work as "social experimenters" (Martin and Schinzinger, 1989), engineers must pay greater attention to the social impact of their decisions. The product designer in the United States may affect life in Russia as never before. The Swedish traffic engineer may influence the environment in a remote section of Africa, and the chemical engineer from Australia may be making life-and-death decisions in drafting facility designs for India. In this section, we address some of these increasingly wide social dimensions.

The political environment in which some engineers work may provide challenges with respect to the *distribution of public services*. The city engineer or public works director may make decisions, or may be subject to the decisions of others, regarding public works improvements for certain areas to the exclusion of other areas. There are laws that provide equal protection under some circumstances, and these must be respected. Beyond that, tendencies toward some forms of subtle discrimination may be part of the environment in which persons in these positions must work.

We have already noted the increasing recognition of *environmental ethics* as a legitimate basis for decision making. It is not our purpose here to take a position on how badly our earth has suffered from development and industrialization or to predict the future of ecological trends (or to speculate, as some do, over how many decades we have to live). Certainly environmental politics can lead to the most heated of controversies and debates, and often engineers find themselves in the middle. Disastrous flooding in America's Midwest in 1993 led, among other things, to charges that the U.S. Army Corps of Engineers had overbuilt a levee system along the Mississippi River that contributed to greater flood damage than there would have been if no such protection had been in place.

Issues include acid rain (from the sulfur dioxide and nitrogen oxides produced in the burning of fossil fuels), nuclear power generation, chemical production, asbestos from mining operations, water reservoirs that flood nesting areas, and oil spills that damage animal and plant life, as well as the view as seen by the human eye. The code of ethics that stipulates that the engineer must hold paramount the health, safety, and welfare of the public will be interpreted to include the environment as well, and some codes, such as that of the Accreditation Board for Engineering and Technology, already contain such provisions. Surely the traditional benefit–cost analysis will need to recognize the benefits and costs of environmental impacts, however difficult such recognition may be.

While the threat of global war seems to have lessened in the early 1990s, *national defense and weapons development* are still major industries, and engineers continue to play major roles. Oh that there might be an easy answer to this dilemma! The duty ethicist may say either "Thou shalt not kill!" or "Protect the homeland and innocent peoples at any cost!" And the utilitarian may justify either war or no war on the basis that the respective results are best for the overall good. Martin and Schinzinger (1989) pose six hypothetical situations in the weapons development field, of which two are quoted here.

> Ron is a specialist in missile control and guidance. He is proud to be able to help his country through his efforts in the defense industry. The missiles he works on will carry single or multiple warheads with the kind of dreadful firepower which, in his estimation, has kept any potential enemy in check since 1945. At least there has not been another world war—the result of mutual deterrence, he believes. . . .

> Joanne is an electronics engineer whose work assignment includes avionics for fighter planes which are mostly sold abroad. She has no qualms about such planes

going to what she considers friendly countries, but she draws the line at their sale
to potentially hostile nations. Joanne realizes that she has no leverage within the
company, so she occasionally alerts journalist friends with news she feels all
citizens should have. "Let the voters direct the country at election time"—that is
her motto.

The activities of engineering firms who do *business in other countries* and
the influences of large, multinational corporations also testify to an increasingly
interconnected world. "When in Rome do as the Romans do" suggests that a
company operating in a region with different ethical standards should drift with
the local conventions. However, one can choose not to participate, as shown in the
following example.

An American design firm submitted a proposal to a European company, and the
officers were invited for an interview. The firm was ranked first, whereupon it
submitted a fee proposal for a detailed scope of services. That proposal was
accepted, and the principals worked with the owner to draft a professional
services contract. When the document was finalized, members of the firm were
invited to a contract signing at the owner's headquarters. At that meeting,
the owner presented, for the first time, an addendum that included four new
requirements: (1) that professional liability insurance, to be paid by the design
firm, was to be carried by a firm headed by the contract officer's wife; (2) that
all printing and reproduction services would be performed by a company headed
by the son of the head of the owner company; (3) that travel arrangements for
the contract would be handled by a travel agency owned by the same son; and
(4) that 2.5% of the proposed fee of just over $4 million was to be rebated to
the owner firm's "Employee Morale Fund." Upon receiving these conditions,
the American firm ended the negotiations and came home, "empty-handed but
wiser" ("Enterprise Ethics and the International Marketplace," 1990).

One is moved by such a story to suggest that it may be time to develop an
international code of ethics. Could this be done? Would such a document be
respected? Could it be enforced?

13.4.7   A Comment and a Concluding Case

In the interest of a "clean" outline, the classifications presented in Sections 13.4.1
to 13.4.6 have been somewhat arbitrary, and a reading of them makes it clear that
several of the cases might be placed in more than one category. Nevertheless, the
classifications demonstrate the diverse kinds of cases and their complicated and
many-faceted nature.

We conclude this section with a case involving an environmental issue that
entails matters of loyalty to employer, professional conduct, and perhaps conflict
of interest. NSPE's Board of Ethical Review dealt with this case in late 1992.
We provide the facts and questions as presented in the BER report, plus selected
portions of the discussion and conclusions (BER Case 92-4, 1992).

An environmental engineer employed by a state environmental protection division was directed to prepare a construction permit for a manufacturing facility power plant. He was told by a superior to move quickly and to "avoid any hangups" on the technical issues. He believed the plans as presented were inadequate because outside scrubbers to reduce sulfur dioxide emissions had been omitted. Without such devices, plant operation would result in air pollution under standards set forth in the 1990 Clean Air Act. Thus it would be wrong to issue a permit. The engineer's superior believed that an alternative method would allow the plant to meet the regulatory requirements. The engineer contacted his state registration board and was told that suspension or revocation of his license was a possibility if he prepared a permit to operate in a way that violated environmental regulations. He refused to issue the permit and submitted his findings to his superior, whereupon the department head authorized issuance of the permit. The case was reported widely by the news media, and at the time NSPE reviewed it, it was being investigated by state authorities. The questions considered were:

1. Would it have been ethical for the engineer to withdraw from further work on the case?
2. Would it have been ethical for the engineer to issue the permit?
3. Was it ethical for the engineer to refuse to issue the permit?

The BER called the situation

> a classical dilemma faced by many engineers in their professional lives. Engineers have a fundamental obligation to hold paramount the safety, health and welfare of the public in the performance of their professional duties . . . we do not believe it is incumbent upon [the engineer] to bring this issue to the attention of the "proper authorities" [since] such officials are already aware of the situation and have begun an investigation. . . . [W]e believe it would not have been ethical for [the engineer] to withdraw from further work on this project because [he] had an obligation to stand by his position consistent with his obligation to protect the public health, safety and welfare and refuse to issue the permit. Engineers have an essential role as technically-qualified professional to "stick to their guns" and represent the public interest under the circumstances where they believe the public health and safety is at stake. . . . [The engineer's actions in seeking advice from the registration board] constitute appropriate conduct and actions consistent with [the NSPE Code of Ethics]. (BER Case 92-4, 1992)

Thus, the board's answers to the three questions were (1) no, (2) no, and (3) yes.

13.5   ETHICS AND THE ENGINEERING MANAGER

If an engineering organization and its employees are expected to make decisions that are ethical, it certainly is incumbent on the engineering manager to provide the proper setting. This responsibility is both a privilege and an overwhelming challenge. The manager has an obligation to the public, the employee, the organization, and the profession to enforce ethical decisions and behavior.

Primary among the ways the engineering manager can instill ethical behavior is to *provide the good example*. L. L. Lammie, of Parsons Brinkerhoff, Inc., was the 1991 recipient of the ASCE's Management Award. In his acceptance remarks, he asked how one maintains cultural values and ethical standards in an organization of 3000 employees. His answer was "1) Set the example; 2) communicate; 3) set the example; 4) train; 5) set the example; and 6) communicate by all media— written, verbal and video" ("The Dimensions of Ethics," 1992). The engineering manager demonstrates ethical behavior when he maintains high personal standards, when interpersonal relationships within the organization are above reproach, when relations with associates and competitors are conducted in an atmosphere of respect, recognition, communication, and fairness, and when clients are selected because of their ethics, contracted with on a fair basis, and told the truth about project status and problems (rather than what they might want to hear).

As a supervisor, the engineer must *be available* to discuss matters of ethical concern with employees, to assure them that the ethical decision is always the right decision, and to convey an attitude that assures the employee that the manager can be trusted (Frantz, 1988). Beyond setting the example, an active involvement with employees assists them in carrying out their responsibilities.

Finally, the engineering manager can provide opportunities for *education and information exchange*. We have referred to several cases published by the National Society of Professional Engineers. NSPE's Institute for Engineering Ethics publishes *Engineering Ethics Update*, several issues of which are cited in the reference list for this chapter. That institute has also published a resource guide (*Professional Ethics and Engineering: A Resource Guide*, 1990), which includes a bibliography, a list of ethics organizations, a number of films, and a list of resource professionals in engineering ethics, and it has produced videotapes that can be used for employee motivation and training.

The cases appearing at the end of this chapter, with "answers" provided in the Appendix, could form the basis for a lively training and discussion session. Another source of cases is Glenn's *Ethics in Decision Making* (1986).

The Martin Marietta Corporation has produced a board game called *Gray Matters: The Ethics Game* (1992), consisting of a series of 55 mini-cases, in which player teams select the most correct ethical choice. The leader's guide provides the "correct" answer to each case, with an explanation. Scores are recorded, and the winning team rewarded at the end. The name of the game seems appropriate, since we now know, as we approach the end of this long chapter, that most such matters are seldom black and white, despite Bertrand Berube's observation to the contrary.

13.6   IN CONCLUSION

Are there any absolutes anymore? Near the beginning of this chapter, we asserted that there are. It is hoped that the material considered throughout the chapter will provide a modest basis for confronting the ethical questions that are going to arise in the course of an engineer's career. For starters, there will be your own *personal sense of morality.* Then, the various *codes of ethics* and guidelines for professional employment can provide a reference point for considering your choices. Finally, the *experiences of others,* as illustrated by the cases presented above and the code interpretations by such groups as NSPE's Board of Ethical Review, will give some guidance in choosing a response to a particular situation.

13.7   DISCUSSION QUESTIONS AND CASES

1. In Section 13.2, we identify a series of roles that codes of ethics are intended to play. Which three of these roles might have negative consequences? Give an example of each.

2. Suppose you are an engineer in the biotechnology field engaged in developing a new genetically engineered compound. Describe some of the potential conflicting obligations you might encounter, based on the discussion in Section 13.3.

3. A conflict arises between you and your supervisor. You suspect her of taking actions that could endanger the public. You further believe that she is not telling the truth about these actions. In trying to decide what action *you* should take, if any, you remember the five theories listed in Section 13.1. Might you possibly reach a different conclusion depending on which theory you adopt? Explain.

4. How might the adoption of one or another of the five theories in Section 13.1 lead to a different management decision in the case of (a) deciding whether to introduce robotic control into a manufacturing operation, (b) selection of a highway traffic improvement scheme, and (c) choice of a space shuttle launch time during inclement weather.

5. In Section 13.4.1, we describe a case in which an engineer was offered indemnification against claims of poor product performance in exchange for specifying that the product was to be used in a project. Review the NSPE Code of Ethics and identify the sections that might apply to this case. After your review, tell whether you agree with the conclusion reached by NSPE's Board of Ethical Review in this case, and why.

Questions 6 through 15 are taken from a series of articles by Jacob L. Peterson in the *Engineering Management Journal.* The articles appeared in Vol. 1, No. 2, June 1989 (Questions 6 and 7); Vol. 1, No. 4, December 1989 (8 and 9); Vol. 2, No. 2, June 1990 (10, 11, and 12); and Vol. 2, No. 4, December 1990 (13, 14, and 15). They are reproduced by permission of the *Engineering Management Journal.* Summaries of reader responses were printed in subsequent issues of the *Journal* and appear, sometimes slightly shortened, in the Appendix.

6. *Rules*

    You have scheduled a crew to work the weekend to complete a task that cannot be done during normal operating periods. This task is in the critical path of an important larger project; thus, if the task is not completed this weekend, the whole project is likely to be delayed by a week. Management will be aware of this shortfall because of the costs involved in any delay. Unknown to you at the beginning of the shift, the certified crane operator failed to report to work. The crew and their supervisor proceeded with their work using a non-certified operator to run the crane, which is in violation of company rules. While overseeing the work area, you note the situation. Obviously, the crew does not feel this is a safety issue or they would not be working.

You would:

_____ Immediately stop the work with the most likely result being the loss of a week in the schedule.

_____ Call over the supervisor, "slap him on the wrist" for violating the rules, but allow the work to continue.

_____ Walk on by.

Comments: _____

_____

_____

7. *What Is the Worth of an Eagle?*

Your engineering group has let contracts for a major railroad bed repair in a remote area. Following normal procedures, tight controls on costs, specifications, and performance dates are set. An employee then points out that the work is scheduled to take place near the nesting place of a pair of endangered bald eagles during the mating season. You know that the human activity could be disruptive to the nesting eagles. Your contractor indicates that it will cost $10,000 to delay at this point and defer work until after the nesting season. From past experience with your boss, you know that he will not be sympathetic to spending money on that kind of "foolishness."

You would:

_____ Take the decision to your boss who will turn down the overrun in cost and proceed on the original schedule.

_____ Proceed with the original schedule and hope for the best.

_____ Delay the work and cover the additional cost with extras to the contract rather than expose a lack of foresight on your part.

_____ Delay the repair work and simply take the heat for the cost overrun.

Comments: _____

_____

_____

8. *Standard Operating Procedures*

You are managing a substantial project of work to be done for a municipality. The city inspector refuses to allow release of the final payments by continually nit-picking. When confronted, the inspector broadly hints that "a few thousand dollars for his bosses" would solve the problem. You bring the matter to your firm, which is pressing for a resolution because it is costing thousands of dollars for each day that passes. The next morning, you receive an envelope marked "Personal and Confidential" that has in it a few thousand dollars.

Would you pass the envelope onto the inspector? Yes_____ No_____

Comments: _____

_____

_____

9. *Late or Never*

You are completing a project management assignment that has been highly successful for the client and your firm. In writing a summary report and checking a minor discrepancy, you uncover an error made by the Engineering Design Department. It looks like an honest mistake in calculation, as they are normally competent and professional. You realize that you could have caught it earlier had you checked out the discrepancy when it first appeared. You know enough about the design to be able to estimate it will now cost $100,000 to make the correction, making the project over budget and, of course, late. As designed now, there is the likelihood everything will perform successfully. If there would be a failure, with a remote possibility of an operator injury, an investigation could reveal the obvious design calculation error.

You would:

_____ Complete the summary report, close out the project, and keep the information to yourself.

_____ Informally notify your organization and let them decide what is appropriate.

_____ Formally notify your organization with a memo to be retained in the files.

If no action is taken, would you notify the client to work out a solution with them? Yes_____ No_____

Comments: _____

_____

_____

10. *Three Bids*

You are required to secure three bids on work to be completed by outside contractors. Unless there are obvious and compelling reasons, the bid must be awarded to the lowest bidder. One firm has been very helpful in suggesting ideas and new design concepts for the work being considered. They are competent and price their services in the middle of the price range. From your experience you know capable contractors who are likely to submit lower prices and contractors who do excellent work and will almost always be at the top of the price range.

Is it ethical to secure two bids from the contractors who price their services at the high end and the third bid from the contractor you favor?

Answer/Comments: _____

_____

_____

11. *Equity*

There is a budget cut and austerity measures are announced: a hiring freeze, no overtime pay, no salary increases in the coming year. Your promotion in grade level, which you were led to believe was imminent, appears to be a dead issue even though your contributions were acknowledged to be outstanding.

You have worked hard to develop a new project which, if implemented, will make the operation significantly more efficient. Because of the obvious merit, the project is approved but there is no additional staffing available. Therefore much of the added new work will fall on your shoulders for at least six to eight months.

Would you continue working fifty to sixty hours per week to make the project a success in spite of the pay situation?

Would you just "put in" your forty hours and take it easy because of the short-sighted decisions that have been made and the unfairness of your pay situation?

Answer/Comments: _____

_____

_____

12. *The Bad News*

Recalling how in ancient times the messenger, bringing bad news from the war, might have been run through with a javelin, you fear reporting some bad information to your boss. Press releases have been issued, the directors have been notified, and a letter has been sent to the whole organization. Your boss looked at the PERT chart you prepared and thought the mean expected time represented the not-to-exceed rather than a 50% probability completion date. To be realistic, four more months need to be added to the date announced.

You are confident that pointing out the error your boss made will result in your being blamed.

You decide to try to protect yourself by writing a memo to the public relations executive explaining the correct technical interpretation of the analysis. You are confident that there will be no retraction of the press releases and the bad news will not be passed on.

If trouble occurs later, your memo will serve as a defense that you tried to correct the misinterpretation and so you cannot be blamed.

Is this a politically savvy tactic? Is it ethical?

Answer/Comments: _____

_____

_____

13. *Consequences or Truth*

You are working on maintenance and retrofit work for a large utility. The work is performed under a continuous support agreement. Projects are assigned on the basis of a sole-source negotiated agreement. The client's engineer provides a scope of work that is then reviewed, and a mutually agreed-upon man-hour budget is developed. Authorization is then given on a not-to-exceed basis.

On one such project request, the client's engineer reviews the submitted budget and asks you to increase the man-hours by 15%. He advises you that he has some unauthorized work that he wants performed in the same plant. That work scope was rejected by his management but he nevertheless wants it performed. He requests that you perform the authorized work within the original budget and provide the unauthorized service using the additional 15% he has approved.

Your management, upon being informed about the matter, tells you to do the unauthorized work but be sure to have any reference to it hidden in the status and time reporting. They also advise you that they hold you responsible for seeing that the effort is paid for. If you bring the matter to the attention of the client engineer's superiors, you may find that they agreed with the engineer and directed his efforts to obtain the unauthorized work. If so, you could lose your client. If the engineer was acting unilaterally, he would be reprimanded by his supervisor but probably not be replaced or transferred. Therefore, you would probably experience difficulty in obtaining additional work from this engineer or at least find him less than cooperative on future efforts. If you perform the unauthorized work and a subsequent audit uncovers the facts, your company will be back-charged for the amount and the client may very well cancel the continuing services agreement.

Do you:

1. Keep quiet, perform the work and hope it is never uncovered?
2. Inform the engineer's supervisor?
3. Inform the client's vice president of engineering?
4. Get your resume out on the street?

Answer/Comments: _____

_____

_____

14. *Need-to-Know Basis*

One of your work crews accidentally shorts a main feeder; the computer fails, and other electrical damage occurs to the system. The computer service contract covers only normal usage and operation. The computer repair person who comes to service the system is inexperienced and cannot determine the cause of the extensive damage. You are asked about the probable causes.

If you mention the work crew incident, you run the risk of being billed the full amount of the repair. This would come at a bad time, when management is exercising tight scrutiny over expenditures, especially any unnecessary expenses.

Would you tell the service person about the work crew incident and expect a budget overrun? Would you answer any direct questions but avoid mentioning the work crew incident as being speculative? Would you divulge the work crew information only if required in a legal proceeding?

Answer/Comments: _____

_____

_____

15. *Whistleblowing*

While the firm for which you work supplies for industry, some items are usable for both military and industrial applications. An order comes to your attention involving specifications and a mix of items that you conclude are likely to be for military use. The documents have been issued by a European representative, but you learn the actual destination is a Middle Eastern country. The negotiated price is unusually high for these items.

The agent handling the sale assures you that even though this is a new customer, there is nothing to be concerned about. The firm's top manager tells you to stop asking questions, indicating it is none of your business to whom sales are made.

Would you alert a federal agency about a possible illegal military shipment or take some other provocative action?

Answer/Comments: _____

_____

_____

13.8   REFERENCES

Accreditation Board for Engineering and Technology. 1985. *Code of Ethics of Engineers* and *Suggested Guidelines for Use With the Fundamental Canons of Ethics.* New York: ABET.

American Society of Civil Engineers. 1991. "Procedures for Professional Conduct Cases." New York: ASCE, October.

Babcock, D.L. 1991. *Managing Engineering and Technology.* Englewood Cliffs, NJ: Prentice-Hall.

BER Case 90-2. 1990. "Expert Witness—Accusation of Unprofessional Conduct." Alexandria, VA: National Society of Professional Engineers Board of Ethical Review.

BER Case 87-5. 1987. "Gift—Complimentary Seminar Registration." Alexandria, VA: National Society of Professional Engineers Board of Ethical Review.

BER Case 88-5. 1988. "Signing of Drawings by Engineer in Industry." Alexandria, VA: National Society of Professional Engineers Board of Ethical Review.

BER Case 90-5. 1990. "Failure to Report Information Affecting Public Safety." Alexandria, VA: National Society of Professional Engineers Board of Ethical Review.

BER Case 90-6. 1990. "Use of CADD System." Alexandria, VA: National Society of Professional Engineers Board of Ethical Review.

BER Case 91-6. 1991. "Conflict of Interest—Hazardous Waste Services." Alexandria, VA: National Society of Professional Engineers Board of Ethical Review.

BER Case 91-7. 1991. "Indemnification—Product Specification." Alexandria, VA: National Society of Professional Engineers Board of Ethical Review.

BER Case 92-4. 1992. "Public Welfare—Duty of Government Engineer." Alexandria, VA: National Society of Professional Engineers Board of Ethical Review.

Berube, B.G. 1988. "A Whistle-blower's Perspective of Ethics in Engineering." *Engineering Education,* Vol. 78, No. 5, February, 294–295.

Blinn, K.W. 1989. *Legal and Ethical Concepts in Engineering.* Englewood Cliffs, NJ: Prentice-Hall.

Boisjoly, R.J., and E.F. Curtis. 1987. "Roger Boisjoly and the Challenger Disaster: A Case Study in Engineering Management, Corporate Loyalty and Ethics." *Proceedings of the ASEM Eighth Annual Meeting,* American Society for Engineering Management, October, pp. 8–15.

Bok, S. 1988. "Whistleblowing and Professional Responsibilities." In Callahan, J. C., *Ethical Issues in Professional Life*. New York: Oxford University Press, pp. 331–340.

Callahan, J. C. 1988. *Ethical Issues in Professional Life*. New York: Oxford University Press.

Code of Ethics and Business Conduct Guidelines. 1993. Foxboro, MA: Maguire Group.

Committees on Standards of Practice and Employment Conditions. 1990. "Whistle Blowing—Responsible Action vs. Conflicting Interests." American Society of Civil Engineers National Convention, San Francisco, November 5.

"The Dimensions of Ethics." 1992. *Journal of Management in Engineering*, American Society of Civil Engineers, Vol. 8, No. 4, October, 313–315.

"Enterprise Ethics and the International Marketplace." 1990. *Engineering Ethics Update*, National Institute for Engineering Ethics, Vol. 1, No. 2, Winter, 1–2.

"Ex-GSA Employee Gets $560,000 for Discharge." 1988. *Engineering News-Record*, Vol. 221, No. 11, September 15, 23.

Firmage, D. A. 1980. *Modern Engineering Practice*. New York: Garland STPM Press.

Frantz, L. R. 1988. "Engineering Ethics: The Responsibility of the Manager." *Engineering Management International*, Vol. 4, No. 4, January, 267–272.

French, P. A. 1982. "What is Hamlet to McDonnell-Douglas or McDonnell-Douglas to Hamlet: DC-10." *Business and Professional Ethics Journal*, Vol. 1, No. 2, 1–13.

Glazer, M. 1988. "Ten Whistleblowers and How They Fared." In Callahan, J. C. *Ethical Issues in Professional Life*. New York: Oxford University Press, 322–331.

Glenn, J. R., Jr. 1986. *Ethics in Decision Making*. New York: Wiley.

Gluck, S. E. 1986. "Ethical Engineering." In Ullmann, J. E., Ed., *Handbook of Engineering Management*. New York: Wiley, p. 176.

Gorlin, R. E., Ed. 1990. *Codes of Professional Responsibility*, 2nd ed. Washington, DC: Bureau of National Affairs.

Heilbroner, R. L., et al. 1972. *In the Name of Profit*. Garden City, NY: Doubleday.

"I Gave Up Ethics—To Eat!" 1975. *Consulting Engineer*, Vol. 45, No. 6, 39–41.

Kemper, J. D. 1990. *Engineers and Their Profession*, 4th ed. Philadelphia: Saunders College Publishing.

Lewis, B. J. 1977. "The Story Behind the Recent National Scandals Involving Engineers." *Engineering Issues*, American Society of Civil Engineers, Vol. 103, No. EI2, April, 91–98.

Maguire Group Ink. (Foxboro, MA) 1992. Vol. 5, No. 9, September 15, 2.

Maguire Group Ink. (Foxboro, MA) 1993. Vol. 6, No. 3, March 15, 1.

Martin Marietta Corporation. 1992. *Gray Matters: An Ethics Game*. Orlando, FL: Martin Marietta.

Martin, M. W., and R. Schinzinger. 1989. *Ethics in Engineering*, 2nd ed. New York: McGraw-Hill.

"MCC Report in the Matter of Virginia Edgerton." 1983. In Schaub, J. H., and K. Pavlovic, Eds., *Engineering Professionalism and Ethics*. New York: Wiley, pp. 499–504.

McCarthy, K. A. 1988. "What It Takes to Blow the Whistle." *Engineering Education*, Vol. 78, No. 9, October, 26.

Morrill, R. L. 1980. *Teaching Values in College*. San Francisco: Jossey-Bass.

National Institute for Engineering Ethics. 1990. *Professional Ethics and Engineering: A Resource Guide.* Alexandria, VA: NIEE.

National Society of Professional Engineers. 1993. *Code of Ethics for Engineers.* Alexandria, VA: NSPE.

Nelson, C., and S. Peterson. 1979. "Ethical Decisions for Engineers: Systematic Avoidance and the Need for Confrontation." *Civil Engineering Education,* American Society of Civil Engineers, Vol. 1, pp. 628–634.

Oldenquist, A. G., and E. E. Slowter. 1979. "Proposed: A Single Code of Ethics for All Engineers." *Professional Engineer,* Vol. 49, No. 5, May.

Peterson, J. L. 1989, 1990, 1991. "What Would You Do?" *Engineering Management Journal,* Vol. 1, No. 2, June 1989, 3–6; Vol. 1, No. 4, December 1989, 3–8; Vol. 2, No. 2, June 1990, 37–42; Vol. 2, No. 4, December 1990, 22–28.

"Professional Responsibility and the Dispatching of Police Cars — A Case Study." 1983. In Schaub, J. H., and K. Pavlovic, Eds., *Engineering Professionalism and Ethics.* New York: Wiley, pp. 491–498.

Sommers, C. H. 1991. "Teaching the Virtues." *Imprimis,* Hillsdale College, Vol. 20, No. 11, November.

"Survey Finds Ethics a Concern." 1990. *Engineering Ethics Update,* National Institute for Engineering Ethics, Vol. 1, No. 3, Spring, 2.

Vandiver, K. 1972. "Why Should My Conscience Bother Me?" In Heilbroner, R. L., et al., *In the Name of Profit.* Garden City, NY: Doubleday.

Vesilund, P. A. 1988. "Rules, Ethics and Morals in Engineering Education." *Engineering Education,* Vol. 78, No. 5, February, 289–292.

"The Whistleblowing Dilemma." 1992. *Executives' Digest,* Herman Goldner Company, December, p. 2.

Wright, P. H. 1994. *Introduction to Engineering,* 2nd ed. New York: Wiley.

Zinckgraf, G. 1992. "Ethics and Engineering." *California Professional Engineer,* California Society of Professional Engineers, Vol. 26, No. 5, September/October, 5.

____14

The Engineering Professional

We began Chapter 1 by describing engineering as both a profession and a career. In this last chapter, we emphasize several more professional aspects of engineering. Throughout the book, and especially in the chapters on liability and ethics, we have stressed the professional nature of the work of the engineer and the engineering manager.

Just what do we mean by a profession? Many years after he retired from the position of dean of law at Harvard, Roscoe Pound defined a profession as "the pursuit of a learned art in the spirit of public service" (Pound, 1953). The American Society of Civil Engineers adopted this definition and then amplified it as follows:

> A profession is a calling in which special knowledge and skill are used in a distinctly intellectual plane in the service of mankind, and in which the successful expression of creative ability and application of professional knowledge are the primary rewards. There is implied the application of the highest standards of excellence in the educational fields prerequisite to the calling, in the performance of services, and in the ethical conduct of its members. Also implied is the conscious recognition of the profession's obligation to society to advance its standards and prescribe the conduct of its members. (American Society of Civil Engineers, 1992)

To be fair, we should note that the Taft–Hartley Act (Federal Labor–Management Relations Act of 1947) ignores the public service aspect. Rather, it defines a professional employee, for collective bargaining status, as

> any employee, engaged in work (1) predominantly intellectual and varied in character as opposed to routine mental, manual, mechanical or physical work; (2) involving consistent exercise of discretion and judgment in its performance; (3) of such a character that the output produced or the result accomplished cannot be standardized in relation to a given period of time; (4) requiring knowledge of an advanced type in a field of science or learning customarily acquired by a prolonged course of specialized intellectual instruction and study in an institution of higher learning...as distinguished from a general academic education or from an apprenticeship or from training in the performance of routine mental, manual or physical processes.... (quoted in Blinn, 1989)

Many writers have tried to describe the characteristics of a profession (Callahan, 1988; Firmage, 1980; Schaub and Pavlovic, 1983). Among the prevalent attributes often mentioned are the following:

1. There is a requirement for specialized education leading to knowledge and skills not commonly possessed by the general public.
2. The work requires the exercise of discretion and judgment and is not subject to standardization.
3. The profession has legal status and requires well-formulated standards for admission, such as registration or licensing.
4. Standards of conduct are set forth in codes of ethics.
5. Group consciousness promotes knowledge and professional ideals through societies and associations.
6. The professional community exercises oversight and control of member conduct.
7. Activities satisfy indispensable and beneficial needs, with service to the public as the foremost motive. (based in part on Wright, 1994)

Wisely (1978) states the case for the dominance of the service ethic quite eloquently, as follows:

> The obligation to give primacy to the public interest is the very essence of professionalism. Without this commitment, the effort of a group to seek professional status as exponents of a body of specialized knowledge is but a shallow and selfish charade, no matter how sophisticated that body of knowledge may be or how rigorously it may be pursued.

The definition of the *engineering* profession as given by the Accreditation Board for Engineering and Technology (ABET), which we quoted in Chapter 1, bears repeating here:

> Engineering is the profession in which a knowledge of the mathematical and natural sciences gained by study, experience and practice is applied with judgment to develop ways to utilize, economically, the materials and forces of nature for the benefit of mankind. (quoted in Wright, 1994)

In this chapter, we consider professional registration of engineers, including a short history and current status, organization of registration boards, and registration requirements. We then discuss continuing education as an essential element of the professional engineer's life. Professional societies—their organization, purposes, and activities—are presented next, followed by some information about the role of the engineer as an expert witness. We end where we began, with a plea that professionalism is, above all, the keystone of all that we do as engineers.

14.1 PROFESSIONAL REGISTRATION

Professional registration of engineers began in the United States in 1907, when the state of Wyoming enacted a statute governing the practice of engineering and land surveying. Prior to the time, Wyoming had experienced problems with inaccurate plans for waterworks, canals, and streams prepared by lawyers, real estate brokers, insurance brokers, and others posing as engineers or land surveyors. The legislature enacted a statute that restricted registration to those who met defined standards of competence; to them alone the right to practice engineering and land surveying was granted. In addition, the state Board of Examiners was required to deny registration to any person whose moral character, as judged by the board, made him or her "an unsafe employee of any citizen, state corporation or association."

In 1908 Louisiana passed a similar law; at the present time, all 50 states, the District of Columbia, the territories of Guam and the U.S. Virgin Islands, and the commonwealths of Marianas Islands and Puerto Rico have registration boards that set minimum standards of competence and conduct for regulating engineering, architecture, and land surveying (Kimberling, 1988; Rogers, 1980).

Regulation of the professions in the United States is reserved to the individual states under the Tenth Amendment to the Constitution. All state engineering registration laws have the same primary purpose: to safeguard the safety, health, and welfare of the public.

Other nations, such as Canada, have similar requirements, while Japan and Mexico are among those using the U.S. model to incorporate professional registration into their laws. As the European Community matures, it is likely that an engineer who is registered once will be able to work in any of the member countries. The North American Free Trade Agreement poses interesting questions for engineering registration, since traditionally registration in the United States is by individual state; to move to a single registration for all of North America would be a major step, to say the least. In late 1993, however, it did not appear that existing state registration laws would be superseded ("U.S. Officials Say NAFTA Won't Nullify States' Laws," 1993).

The National Council of Examiners for Engineering and Surveying (NCEES) is an umbrella organization that strives for uniformity of requirements and communication among the registration boards of the several states and territories. All boards are currently members of the council. One of the council's prime services is the preparation of semiannual examinations for licensure. NCEES keeps track of registration statistics nationwide. Table 14.1 shows the widely varying propensity toward registration among the various disciplines, with civil engineering by far the most "registered" engineering discipline and chemical and industrial engineers least likely to become registered. Overall, approximately 18% of engineers in the United States are licensed under the registration statutes of at least one state (Taylor, 1993).

Each state statute provides for a registration board, usually appointed by the governor, that is charged with judging applicants' qualifications and otherwise administering the law. Some states have separate boards for engineers, architects, and land surveyors; others have combined boards for two or all three. In Alaska,

TABLE 14.1   Proportion of Engineers Who Are Registered, by Discipline

Engineering Discipline	Approximate Number of Engineers	Approximate Number Licensed	Percent Licensed
Civil	360,000	160,000	44%
Mechanical	395,000	91,000	23%
Electrical	803,000	73,000	9%
Chemical	180,000	15,000	8%
Industrial	133,000	11,000	8%
Agricultural	40,000	5,000	13%
Mining/ Metallurgy	30,000	5,000	17%
Other Disciplines	259,000	40,000	15%
Total	2,200,000	400,000	18%

SOURCE: Taylor (1993). Reprinted by permission of National Society of Professional Engineers.

for example, there is a single Board of Registration for Architects, Engineers, and Land Surveyors (AELS).

Membership on registration boards generally reflects the population being regulated, with several engineering disciplines usually represented. About three-quarters of state statutes currently provide for public, nonengineer members, as well. In Alaska, the AELS board consists of two civil engineers, one land surveyor, one mining engineer, two engineers from other branches of the profession, two architects, and one public member (Alaska Department of Commerce and Economic Development, 1993).

Since individual state boards specify registration standards, we cannot give an all-encompassing list of requirements. Four steps are generally required for professional engineering; they are based on what NCEES calls the tripartite model of experience, education, and examination ("P.E. Registration: The Smart Choice," n.d; Spell, 1993):

1. Graduation from a 4-year engineering degree program accredited by the Engineering Accreditation Committee of the Accreditation Board for Engineering and Technology, Inc. (ABET). Substitution of work experience for some of the education may be allowed. Partial credit may be allowed for non-ABET-accredited programs.

2. Passage of the fundamentals of engineering (FE) examination. This 8-hour, multiple choice test prepared by NCEES is given twice a year. Most states allow engineering seniors to take this exam provided they have completed a specified number of credits. Recently the exam was changed from an open-book test to one in which only a single supplied reference volume, containing formulas and tables, is permitted. As the name implies, the test examines knowledge of the fundamentals in science, mathematics, and engineering science.

3. Completion of a 4-year internship, after passing the fundamentals examination, under the direct supervision of a professional engineer. As an engineering intern, the applicant must demonstrate that he or she has taken on increasing levels of responsibility; in some states, a stated minimum amount of time must be "in responsible charge" of engineering work. This "responsible charge" must generally be certified by an engineer registered in the discipline for which the intern is working. An amount of time greater than 4 years may be required for those from non-ABET-accredited programs, for those from engineering technology programs, and for those with no formal engineering education. Some states allow graduate degrees in engineering to count for one year of the engineering experience.

4. Passage of the principles and practice of engineering (P.E.) examination. This 8-hour test, also prepared by NCEES, consists of both multiple-choice and essay-type questions. Currently, NCEES prepares P.E. exams in the following 16 disciplines: aeronautical/aerospace, agricultural, chemical, civil, control, electrical, environmental, fire protection, industrial, manufacturing, mechanical, metallurgical, mining, nuclear, petroleum, and structural.

Some states have special registration requirements because of special conditions within the state. States in populated, earthquake-prone regions such as California tend to require demonstration of competence in seismic design. In Alaska, all registered engineers and architects must take and pass an approved course in arctic engineering.

An engineer registered in one state cannot automatically practice in another state. However, all states provide for *registration by comity*, under which a registered engineer from one state can become registered in another by simply paying a fee and completing an application, provided the requirements of the original registration are at least as stringent as those of the new state. If you are registered in Florida and wish to become registered in Alaska, you must apply and pay a fee, but you must also demonstrate competence in arctic engineering.

In addition to responsibilities for administering the exams, deciding which applicants are qualified to take them, setting passing standards, issuing certificates, maintaining rosters of registrants, and other matters related to technical competence, boards of registration are also usually charged with issuing and enforcing codes of professionalism and ethics. This activity is an example of the means by which professions police their own members' conduct. As a result of such enforcement activity, registrants' licenses to practice may be suspended or revoked and/or fines may be imposed. Registration boards often publish the results of such investigations and may include the names of the parties. Three examples of such published reports are given:

A professional engineer was found guilty of negligence for signing and sealing building plans that were inaccurate, incomplete and below minimum standards. Penalty: Indefinite revocation of license. May apply for relicensure no sooner than three years from date of revocation. (An Ohio action, from National Council of Engineering Examiners, 1988)

[Name omitted, although it was included in the original report] Civil Engineer Registration Number C [...], Revoked effective April 15, 1988. On April 17, 1992, Mr. [...] petitioned the Board for reinstatement of his revoked registration. The Board determined that Mr. [...] displayed ignorance of the statutes and regulations which govern the licensed activities of civil engineers and land surveyors. Evidence also established that Mr. [...] shows no remorse for his past actions and demonstrates no rehabilitation. Based on these determinations, the Board denied the petition. (California Society of Professional Engineers, 1992)

[Name omitted], Conviction of three counts of bribing public officials, 3 years suspension starting 4/17/92; upon relicensure to be prohibited from participating in any manner in a state or municipal project in the state of Alaska for a period of three years, $2,000 civil fine. (Armstrong, 1993)

Two controversial topics related to professional engineering registration should be mentioned. One has to do with *industrial exemptions*. One might be inclined to ask, when viewing the data in Table 14.1, why all practicing engineers are not registered, rather than just 18% of the total. The answer is that registration laws permit certain groups to be exempt from the regulations. Included, typically, are those who build projects designed by professional engineers, those employed as engineers by the federal government (and, less commonly, by state and local governments), those working under the direct supervision of a registered engineer, those designing projects of limited size and/or use, and those employed in industry. It is this last exemption that has received the greatest attention of late. Those who favor eliminating the exemption argue that if this were to happen, industry employers would be assured of commitment to high ethical standards, would be provided good public relations, and would have engineers authorized to sign and seal design documents, a growing need in an increasingly litigious society (Taylor, 1993). They cite surveys that indicate, for example, that 85% of registered engineers believe that public health and safety would be protected much better or somewhat better if engineers in industry were subjected to the same licensing requirements in place in private practice. ("Reader Survey: January Survey Results," 1993).

On the other side, it is argued that engineers in industry do not perform work related to public safety. In a letter to the editor of *Engineering Times*, one writer said:

[A]rguments for registration made by individuals personally (and professionally) involved in that process are totally self serving. Registration only creates an elitist club of individuals to serve its own limited needs. Industry does not necessarily benefit; it makes no one a better or more ethical engineer. It's merely a formality. It also does not serve to reduce the litigiousness of our society but simply helps protect the individual.... (Price, 1993)

Those two contrasting views indicate the controversial nature of this topic! A resolution is unlikely to come in the near future.

The other issue that has sparked considerable discussion and debate is the matter of *continuing professional competence* (CPC). Should an engineer be required to show participation in some minimum level of professional development and education during the past time period in order to renew his or her professional license? Some states have adopted such a rule, and others may follow suit.

14.2   CONTINUING PROFESSIONAL DEVELOPMENT

Elsewhere in this book, we have stressed the importance of training and development. We talked about the responsibility of the individual to maintain competence in Section 1.4, and we stressed the engineering manager's role in encouraging and providing such opportunities for employees in Section 5.4. We confront that topic one more time in this chapter in professionalism. Indeed, if the engineering professional conducts his or her affairs under the prime motive of service to the public, then a primary individual responsibility of such a professional is to assure that such service is based on up-to-date knowledge of current practice. *Professional persons are those who maintain their competence throughout their careers!*

Continuing professional development includes structured development programs, formal and informal coursework, and such individual activities as reading and conference attendance. In a discussion of professional development of engineers in the heavy construction industry, Tatum (1980) argues that "Training is important; however, progression through a series of varied assignments is critical."

Early assignments, even in a project setting, can involve a planned rotational program that gives the new employee exposure to the responsibilities and activities of several functional departments, before the person is assigned in any one area. Some tactics for the successful use of such an approach include assigning a counselor or mentor to the new employee, making sure the assignments have the potential for making a genuine contribution, and conducting periodic reviews at, say, 6 months and a year, to give performance evaluation and obtain employee reactions and suggestions.

As employees progress through their careers, continuing development must not be neglected. Each employee will have different needs and expectations for specific development opportunities, but the importance of assignments that provide a variety of responsibilities and experiences cannot be neglected. Tatum (1980) makes the bold suggestion that "work assignments and their duration must be made with career paths, not job requirements, as the primary consideration."

Kimmons (1980) provides the following listing of possible forms of training and development for the engineering professional:

1. In-house training programs—especially effective as they deal directly with the specific environment and conditions that exist on the job.

2. Selected off-premises courses—these must be carefully chosen and used where special expertise is not available within the organization.

3. Continuing education programs—an excellent enticement to advance the knowledge of the employee. The subsidy of advanced education has become an important fringe benefit.

4. On the job training—generally conceded to be the most effective per dollar invested. This is true only when time is taken to plan the program adequately.

5. Rotational training—this is especially important in smaller offices where generalists are necessary to cover varying workload situations adequately.

6. Technical society participation and professional registration.

Many engineers choose to pursue advanced degrees in management. Traditionally, the engineering undergraduate degree plus the master of business administration (MBA) have been a typical path to a successful career in the management of engineering. Over the past few decades, however, an alternative to the MBA has emerged for technical personnel wishing to pursue advanced degrees in management. Most such programs call their offerings "engineering management," although other terms (e.g., engineering operations, management engineering, engineering administration, management science) are also used. The worldwide proliferation of degree programs in engineering management and related areas since 1955 is shown in Figure 14.1. In 1990, 94 U.S. universities offered a total of 121 degrees in engineering management: 28 bachelor's, 74 master's, and 19 doctoral (Babcock, 1993; Kocaoglu, 1990; Sarchet, 1993). The usual master's program requires a combination of quantitative, management, legal, financial, and technical courses totaling between 30 and 36 semester hours. Generally, a capstone project or thesis is required.

Although most engineering professionals agree that continuing development is an essential aspect of the professional's career, opinions are honestly divided over whether such activities should be *required* for professional licensure. Proponents point to other professions, such as medicine and accounting, which have such

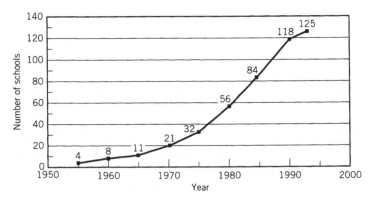

Figure 14.1. Growth of engineering management programs worldwide: number of schools offering engineering management versus year. Schools offering more than one EM degree are counted only once.

requirements. They say that any profession "worth its salt" should have a formal structure for helping its members stay on the "cutting edge" technically. On the other side, the argument goes, engineers keep themselves up to date anyway, so it is "a question of letting integrity and ethics play the policeman." New rules would create an unnecessary bureaucracy to monitor an area of performance that is already satisfactory. Furthermore, market forces eventually drive out of business those who do not maintain their competence ("Continuing Professional Development," 1992).

Since individual states in the United States set the laws and regulations for professional registration, the requirements for continuing professional development (CPD) are set by those states. By late 1993, two states, Alabama and Iowa, had legislated compulsory continuing education. The Alabama rules, effective for 1994 license renewals, require registrants to earn a minimum of 15 "credit hours" per year. One semester hour of a university course converts to 45 "credit hours," while one hour of professional development education gives one "credit hour" ("Stateline," 1992).

The American Institute of Architects, beginning in 1996, will require periodic continuing education as a condition for maintaining membership eligibility. AIA members who fail to meet the requirement will be subject to termination ("AIA Requires CPD," 1992).

The National Council of Examiners for Engineering and Surveying developed a set of continuing professional competency model rules. Since NCEES has no power to mandate state requirements, the purpose of the rules is to provide state boards with a *suggested* program, in case continuing education is mandated, to facilitate the advent of consistency nationwide ("NCEES Approves Model Rules," 1992). There are many ways to demonstrate continuing competence besides taking the P.E. exam again! An anecdote attributed to Hugo Black quotes the late Supreme Court Justice as follows: "I have this terrible nightmare; it is always the same. I am required to sit for the bar exam again" (Ingersoll, 1977). The NCEES Model Rules suggest the following ways to earn professional development hours (PDHs) for registration renewal, with various equivalent credits for activities of different kinds:

1. Successful completion of approved college courses.
2. Successful completion of approved continuing education courses.
3. Successful completion of approved correspondence, televised, videotaped, audiotaped, or other short courses and/or tutorials.
4. Active participation in seminars, in-house courses, workshops, and professional conventions.
5. Teaching or instruction in (2) through (4) above.
6. Publishing papers, articles, or books.
7. Active membership in professional or technical organizations, societies, or boards.
8. Approved active professional practice.

9. Active participation in professional peer review as a member of a board, committee or commission charged with evaluation. ("Continuing Professional Competency Model Rules," 1993)

14.3 PROFESSIONAL ENGINEERING SOCIETIES

More than 200 organizations represent the technical and professional interests and needs of engineers in the United States. They provide a range of services from publications and conferences to various political activities and investigations of member misconduct.

It is helpful to categorize these organizations into four groups, as suggested by Weinert (1986). One group consists of associations formed on established emerging disciplines. Among these are the so-called founder societies, which are five of the oldest and perhaps most prominent engineering organizations. These societies and their dates of founding are as follows:

American Society of Civil Engineers (ASCE, 1852)

American Institute of Mining, Metallurgical, and Petroleum Engineers (AIME, 1871)

American Society of Mechanical Engineers (ASME, 1880)

Institute of Electrical and Electronic Engineers (IEEE, 1884)

American Institute of Chemical Engineers (AIChE, 1908)

A second group comprises societies that focus on a broad occupational field. Examples are the Society of Automotive Engineers (SAE), the Society of Manufacturing Engineers (SME), and the Society of American Military Engineers (SAME).

A third category includes organizations focused on a specific technology or group of technologies, including the American Society of Heating, Refrigeration, and Air-Conditioning Engineers (ASHRAE) and the Society of Plastics Engineers (SPE). In his fourth group, Weinert places organizations "formed either by individual engineers or by groups of societies to accomplish a specific purpose." Examples include the following list:

Accreditation Board for Engineering and Technology (ABET)

American Association of Engineering Societies (AAES)

American Society for Engineering Education (ASEE)

American Society for Engineering Management (ASEM)

National Council of Examiners for Engineering and Surveying (NCEES)

National Society of Professional Engineers (NSPE)

The second organization on the preceding list, the American Association for Engineering Societies, deserves special mention. AAES is an umbrella group that represents the engineering profession as a whole. Over the years, several versions of a unity organization for the engineering profession have existed, with AAES being formed in 1980. As Figure 14.2 indicates, the association is really an organization *of organizations,* not of individual members. Studies continue on the concept of a unity organization for individual engineer members, with an NSPE task force proposing the formation of the American Engineering Society (AES) (National Society of Professional Engineers, 1991).

Engineers also participate in other professional organizations whose membership includes both engineers and nonengineers. Examples of these interdisciplinary associations are the American Society for Quality Control (ASQC), the International

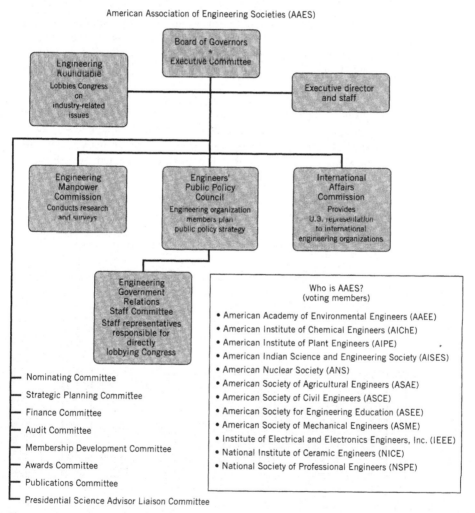

Figure 14.2. Organization chart of the American Association of Engineering Societies. (Reprinted with permission of AAES.)

Conference of Building Officials (ICBO), and the Project Management Institute (PMI).

Figure 14.3 displays the results of a survey of a random sample of 1000 engineering graduates conducted by NSPE (National Society of Professional Engineers, 1991). It shows the percentage of respondents who are members of various professional societies, and similar percentages for the registered engineers in the survey. The percentages total more than 100% for both groups because some individuals belong to more than one society. It seems significant, and disappointing, that more than one-third of all those surveyed, and 28.9% of the P.E.s, are members of *no* engineering society.

The American Society of Mechanical Engineers (1993) states its overriding goal as follows:

> To move vigorously from what is now a society with essentially technical concerns to a society that, while serving the technical interests of its members 'ever better, is increasingly professional in its outlook, sensitive to the engineer's responsibility to the public's interest, and dedicated to a leadership role in making technology a true servant of man.

The various professional societies are generally organized into practice divisions or interest groups, plus several administrative committees. Each has a national headquarters served by full-time staff members. Most activity, however, takes place at the local level, where chapters, sections, or branches are located. ASME member

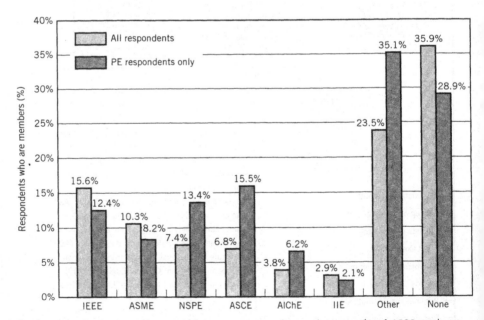

Figure 14.3. Percentage of respondents to a survey of a random sample of 1000 graduates who are members of indicated engineering societies. (Based on National Society of Professional Engineers, 1991.)

activities are subdivided into councils on public affairs, codes and standards, member affairs, engineering, and education. Figure 14.4 shows the ASME organization chart and distinguishes between member and staff responsibilities.

Activities conducted by the professional engineering societies are numerous; all are intended to assist the individual engineer, and the profession collectively, to do a better job of serving the public. A major effort involves publications, which provide members with both general and specialized information through magazines, transactions, journals, monographs, and conference proceedings. National meetings, held annually or more frequently, usually feature specialized paper presentations, often in several concurrent sessions, general lectures, business meetings, tours of professional interest, and meals and social events. Specialty conferences sponsored by practice divisions feature, as the name implies, topics with a narrow range of interest, such as reliability in manufacturing, radiowave probing of the high latitude ionosphere, or cold regions engineering. National committees conduct various investigations, sponsor special events, and prepare special publications.

For most members of professional organizations, the greatest amount of activity takes place at the local level, where meetings are held monthly or more frequently, often with a meal, a business meeting, and a speaker. Local sections sponsor public service projects, provide information about engineering to schools and the public, organize such events as Engineers' Week and the MATHCOUNTS competition for junior high students, and support such professional activities as student chapters at colleges and universities and the proctoring of licensure examinations. Regional and statewide activities add to the many opportunities provided by the societies to serve the public and the profession.

Continuing education is another important aspect of most professional society activity. Most societies sponsor structured programs of face-to-face classes in several cities each year, with subject matter experts as instructors. In addition, many have developed video-based and other self-paced instructional materials to assist working engineers in maintaining currency in the field.

In the political arena, some societies have organized independent political action committees that raise their own funds and act as advocates for certain issues and candidates. Also, representatives of the societies often testify, on behalf of their members, before local, state, and national hearings, bringing their special expertise to bear on legislation and other matters of public concern. Several organizations have developed systems whereby individual members are notified of topics of current national interest, with the suggestion that key public officials be contacted. ASCE's Key Contact program and NSPE's Engineer Ambassador program are examples.

Like the state registration boards, professional societies have become involved in so-called professional conduct matters. Upon a report of a possible member violation of a society's code of ethics, a process can be engaged to investigate the charge and take appropriate action, which can include removal from membership. We briefly described this process, as implemented by ASCE, in Section 13.2. During the decades between 1972 and 1992, 298 cases were brought to ASCE's Committee on Professional Conduct. Of these, 220, or 74%, were dropped as a result of the committee's investigation. The other 78 cases were ultimately the

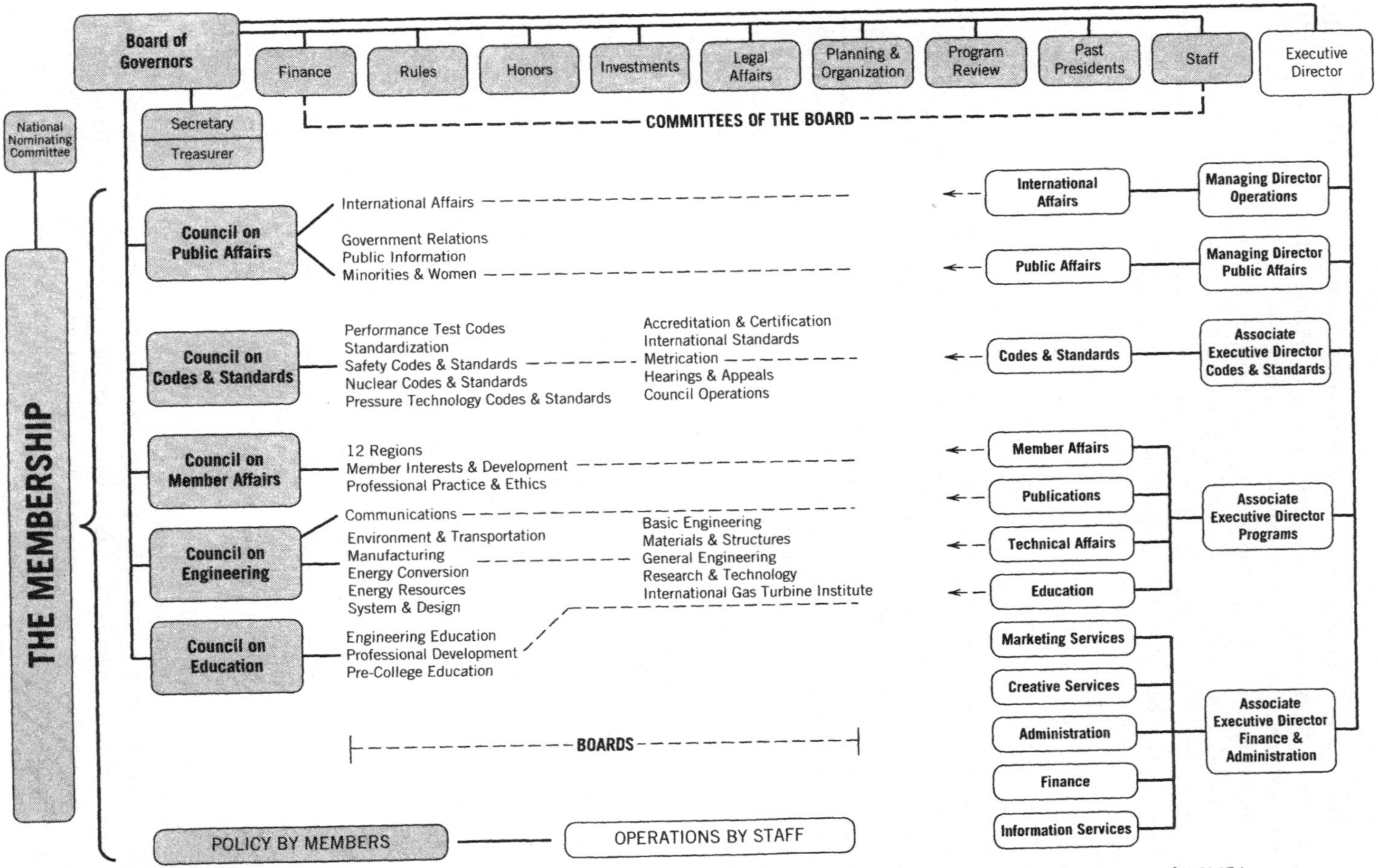

Figure 14.4 Organization chart of the American Society of Mechanical Engineers (1993). (Used by permission of ASME.)

subjects of hearings before ASCE's Board of Direction. These 78 cases led to the following actions ("Summary of Professional Conduct Cases," 1992):

Member admonished	18	(23% of 78)
Member suspended (for between 1 and 5 years)	35	(45%)
Member expelled	9	(12%)
Member resigned with prejudice	16	(20%)

A very small proportion of ASCE's 110,000 members are investigated for professional misconduct, as the preceding tabulation indicates. The important point is that the organization seeks to protect the public from violators of its code of ethics and has institutionalized a process to handle such situations.

14.4 THE ENGINEER AS EXPERT WITNESS

A specialized role the engineering professional is often called upon to fulfill is that of expert witness. Such a person is so well-versed in the subject under dispute that the court (or other forum) permits him or her to offer an opinion about the facts of the case.

As a general rule, witnesses in court proceedings provide evidence about the facts of the case—what happened, who was involved, the timing of the events, and the like—but they are not permitted to offer opinions about the causes of the events or future consequences of conditions or events. An important exception is made in the case of expert testimony; an expert witness can offer analysis and opinion that go beyond the presentation of facts. This exception is needed because such testimony often is essential in assisting the trier of the facts (jury or judge) when the subject matter is beyond the knowledge of the ordinary layperson (Blinn, 1989; Sweet, 1989).

Tozer (1967) states well the need for, and importance and challenge of, expert testimony:

> How can we hope to communicate sophisticated knowledge of technical facts to a jury of six retired mailmen and six bored housewives? How can we hope to make instant chemists or physicists out of the jurors? How can we hope to get across to a jury the intricacies of electronic production–control equipment, or structural principles on which a design is based, or the knowledge of biochemistry necessary to judge whether a particular drug could have caused the condition from which the plaintiff suffers?

In a case in which the plaintiff's house had been destroyed in a flood, the California Supreme Court, in Miller v. Los Angeles Flood Control District [8 Cal. 3d 689, 505 P. 2d 193 (1973)] found in favor of the defendant. The court based its decision in part on the failure of the plaintiff to introduce expert testimony as to whether due care had been exercised in the construction of the home. Without the aid of expert testimony, the jurors could not have made a rational determination of whether such care had been violated. The court wrote:

> Building homes is a complicated activity. The average layman has neither training nor experience in the construction industry and ordinarily cannot determine whether a particular building has been built with the requisite skill and in accordance with the standards prescribed by law or prevailing in the industry. In the instant case, then, issue as to whether or not the Miller home had been negligently constructed involved a multitude of subsidiary questions bearing not only upon the erection of the structure itself but also upon the location of the house on the particular lot, the elevation of the lot, the influence of the surrounding terrain, the possibility of run-offs and floods, and the existence of the debris dam. These were not questions which the jury could have resolved from their common experience.... (quoted in Sweet, 1989)

Thus, for the use of expert testimony to be warranted, two primary conditions must be satisfied:

1. The subject of the testimony must be so specialized in some science, business, profession, or occupation that it is beyond the understanding of the ordinary layperson.
2. The expert witness must possess sufficient knowledge, skill, expertise, training, and/or education in that subject as to make it appear that the expert's opinion will assist the trier of the facts in the search for truth. (Blinn, 1989)

In what kinds of situation might the engineering professional be invited to serve as an expert witness? The most obvious is a trial in a court of law. But there are several others. The author has served in this capacity in a master's hearing for the determination of the value of property to be taken for a highway, in several arbitration proceedings involving construction contract disputes, and in claims hearing board proceedings involving highway construction contracting, as well as in court trails (property dispute, building failure, construction injury, and subcontract cases.) Other settings in which the engineer might serve as an expert witness are commission hearings, legislative committee hearings, hearings before zoning boards, and meetings relating to building codes (Bockrath, 1986).

While the law does not require engineer expert witnesses to be professionally registered, certainly a license will fortify the testimony with additional credibility in most cases. Beyond the license, the expert must show expertise that is directly relevant to the subject of the dispute. A general knowledge of engineering will not suffice. This special expertise must be established in the courtroom (or wherever the testimony is to be given) before the witness may testify as an expert. Be prepared to answer questions about your education, your general and specialized experience, your publications, your professional activities, and any other background that might relate to whether you are qualified for a particular assignment. (Bockrath, 1986; Hough, 1981; Sunar, 1989). And be prepared to have the opposing attorney attempt to show how poorly qualified you are!

The functions performed by the expert witness range from the highly visible to the less obvious but just as vital:

1. The expert witness may provide actual testimony at the proceedings regarding a particular point at issue. In this role, the expert is assisting those who will decide the case to understand the technical details.

2. He or she may perform background investigations, visit the site, write reports, prepare exhibits, and otherwise assist in the development of the client's case, regardless of whether an appearance at the proceedings is required. In this role, the expert may advise the client on the strength of the case; in some situations, the case may be dropped as a result of the expert's study.

3. The expert may act as an advisor to the attorney during the proceedings, suggesting questions to be asked during cross-examination and otherwise helping to manage the case.

Despite what one might gather from television, the vast majority of the effort involved in any dispute resolution does *not* occur in the court chambers or hearing room! The work of *preparation* is essential, and it is in this task that the expert witness is likely to play a major role. In one recent assignment, the author spent about 15 hours preparing for a hearing at which his testimony lasted less than 30 minutes. In another, 53 hours were devoted to visiting a construction injury site, reviewing documents, preparing an elaborate model, and holding numerous meetings with the attorney and others. Just prior to the scheduled testimony, however, the case was settled out of court!

We discussed the various steps in a formal trial process in Section 8.5. The expert witness can be involved in all those steps. Generally, the engineer's involvement begins with a call from the attorney who represents one of the parties to a dispute. The prospective witness must then decide whether he or she has sufficient expertise, time, and interest to accept the assignment; the appropriate time to say "no" is very early in the process. If the answer is "yes," the prospective witness will want to become thoroughly familiar with the background of the case before beginning actual preparation for testimony. In the early phases of the investigation, the expert may conclude that the case lacks sufficient merit; the expert will render a major service by so informing the client immediately.

One of the ways the engineer can assist in preparing effective testimony is to develop visual aids that can explain the case to a jury of laypersons or a judge who may be unfamiliar with the technical aspects. Such simple but complete mechanisms can require substantial effort, but they can be very effective in conveying the message. Figure 14.5 depicts a partial building model prepared by the author and a colleague to illustrate the conditions in a construction injury case. Figure 14.6 shows two pages of a photograph album the author prepared for a case involving a building condition dispute.

During the discovery process, each side has the opportunity to obtain information about the case from the other side. While such a process may seem

Figure 14.5. Building model for a construction injury case. (Photo by permission of Call, Barrett, and Burbank.)

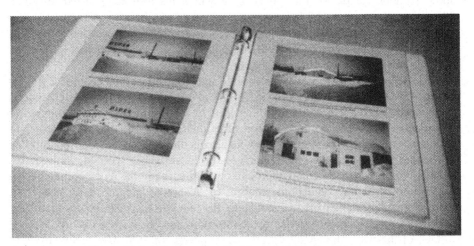

Figure 14.6. Photograph record for a building condition case.

strange, and the particular rules vary from jurisdiction to jurisdiction, discovery is very much an accepted part of American law. For starters, discovery serves to eliminate the element of surprise (as in *Matlock!*), give each side a "level playing field," and provide information that may lead to settlement before the trial. Two parts of the discovery process may involve the expert witness: depositions and interrogatories.

In a *deposition*, the witness is examined orally, under oath, by the opposing attorney. The purpose is essentially to learn the facts in the witness's possession and the individual's opinion about the facts. This testimony will be available, and will likely be used, in the eventual court trial. If the deposition testimony and the trial testimony are not consistent, the witness must be prepared to explain any discrepancies. At the deposition, the witness's image and demeanor will become apparent and may set the pattern for the entire balance of the proceedings. Thus careful preparation is important for the deposition, as it is for the trial.

An *interrogatory* is a list of questions prepared by one party to be answered by the other party. The expert witness may be involved in two ways. First, questions posed by the other side need to be answered; in this connection, the expert may be asked to provide technical assistance, or even to develop rough draft responses for review and editing by the attorney who will have to answer. Second, the witness may assist with the preparation of interrogatories to be directed at the other side (Hough, 1981; Sunar, 1989).

A great deal has been written about proper conduct by the expert witness while on the stand. Remember, first, the data presented in Section 6.3 about the importance of the *vocal* and the *visual* in oral communication. How you sound and what the audience sees can be even more crucial to the believability of your message than the information itself. What you say needs to be understandable without talking down to the jury; a description of bending a paper clip several items to failure might illustrate fatigue distress of a metal member, for example. The way you answer—hurriedly or slowly, enthusiastically or disinterestedly, seriously or offhandedly, with a steady or a wavering voice, looking at the jury or at the floor—will have an impact on the confidence placed in your answers. The way you dress will also influence the believability of your message. Visual materials— models, drawings and sketches, pictures, physical components—will also tend to strengthen your testimony, provided you are able to explain them adequately.

Many authors have provided checklists of "hints for the expert witness" (Bockrath, 1986; Hough, 1981; Sunar, 1989). We present, in Table 14.2, a list of 41 "don'ts" for expert witnesses, prepared for the Cornell Law School.

14.5   IN CONCLUSION

We end with a quotation, a creed, and a quotation. The first was penned by an engineer who later became president of the United States. This quotation, from the memoirs of Herbert Hoover, uses masculine language consistently, as was commonly done in 1951.

> The great liability of the engineer compared to men of other professions is that his works are out in the open where all can see them. His acts, step by step, are in hard substance. He cannot bury his mistakes in the grave like the doctors. He cannot argue them into thin air or blame the judge like the lawyers. He cannot, like the architect, cover his failures with trees and vines. He cannot, like the politicians, screen his shortcomings by blaming his opponents and hope that the people will forget. The

TABLE 14.2   Advice for Expert Witnesses

Don'ts for Expert Witnesses

Prepared by Lyman P. Wilson for use in the Cornell Law School

1. Don't go into any case against your own best judgment.

2. Don't go into court if you are not sure your lawyer knows *your* stuff.

3. Don't forget that it is part of your job to educate the lawyer who calls you, if he needs it (and he usually does).

4. Don't fail to help the lawyer who is examining you. (There may be occasions when you can say: "I think you did not state just what you mean. You mean, do you not...?" You may even score a point on a cross-examiner who has asked a misleading question or who does not know his stuff.)

5. Don't speak in low tones or indistinctly; you are supposed to have something worth hearing.

6. Don't take your technical terms to court. (Language was invented to convey, not to conceal, ideas.) Use the simplest language which is capable of conveying your thought.

7. Answer the questions as asked (if you can). Don't enlarge upon the answer demanded by a question, BUT.

8. Don't be content with a mere answer; give your reasons for reaching your conclusions.

9. Don't exaggerate (here is a favorite trap).

10. Don't hurry; never let the lawyer crowd you into hasty conclusions.

11. Don't hesitate if the answer to the question is obvious. It is highly effective if you can answer promptly and safely; it may leave the jury waiting on the lawyer and not on you.

12. Don't let the lawyer get your goat; just grin at him, however angry you may be. (The jury loves to see a badgering lawyer fail.)

13. Don't forget that it is possible for you to compel a lawyer to get deeper into the subject. (If you answer "sometimes," "usually not," "under certain circumstances," etc., he is almost forced to ask you to complete the answer, and the door is then opened for a complex statement.)

14. Don't miss a clear opportunity to show that a cross-examiner is not familiar with the field covered by your testimony. If he hesitates and fumbles for his next question, ask: "Is there some question which you have asked that I have not answered completely?"

15. Don't try to answer several questions at once; make the lawyer choose the one you are to answer.

16. Don't let him confuse you with rapid-fire questions; be deliberate. (He gets nowhere by asking questions which are not answered.)

17. Don't guess; if you don't know, say no.

18. Don't be afraid to make a mistake or to qualify an answer; a reputation for honesty and sincerity is valuable.

TABLE 14.2   continued

19. Don't be cocksure; undue positiveness is the refuge of liars. (And the prophet said: "Pride goeth before a fall and a haughty spirit before destruction.")

20. Don't take the stand if you have not carefully prepared your stuff; you will have to be ready to face skilled, and often unfair, cross-examination.

21. Don't try to be cute; the lawyer is playing on his own home grounds and he knows a few tricks that you have not heard about. He has the advantage. If you forget, he'll show you.

22. Don't forget that the opposing lawyer may be loaded for bear, and that for the moment he may be better informed upon some point than you are. However dumb he may seem, don't underrate him.

23. Don't be too ready to claim agreement with the authorities in your field, and don't accept a book as authoritative unless you know what it contains. (The lawyer may have a copy under the table.)

24. Don't be surprised if certain stock questions are sprung on you, such as:

    (a) Have you talked with anyone about this case? ("Certainly. I talked it over at length with the lawyer who called me here.")

    (b) How much are you being paid to testify? (The judge is now quite likely to intervene to protect you. If he does not, state the amount frankly and say: "That is what my time is worth." If the lawyer is particularly offensive you may even say: "I am usually paid..., but some lawyers don't pay their bills.")

    (c) You knew what you were going to say before you took the stand, did you not? ("If the evidence disclosed certain facts, yes, for then there would be only one reasonable conclusion—the one I have stated.")

    (d) Have you not frequently differed from other experts? ("Perhaps, but in the present instance I see no basis for any difference of opinion" or "Perhaps, but I was then and still am convinced that my opinion was correct.") In close cases there may well be room for such differences.

25. Don't fail to be as courteous to the opposing lawyer as you are to the one who engaged you.

26. Don't try to answer by "Yes" or "No" when such an answer cannot possibly be correct; if the insists, appeal to the judge.

27. Don't risk your reputation in the hands of a careless lawyer. (An ignorant lawyer is bad, but a careless lawyer is an unspeakable menace. Avoid employment by the latter.)

28. Don't try to reform the rules of evidence or court procedure. (They do need reforming, but few laymen know where or why.) Just do your best to tolerate the rules and leave it to your lawyer to guide you.

29. Don't forget the answer of a student in medical jurisprudence: "Answer the question only; volunteer nothing; give only your opinions. When questions are objected to by attorneys, shut up! Wait until the learned judge decides whether you may answer or not. Don't get mad at the attorney; it is part of his business to get your goat.

TABLE 14.2   continued

More Don'ts for Expert Witnesses

Through the kindness of Hon. Alvah W. Burlingame, Jr.
(Judge of the Court of Special Sessions, New York City)

30. Don't talk about the case, or its facts or your opinion, in the corridor, or the court-room. Remember you are being paid to give your expert opinions to the jury (or judge) from the witness stand. Some lawyers post clerks alongside of or near opposing witnesses and use what is overheard in the hall to trip up witnesses on cross-examination.

31. Don't attract everyone's attention by boisterously greeting the opposing expert, even if he is an old college friend. By so doing you are making much of him. You are being paid to make little of him and to destroy his opinion (rightfully) by your superior opinion. Why give him a buildup?

32. Don't go into a consultation with the opposing expert or experts. If you know your stuff why should you prime your opposing experts to take the shine off your testimony? You have not been hired to do that.

33. Don't forget jurors have eyes as well as ears. As an expert, look the part. Top-notched experts command high fees because they are leaders in their field or profession. They thus have the means to dress well. Therefore, look well to the shave, the shine, and the collar and tie, the suit, hat and coat!

34. Don't forget, from the time you enter the courtroom until you leave the witness stand, you are "on parade." You may become the star or chief actor on which your client's case succeeds or fails and your own personal future as an expert rises or falls. Remember that you may have, sitting next to you before the case is called for trial, a man or woman who will very soon be seated in the jury box. If you have stepped on his feet, coughed in his face, etc., the juror has a bad impression from the start.

35. Don't slump down in the witness chair or cross your legs. Don't support your head on your hand with your elbow on the arm of the chair and don't keep your hand over your mouth. Sit straight up with the base of your spine against the back of the chair and at attention and alert at all times. Don't appear dumb. Remember you are "on parade." The place to sleep is at home, and to lounge, at your fireside in the easy chair.

36. Don't take your eyes off the lawyer when he is asking you a question. Listen carefully from start to finish, then don't tell the lawyer the answer (if he asked you the question he should know the answer), BUT turn your head, look straight at the jury, and tell them the answer. The lawyer is on the sending end and the jury on the receiving end. Lose the attention of the jury and you lose the case.

37. Don't take any chance with a long, involved, hypothetical question. For your own safety and reputation, if in any doubt, ask that the stenographer read the question and then be ready to answer promptly and deliberately. It won't hurt the jury to hear the question twice; it may wake some of the jurors up to the importance of the forthcoming answer. Impress the jury at all times of your care and desire to be accurate.

TABLE 14.2 continued

---

38. Don't be fooled by compound questions. If the council asks you two questions, in one, say "as to the first questions "Yes" and the second question "No," or just say "Yes" and "No". You will perforce, be asked to explain. Mr. Lawyer is not likely to ask many more such questions.

39. Don't let an opposing lawyer get away with a trick question. You may properly be compelled to answer "Yes" or "No" as the trick question indicates, but meet it this way: "Yes I can explain that." If the opposing lawyer ducks the explanation, the jury is immediately awakened to the trick attempted. When your own lawyer hears you say "I can explain that," he will take note. You rely on him to bring out the explanation on redirect examination.

40. Don't hesitate (if it is a fact) to admit you have been called upon to testify as an expert many times. The fact that your opinion is much sought after is proof of your knowledge of subject which is not common knowledge. Add, if true, "And I am called in consultation very often." The dumb lawyer will ask "How often?"; give him the works.

41. Don't hesitate to question statements in text books, by alleged authorities, if you disagree. If you know it's an old book and out of date, ask "When was that book written?" or "What edition do you find that statement in?" or say "I'm afraid the author had no practical experience when he wrote that," or "You can find authority for almost anything in books that won't stand up under cross-examination," or fall back on "I can explain that." If called upon to explain, tell what books are the real authority that support your "expert opinion."

---

engineer simply cannot deny that he did it. If his works do not work, he is damned. That is the phantasmagoria that haunts his nights and dogs his days. He comes from the job at the end of the day resolved to calculate it again. He wakes in the morning. All day he shivers at the thought of the bugs which will inevitably appear to jolt its smooth consummation.

On the other hand, unlike the doctor, his is not a life among the weak. Unlike the soldier, destruction is not his purpose. Unlike the lawyer, quarrels are not his daily bread. To the engineer falls the job of clothing the bare bones of science with life, comfort, and hope. (Hoover, 1951)

The *Engineer's Creed* was adopted by the National Society of Professional Engineers in 1954. We present it in Table 14.3 as a fitting illustration of the professionalism for which all engineers strive.

We leave the final word to William H. Wisely, longtime executive director of the American Society of Civil Engineers, who wrote,

The engineering ethic is service of the public interest with integrity and honor. (Wisely, 1977)

TABLE 14.3   Engineers' Creed

As a Professional Engineer, I dedicate my professional knowledge and skill to the advancement and betterment of human welfare. I pledge:

  To give the utmost of performance;

  To participate in none but honest enterprise;

  To live and work according to the laws of man and the highest standards of professional conduct;

  To place service before profit, the honor and standing of the profession before personal advantage, and the public welfare above all otehr considerations.

In humility and with need for Divine Guidance, I make this pledge.

SOURCE: (Reprinted by permission of the National Society of Professional Engineers.)

14.6   DISCUSSION QUESTIONS

1. Why are engineers who are employed by the federal government exempt from state professional registration laws?

2. Study the arguments in Section 14.1 for and against requiring engineers in industry to be professionally registered. Add any other points that you can. Which side do you favor? Why?

3. What plans do you have, personally, for continuing professional development during your career as an engineer?

4. Suppose the American Institute of Chemical Engineers (or your own discipline's professional society) gave members certain mandatory continuing professional development activities to fulfill every year as a condition of renewing their membership. What effects might such a requirement have on the individual, the professional society, the profession, and society in general? What kinds of activity might be appropriate?

5. The American Society of Mechanical Engineers 1993 goal statement, quoted in Section 14.3, calls for ASME to move away from a single focus on technical concerns. In what kinds of new activity do you foresee ASME becoming involved? Give some examples.

6. Among the different activities provided by engineering's professional societies, which are likely to be most important during the early phases of an engineer's career? In what ways might this list change during the later career phases? Why?

7. The American Society of Civil Engineers has considered, at various times in its history, changing its name to the American Society for Civil Engineering. In what ways do you think this new ASCE would be different from the old?

8. Suppose you are engaged by the city engineer's office as an expert witness in an arbitration case involving a sewer line installation contract. The city

accuses the contractor of breaching the contract by failing to complete work on time and using substandard piping material. Furthermore, the contractor has submitted a claim for extra compensation, claiming that actions by city personnel caused delays and necessitated inefficient work practices. Describe some of the activities you might be asked to work on prior to the arbitration hearing and some of your responsibilities during the hearing.

9. Consider a product liability court case involving personal injury that is alleged to be due to a piece of improperly manufactured agricultural equipment. Recall the discussion of product liability in Chapter 12. If you are engaged as an expert witness by the manufacturer, what might be some of your tasks prior to and during the trial?

10. Suppose you are an expert witness testifying before a jury on a complex matter in which the locations of the boundaries of several parcels are in dispute. What are some means you might use to try to assure that you communicate your story effectively?

11. Based on your study of this book and any other contacts you may have had with the profession, list five characteristics you deem essential for a person who claims to be an engineering professional.

12. List and describe four things the engineering manager can do to help the engineers in his or her organization to "be professional" (however you might define that term).

14.7 REFERENCES

"AIA Requires CPD." 1992. *Engineering Times*, Vol. 14, No. 10, October, 8.

Alaska Department of Commerce and Economic Development. 1993. *Statutes and Regulations: Architects, Engineers and Land Surveyors*, July.

American Society of Civil Engineers. 1992. *Official Register 1993*. New York: ASCE.

American Society of Mechanical Engineers. 1993. "Organizational/Activities MM-2." New York: ASME.

Armstrong, R. 1993. "Additional Meeting Highlights." *Alaska Designs,* Alaska Professional Design Council, Vol. 16, No. 4, April.

Babcock, D. L. 1993. Personal correspondence, November.

Blinn, K. W. 1989. *Legal and Ethical Concepts in Engineering*. Englewood Cliffs, NJ: Prentice-Hall.

Bockrath, J. T. 1986. *Dunham and Young's Contracts, Specifications and Law for Engineers*. New York: McGraw-Hill.

California Society of Professional Engineers. 1992. "Board of Registration." *California Professional Engineer,* Vol. 26, No. 6, November–December, 8.

Callahan, J. C. 1988. *Ethical Issues in Professional Life*. New York: Oxford University Press.

"Continuing Professional Competency Model Rules." 1993. *Industry Engineer,* National Society of Professional Engineers, Vol. 10, No. 2, December 1992–January 1993, 3.

"Continuing Professional Development." 1992. *Engineering Times,* Vol. 14, No. 8, August, 4.

Firmage, D. A. 1980. *Modern Engineering Practice.* New York: Garland STPM Press.

Hoover, Herbert. 1951. *Memoirs of Herbert Hoover,* Vol. 1, *Years of Adventure.* New York: Macmillan.

Hough, J. E. 1981. "The Engineer as Expert Witness." *Civil Engineering,* Vol. 51, No. 12, December, 56–58.

Ingersoll, A. C. 1977. "Qualifications for Continued Practice." *Ethics, Professionalism and Maintaining Competence.* American Society of Civil Engineers Specialty Conference, March, 235–247.

Kimberling, C. L. 1988. "Professional Registration: You Owe It to the Public, the Profession and Yourself." *The Bulletin of Tau Beta Pi.* March, 1–3.

Kimmons, R. L. 1980. "Coordinated Advancement for Professionals." *Issues in Engineering,* American Society of Civil Engineers, Vol. 106, No. EI3, July, 227–239.

Kocaoglu, D. F. 1990. Personal correspondence to D. L. Babcock, March 5.

National Council of Engineering Examiners. 1988. *Registration Bulletin,* May–June, 8.

National Society of Professional Engineers. 1991. "The American Engineering Society (AES): A Vision for the Future of U.S. Engineers." Report of the 2000 Task Force.

"NCEES Approves Model Rules for Continuing Development." 1992. *Engineering Times,* Vol. 14, No. 10, October, 8.

"P.E. Registration: The Smart Choice." n.d. The NCEES Guide to Engineering Registration. Clemson, SC: National Council of Examiners for Engineering and Surveying.

Pound, R. 1953. *The Lawyer from Antiquity to Modern Times.* St. Paul, MN: West Publishing.

Price, C. 1993. "Letters: Industry Exemptions." *Engineering Times,* Vol. 15, No. 3, March, 4.

"Reader Survey: January Survey Results." 1993. *Engineering Times,* Vol. 15, No. 3, March, 5.

Rogers, F. H., Sr. 1980. "Engineering Registration Trends." *Issues in Engineering,* American Society of Civil Engineers. Vol. 106, No. EI4, October, 397–403.

Sarchet, B. R. 1993. Personal correspondence, November.

Schaub, J. H., and K. Pavlovic, Eds. 1983. *Engineering Professionalism and Ethics.* New York: Wiley.

Spell, B. D. 1993. Personal communication, November 12.

"Stateline." 1992. *Engineering Times,* Vol. 14, No. 10, October, 7.

"Summary of Professional Conduct Cases from 1972 to the Present." 1992. New York: American Society of Civil Engineers, May.

Sunar, D. G. 1989. *The Expert Witness Handbook: A Guide for Engineers,* 2nd ed. Belmont, CA: Professional Publications.

Sweet, J. 1989. *Legal Aspects of Architecture, Engineering and the Construction Process,* 4th ed. St. Paul, MN: West Publishing.

Tatum, C. B. 1980. "Professional Development in Heavy Construction." *Issues in Engineering,* American Society of Civil Engineers, Vol. 106, No. EI3, July, 189–203.

Taylor, P. 1993. "Doing Away with Industry Exemptions." *Engineering Times,* Vol. 15, No. 1, January, 5.

Tozer, F. L. 1967. "Preparation and Use of Technical Evidence in Product Liability Cases." *Defense Law Journal,* Vol. 16, 669.

U.S. Officials Say NAFTA Won't Nullify States' Laws." 1993. *Engineering Times,* Vol. 15, No. 9, September, 2.

Weinert, D. W. 1986. "The Structure of Engineering," in Ullman, J. E., Ed., *Handbook of Engineering Management.* New York: Wiley, 28–30.

Wisely, W. H. 1977. "The Influence of Engineering Societies on Professionalism and Ethics." *Ethics, Professionalism and Maintaining Competence,* American Society of Civil Engineers Specialty Conference, March, 51–62.

Wisely, W. H. 1978. "Public Obligations and the Ethics System." American Society of Civil Engineers, Preprint No. 3415.

Wright, P. H. 1994. *Introduction to Engineering,* 2nd ed. New York: Wiley.

APPENDIX

Summaries of Reader Responses to Ethics Questions

Summaries of reader responses to a series of ten ethics questions in Chapter 13. Reprinted from J. L. Peterson, 1989, 1990, and 1991. "What Would You Do?" *Engineering Management Journal,* Vol. 1, No. 4, December 1989; Vol. 2, No. 2, June 1990; Vol. 2, No. 4, December 1990; Vol. 3, No. 2, June 1991; by permission.

6. Rules

You would:

- 58% Immediately stop the work with the most likely result being a loss of a week in the schedule.
- 25% Call over the supervisor, "slap him on the wrist" for violating the rules, but allow the work to continue.
- 16% Walk on by.

Most comments indicated that safety was the concern in spite of what was claimed, and, therefore, there was no contest among the choices. A supplementary argument was that in today's circumstances, with litigation and organizations full of second-guessers, the risks associated with not complying with the rules dictate compliance. The following comments were received:

"This is an easy call; human life is at stake."

"Safety is paramount."

"Violation of crane safety rules is potentially too significant to tolerate."

"The exposure to risk is small, but the magnitude of liability is so great as to be overriding."

"The crew may feel it's safe, but the company rules are rules."

"You cannot expect others to follow rules and procedures if the boss doesn't."

452

The responses to this case highlight one of the ethical obligations that seems to fall naturally [to] and is readily accepted by engineering management. It is the responsibility for safety. Certainly in the design stage, as well as during implementation, engineers are looked to as guardians of safety. With increasingly sophisticated technology, only the technically trained can responsibly discharge the obligation to maintain safe conditions. Other professions play modest roles in safety. In your experience, how many times have accountants been heard to say, "Spend more money so that it is safe"? Even lawyers seem to confine their roles to responses to accidents rather than prevention. Perhaps explicitly recognizing the vital role engineering management plays in safety would be very valuable. Credit should be given where credit is due.

7. What Is the Worth of an Eagle?

You would:

27% Take the decision to your boss, who will turn down the overrun in cost and proceed on the original schedule.

12% Proceed with the original schedule and hope for the best.

3% Delay the work and cover the additional cost with extras to the contract rather than expose a lack of foresight on your part.

46% Delay the repair work and simply take the heat for the cost overrun.

12% None of the above.

Interestingly, the responsibility for safety is more strongly perceived than that for the environment. The professional responsibility for the environment seems to be less obvious or certain. Comments on "What Is the Worth of an Eagle?":

"Two eagles are hardly worth $10K."

"$10,000 is nothing compared to the potential liability of violating the Endangered Species Protection Act."

"Some things are obviously more important than money."

Another aspect to keep in mind in this discussion is the sphere of competence highlighted by the comment, "Advise the local Audubon Society.... This is out of your area of expertise."

A provocative comment was, "This is not an 'engineering' decision but a management decision." If so, then the question of engineering management's role in environmental issues is problematic. Perhaps the responsibility is an individual one based on personal knowledge and concern. Again, this seems to be a necessary and fruitful area of consideration for the profession.

8. Standard Operating Procedures

Yes 0%   No 100% $N = 12$

The prohibition of the use of bribes as a means of doing business is generally believed to be one of those easily identifiable values that society and the profession readily affirms. But this is neither obvious nor trivial. In some places, underpaid public employees expect to augment their income with payoffs or kickbacks. Others use their public position and power to become wealthy.

This may be a commonly accepted practice which "everybody" would follow if they were in such a favored position. In a poll of 30 graduate students, 10% answered yes, they would pass on the envelope. Their reasoning was that it gets the job done and it saves money.

Some might argue that bribery is merely inefficient. Higher prices are charged to cover the cost of kickbacks and that cost is borne by someone who would have allocated the scarce resources of money in a more efficient way. However, bribery can be used to cover shoddy, inferior, and unsafe designs and practices. Moreover, firms and individuals who are willing to participate in bribery schemes lose credibility. It becomes difficult to establish trust and contracts become problematic.

Most respondents not only would not participate in the bribery but indicated they would act against the inspector and even the firm by reporting the matter to the authorities.

Aggressive action in this case stems from a high personal standard which was not reflected in the firm. Getting fired is not a penalty commensurate with a possible loss of self-respect and reputation or the penalties that could be imposed by law in the eyes of engineering managers.

9. Late or Never

You would:

   0%   Complete the summary report, close out the project, and keep the information to yourself.

   9%   Informally notify your organization and let them decide what is appropriate.

   91%   Formally notify your organization with a memo to be retained in the files.

Doing nothing can have as important a consequence as consciously making a decision and acting upon it. In the case of late or never, a situation is presented where a manager could do nothing about a problem and chances are nothing would ever happen. To take action means problems are brought to light and solutions requiring time, effort, and money must be implemented. None of the respondents chose the expedient or easy path (i.e., keep the unfavorable information to oneself and overlook the mistake). Their motivations are clear from the comments.

   "To do otherwise would violate the fundamental canon of protecting public safety."

"Honest engineering communications between a supplier and customer is a must."

"We can all benefit from making mistakes but only if we know about the mistakes and a remedy is devised. Covering up mistakes usually causes us to repeat the mistake."

"Not only must you be ethical, but 'Murphy's law' will certainly bite you if you don't."

By a 2 to 1 ratio respondents favored notifying the client if no action was taken by the organization.

"If no action is taken, then it's time to start looking for another job."

10. Three Bids

The uniform response from readers was "No."
Comments:

"Such a situation would unfairly rule out firms who may do excellent work for the lowest price."

"If you did you would be showing blatant favoritism."

"You need to get bids from comparable firms. If the bids are close enough, you may be able to justify giving the bid to the preferred firm. You need to be fair to everyone."

The three-bid system is intended to obtain goods or services at the lowest price, presuming that each bid is of comparable quality in the relevant aspects. It also aims at being fair to bidders. People find ways of adhering to the letter of the law without keeping the intent.

Several principles might be involved in this situation, but the simple notion of honesty seems to be most applicable. The person who solicits the bids in the memo described is misrepresenting the situation to the employer and not treating other bidders creditably.

11. Equity

Why does a person work? A person may work because of the consequences: pay, benefits, security power, status, and the like. When prospects of rewards are diminished, the motivation to work declines. Work can be viewed as a means to an end. That rewards are a valid reason for working is reflected in the comments.
Comments:

"The boss that cares will make sure you get your reward in the long run."

"I would continue to work the 50-to-60 hour week (I have been in such situations).... However, after 8 months, if nothing improves, I'm gone."

Work is also valued by society. Status is accorded to people who work and especially to those who are accomplished workers. Some types of work are prized more than others, but all work is perceived as good.

12. The Bad News

An unfortunate reaction can be expected when pointing out someone's mistake if the embarrassment is great. In the politics of an organization, blame assignment is not always to the culprit. Unfairness may not be the rule, but it is not a rare event. Should a person use a standard political ploy, the letter in the file, as a defense for a political problem? Readers had opposing views on the politics: "Yes, this is politically the expedient course," and "No, ... it's downright dumb...." There was consensus that it was unethical. While the consequences to the individual in the case might have been improved, looking at all the stakeholders and the common good, the usefulness of the ploy is questionable. Public reaction to this kind of circumstance is likely to be most critical. Sins of omission such as not disclosing important information are universally thought to be wrong.

13. Consequences or Truth

One of the issues presented in this case is "bending," but not necessarily "breaking," the rules. In work of this kind, costs are difficult to estimate, and a variance of 15% for contingencies might well be expected. A similar rationale is that this is the way it is commonly done, or "everybody" does it. Rarely is the portion portrayed as a flagrant violation of policy, regulation, or contract.

It is also noteworthy and not unusual in experience that if the arrangement comes apart, you are left with the responsibility for improprieties uncovered. Veteran participants are willing to invoke practicality and expediency—as long as the blame is to be directed elsewhere.

Agreeing to the scheme proposed by the client's representative is aiding a circumvention of the client's control of expenditures and work. This may or may not be an obligation felt by you, but your part is a misrepresentation of the documents. Presumably the work and the costs are in order. The point of examination is the presentation to show unauthorized work as being within the scope of work.

Responses from readers identify the issues:

"You will willingly perform the work but only based on written authorization."

"You may now have to tell your management you have given it extensive thought, and you find you cannot bring yourself to falsify documents."

It was prudently suggested to solve the problem with the client's engineer. If effective, this gets you off the hook. Because your management is pressuring you to be expedient, your next recourse might be to inform the client about the matter. In the short term, painful repercussions are easily envisioned, but that is the essence of an ethical dilemma: to do what is right or to do what is expedient and hope it is never uncovered with attendant consequences.

14. Need-to-Know Basis

A respondent succinctly summed up the issue: "Clearly the act of omission, if you fail to volunteer the work crew incident, is 'wrong,' but it is human to be more tolerant of it than act of commission...." The responses ranged from "I would not tell the service person" to "best to confess immediately." The degree to which you act on behalf of an ethical principle as opposed to self-interest is a reflection of yourself and the standards to which you hold yourself accountable. Of course, your stewardship of the resources entrusted to your care (your budget) is a proper and a good value. Your self-interest may be involved because a favorable budget report is a part of your performance evaluation.

As in most cases in real life, ambiguity exists about the circumstances that allow a rationalization to avoid assigning the cost to your organization. In this case, the cause-and-effect linkage may be too tenuous to accept the blame and the cost. But if in your best technical judgment, costs fairly belong to you because of the work crew's accident, then by inaction do you allow the cost to be assigned to another party? The decision to withhold information is tolerated more than the giving of false information. It would seem quite possible that without the disclosure, a misrepresentation has occurred causing erroneous conclusions to be reached and, finally, cost incurred unfairly. Should all the information become available at a later date, certainly the service vendor and perhaps your own organization will note the breach of trust on your part.*

15. Whistleblowing

There does not appear to be a unique mandate conferred by society on engineering professionals in the waging of wars or international relations, although there is involvement by necessity. Citizenship generally is thought to require duties and obligations as well as privileges. The need to protect national security, even though far-reaching and ill-defined, has traditionally become a part of accepted values for all citizens.

A second important issue is that of whistleblowing. When the normal means of redressing perceived wrongdoing fails to give satisfaction, the issue of going outside of the established channels of responsibility is important. There are cases

*Comment from book author: The response to this "need to know" question is illustrative of the difficulties often encountered in separating ethics from other issues. Is the response given from the perspective of ethics, or personal survival, or risk management, or all three?

where whistleblowing is the right thing to do. It also may be very difficult to realistically determine if the channels are failing to respond appropriately, especially if you are personally and emotionally involved. It is simply a difficult matter. By the same token, it is unfortunate but realistic to expect penalties to be inflected on the whistleblower for threatening the order and stability of the affected organization and for the implied condemnation of the organization. Whistleblowing is a court of last appeals, and this perspective should be kept in mind.

Index

Abnormally dangerous things and activities (in strict liability), 356
Abuse of the legal process, 351
Acceptance, 259, 262, 288, 291
  mail box rule, 263
  mirror-image rule, 263
Accountability (accountabilities), 26, 27, 134
Accounting, 43
Accreditation Board for Engineering and Technology, 2, 389, 412, 426, 428, 434
  Suggested Guide for Use with the Fundamental Canons of Ethics, 389
Act for the Prevention of Fraud and Perjuries, 265
Acting, 57
Activities (in network schedule), 196
  critical, 202
Activity duration estimates, 199
Activity frequencies of engineering managers, 16
Activity-on-arrow networking, 196, 203–206, 207
Activity-on-node networking, 196, 197–203
Activity table, 203, 207, 214
Actual cost of work performed (ACWP), 216, 217, 218
Actual damages, 277
ACWP (Actual cost of work performed), 216, 217, 218
Adequacy of consideration, 264–265, 292
  peppercorn theory, 265
Administrative law, 226, 244
Admissions, 234
Advertising (advertisement), 302–303
Agency, 225, 238–241
  by agreement, 239
  apparent authority, 239, 240
  by estoppel, 239
  mutual, 328
  by necessity, 239
  by ratification, 239
  termination, 240

  triangle, 240
Agent, 238–239, 288, 300, 328
Agnew, S.T., 383
Agreement, 254, 259, 307, 311, 315
  agency by, 239
  partnership, 328, 329
Alaska Board of Registration for Architects, Engineers and Land Surveyors, 428
Alien corporation, 336
Alternatives (in construction contract), 307
Ambiguities, contract, 268, 290
American Arbitration Association, 247, 249
American Association of Engineering Societies, 3, 388, 434–435
American Consulting Engineers Council, 113
American Engineering Society, 435
American Institute of Architects, 76, 287, 290, 397, 433
American Institute of Certified Public Accountants, 388
American Institute of Chemical Engineers, 434
American Institute of Mining, Metallurgical and Petroleum Engineers, 434
American Iron and Steel Institute, 315
American Quality Foundation, 113
American Society for Engineering Education, 434
American Society for Engineering Management, 434
American Society for Quality Control, 53, 54, 435
American Society of Civil Engineers, 55, 78, 383, 388, 397, 404, 425, 434, 437, 439
  Committee on Professional Conduct, 437, 439
  Key Contact program, 437
American Society of Heating, Refrigeration and Air-Conditioning Engineers, 434
American Society of Mechanical Engineers, 434, 436–437, 438
  organization chart, 438
Amos and Sarchet, 14, 121, 127, 134
Anderson, Sweeney, and Williams, 196

Apollo program, 402
Apparent authority, 239, 240, 328, 344
Apparent low bidder, 305
Appeal rights, 251
Appeals court,
  federal, 230
  state, 231
Appellant, 235
Appellate court, 235
Appellee, 235
Appraisal, performance, 140
Arbitration, 142, 247, 249–251, 440
  advantages, 250
  disadvantages, 250
Arbitrator, 249
Arizona Celebration of Excellence Award, 76
Articles of dissolution, 343
Articles of incorporation, 245, 335–336
Articles of partnership, 329
Assault, 349, 350
Assignee, 270
Assignment, contract, 269–270
Assignor, 270
Association, unincorporated, 345
Assumption of risk doctrine, 354
Attorney, 225, 233, 234
Attractive nuisance, 355
Attributes (of project manager), 189–190
Auction, 255
Authority, 25, 42, 134
  apparent, 239, 240, 328, 344
  principle of organizational structure, 25
  project manager, 178
    authority areas, 186

Babcock, D.L., 18, 40, 159–160, 169, 171, 172
Backward pass, 199, 201–202, 205
Badawy, M.K., 25, 28, 40
Baker, E.R., 82
Baldrige, M., 72, 158
  National Quality Improvement Act of 1987,
    72
Baldrige Award, 61, 72–73, 74–75, 98
Bankruptcy, 247
Bankruptcy Reform Act of 1978, 247
Bar chart(s), 193–195, 211, 308
  advantages, 193, 195
  disadvantages, 195
Barker, J.A., 78
Barnard, C., 4
Barriers to delegation, 135
Baseline schedule, 193
Battery, 349, 350
Bay Area Rapid Transit system, 406–407

BCWP (Budgeted cost of work performed), 216,
  217, 218
BCWS (Budgeted cost of work scheduled), 216,
  217, 218
Behavioral traits (required of technical man-
  agers), 11
Bell, A., 245
Benchmarking, 97–98, 99, 115
Bennett, F.L., 207
Bennis, W., 130
Berube, B.G., 409–410, 415
Best evidence rule, 236–237
Bid(s),
  bond, 314
  evaluation and award, 306–307
  form, 313
  mistakes in, 263–264
  opening, 305–306
  preparation, construction, 303
  nonresponsive, 306
  rejection, 305
Bilateral contract, 255
Bill of exchange, 259
Black, H., 433
Blake, R., 24
Blinn, K.W., 251, 399
Board of directors, 336, 337, 338
Board of Ethical Review, National Society of
  Professional Engineers, 400, 401, 402, 410,
  413–414, 416
Bockrath, J.T, 225, 226, 227, 251, 256, 272,
  277, 312, 313, 314
Boilerplate, 312
Boisjoly, R.M., 407–408, 409
Bok, S., 405–406
Bonding company, 314
Bonds, construction, 313, 314
  bid, 314
  payment, 314
  performance, 314
Bonds, corporate, 339
Brainstorming, 27, 92–93
Breach of contract, 254, 273–274, 349
  anticipatory, 273–274
  material, 273
  non-material, 273
  prevention, 274
  remedies, 276–279
  repudiation, 273
  voluntary disability, 274
Breach of legal duty (in negligence), 351
Breach of warranty, 360, 364
  in services liability, 366, 368
Brocka and Brocka, 93

Budgeted cost of work performed (BCWP), 216, 217, 218
Budgeted cost of work scheduled (BCWS), 216, 217, 218
Budgeting, 43
  project, 210, 213–215
Building codes, 362, 440
Business license, 325
Business organization law, 225
Business organizations, legal structures, 325–346
  corporation, 334–343
  franchise, 345
  joint venture, 343–344
  limited liability company, 345
  not-for-profit corporation, 344–345
  partnership, 326–334, 336, 340, 342, 344, 346
  share-office arrangement, 345–346
  sole proprietorship, 325–326
  unincorporated association, 345

C, big, versus little c, 58
C corporations, 339–340, 342
Cable and Adams, 28
Calculations (network scheduling), 209
Campbell, D., 129, 132
Camps, F., 404
Canons, 388–389, 390
  fundamental, 390
  suggested guide for use with, 389
Cardozo, B., 224, 310
Career development (in engineering management), 11
Career progression, 431
Case law, 227
Cause and effect diagrams (Ishikawa or fishbone diagrams), 65, 93, 94
Caveat emptor, 358
Centralization, 26
Certificate of completion, 310
Certificate of dissolution, 343
Certificate of incorporation, 335–336
Certification, 261
Certified payroll, 245
CH2M-Hill, 35, 182, 184, 185
Challenger space shuttle, 407–408
Change, management of, 18
Change order(s), 103, 107, 309
Changes in the work, 309
Channel switching, 163
Characteristics of effective leaders, 131, 132–133
Charles II, 265
Checking, 57
Cheeseman, H.R., 326

Chevron, 138
Citation of court cases, 237–238
Civil law, 228
Civil litigation, flow of, 235, 236
Claims, 105
Claims hearing board, 440
Cleland and King, 178
Cleland and Kocaoglu, 7, 14, 122, 148, 161–162
Clerical errors, 268
Client, duty to, 398–399
Closely held ("close") corporation, 334–335, 338
Closing arguments, 235
Cluster organizational structure, 38–39
Code of Ethics and Business Guidelines, 397, 398
Code of Ethics for Engineers, 389, 390–395, 402
  fundamental canons, 390
  professional obligations, 392–394
  rules of practice, 390–392
  statement by NSPE Executive Committee, 395
Code of Ethics of Engineers, 389
  fundamental principles, 389
Code(s) of ethics, 384, 387–399, 415–416, 426, 429, 437
  core concepts, 389, 396
  corporate, 396–397, 398
  enforcement, 396–397
Collateral income sources, 378
Collective bargaining, 45, 141, 162, 244, 425
Comity, 429
Common causes, 82
Common law, 224, 227–228, 251
Common sense, 266
Common stock, 338
Communicating, 14, 15
Communication(s), 11, 38, 43, 112, 122, 128, 147–163, 375
  adaptive network, 150, 151
  appropriate method, 159–161
  common methods, characteristics, 160
  downward, 148
  grapevine, 152
  improving the process, 163
  informal, 28, 152, 159
  in organization, 148–152
  lateral, 149
  links, 148–151
  listening, 155–156
  meetings, 161–162
  negotiation, 162–163
  one-way, 154–155
  oral, 152–156, 443
  outside the organization, 149

process, 147–148
project management, 149
two-way, 154–155
upward, 148
verbal, vocal and visual, 152, 443
written, 156–159, 161
Comparative negligence, 354
Compensation (and benefits), 127, 141
Compensatory damages, 277
Competent (competency of) parties, 258, 259,
   260–261, 286, 290–291
Complaint (filing), 234
Complaints, employee, 141
Complete performance, 272
Completion,
   certificate of, 310
   date, project, 312
   substantial, 310, 312, 316
   time of, 312, 316–317
Compromise (in law), 225
Computer-aided design and drafting, 371, 411
Computer processing (of project schedule data),
   196, 208
Conditions, 315
Confidentiality and employee loyalty, 385, 401–
   404, 413–414
Conflict (in project organizations), 170
Conflicting obligations, 397–399
Conflict of interest, 385, 399–401, 413–414
Conformance, cost of, 69
Confucius, 385
Connolly, T., 121, 128, 148, 150, 163
Consideration, 260, 264–265, 286, 292
   adequacy of, 264–265, 292
Constancy of purpose, 63
Constitutional law, 226
Construction, design for, 67
Construction contract elements, 290–293
   competent parties, 290–291
   consideration, 292
   form, 293
   meeting of the minds, 291–292
   proper subject matter, 291
Construction contract(s), 247, 249, 287, 290–320
   documents, 311–313
   elements, 290–293
   process, 302–311
   provisions of special interest, 313–320
   types, 293–301
Construction cost estimate, 287
Construction documents, 287, 311–313
   agreement, 311
   bid form, 313
   bond forms, 313

drawings, 311–312
general conditions, 312
instructions to bidders, 313
special conditions, 312
technical specifications, 312–313
Construction manager (management) contract,
   300, 301, 318
Construction phase, 288
Construction schedule, 375
Construction specifications, 159
Construction Specifications Institute, 313
Consultant(s), 45, 287, 290, 300
Consulting contract, 162
Contingent fees, 378
Continuing education, 13, 426, 432
Continuing professional competence, *see* Continuing professional development
Continuing professional development, 431–434
   requirement, for licensure, 432–433
   model rules, 433
Continuous (continual) improvement, 52, 56–58,
   63, 64, 76, 77, 102, 105, 106, 110, 111,
   142
Contract,
   assignment, 269–270
   bilateral, 255
   breach of, 254
   construction contract elements, 290–293
   construction manager (management), 300, 318
   cost plus, 297–298, 306, 308
   definitions, 254
   design-build, 298–299
   discharge, 272–276
   divisible, 257, 308
   elements, 259–266
   end result, 301
   enforceable, 258, 266
   engineering and architectural services, 285–
      290
      contract elements, 286
   entire, 257, 308
   executed, 257
   executory, 257–258
   express, 255
   fast track, 301
   fixed price, 293–294, 299, 306, 308
   formation, 259–266
   function, 301
   implied-in-fact, 255
   implied-in-law, 256
   interpretation, 266–269
   joint, 256
   joint and several, 256–257
   joke, 262

lump sum, 293–294, 299
oral, 298
quasi, 256
remedies for breach, 276–279
severable, 257, 308
several, 256
target price, 294
third-party rights, 269–272
turnkey, 299
two-party, 254
ultra vires, 260, 336
unenforceable, 258, 292
unilateral, 255
unit price, 294–297, 306, 308, 309
void, 258
voidable, 258
Contract ambiguities, 290
Contract drawings, 311
Contract law, 225, 254–279
Contract liability, 328
Contributions and kickbacks, 385, 404–405, 413
Contributory negligence, 353, 355
Control chart(s), 82–86, 87, 88, 103
R-chart, 83–86
x-bar chart, 83–85
types, 83
Control limits, 83
lower, 84–86
upper, 84–86
Conversion of personal property, 351
Conway, W.D., 158
Cook and Hancher, 333
Copyright Act of 1976, 243
Copyright Office of the Library of Congress, 243
Copyrights, 243
Core concepts in engineering ethics, 389, 396
Corporate charter, 260, 335–336
Corporate income tax return, 245
Corporate stock, 245
Corporation, 260, 334–343
alien, 336
articles of incorporation, 335–336
board of directors, 336, 338
C corporations, 339–340
charter, 335–336
closely held ("close"), 334–335, 338
debt securities, 339
dissolution, 343
judicial, 343
voluntary, 343
domestic, 336
financing, 338–339
foreign, 336

formation, 335–336, 342
limited liability of owners, 335, 341, 342
not-for-profit, 344–345
officers, 337
ownership transfer, 340, 341, 342
participation and control, 337–338
piercing the corporate veil, 335
private, 334, 335–343
professional, 340–341, 343
public, 344
publicly held, 334–335
S corporations, 339–340
shareholders, 337
stock, 338–339
taxation, 339–340, 341
termination, 343
voting, 337, 338
Correcting (as an engineering management function), 14, 16
Cost control, 43
project, 207–208, 209, 215–219
Cost estimating, 210, 213, 303
construction, 303
Cost of quality, 52, 62, 69–71
Cost plus contract, 297, 303, 306, 308
fixed fee, 297, 299
percentage of costs, 297
reimbursable costs, 297–298
Cost/schedule control system criteria (C/SCSC), 216
Cost variance, 216, 217
Counterclaim, 233, 234
Counteroffer, 262, 263
Court case citations, 237–238
Court system, 225, 229–233
federal, 229–230
jurisdiction, 231–233
state, 231
Coventurer, 343
Covey, S., 130
CPM (Critical Path Method), 195–207
CPS (Critical Path Scheduling), 195
Credibility, 189
Creditor beneficiary, 271
Cridge, E.S., 155–157
Crime, 261
Crimes (distinguished from torts), 349
Criminal acts, 329
Criminal law, 228
Critical activities, 202
Critical path, 195
calculation of, 202
definition, 196
Critical Path Method (CPM), 195–207

Critical Path Scheduling (CPS), 195
Crosby, P.B., 52, 62, 69, 70
Cross-examination, 234
Cross functional team, 110
Cumulative preferred stock, 338–339
Cumulative progress curve (completion curve),
    174, 215, 217
Cumulative voting, 337
Custom and usage, 267
Customer focus (or orientation), 51, 52, 56, 58–
    59, 65, 105, 110
  customer delight, 58
  internal and external, 52
Customers, 45

Damages, 235, 277–278
  actual, 277
  compensatory, 277
  liquidated, 278, 312, 316–317
  mitigation of, 277
  nominal, 277
  punitive, 277
Damages due to delays, 288
Data processing, 403
Davis, K., 155
Davis-Bacon Act, 246
Debentures, corporate, 339
Debt securities, 339
  bonds, 339
  debentures, 339
  notes, 339
Decentralization, 26
Defamation of character, 350
Defendant, 233
Delay(s), construction contract, 317–318
  by the contractor, 317
  by the owner, 317–318
  compensable, 318
  excusable, 316, 318
  no damages for delay clause, 317-318
Delegation, 11, 26, 120, 128, 134–136
  barriers to, 135
  questionnaire, 135–136
Delighting the customer, 58
Deming, W.E., 51, 56, 60, 61, 63–64, 76, 82,
    91, 100, 101, 102
  14 points "for transformation of American in-
    dustry", 63
Deming Prize, 73
Demurrer, 234
Departmentalization, 26
Department of Energy, 226
Deposition(s), 234, 443

Design,
  for construction, 67
  for manufacturing, 67
Design assurance audit checklist, 374
Design build contract, 298–299, 301
Design development, 287
Design drawings, 311
Design engineer, 365, 369
Design patents, 241
Design phase, 287, 291, 302, 315
Design professional(s), 239, 240–241, 285, 286,
    288, 300, 307, 345
Design review, 374
Destruction of subject matter, 275
Development, personnel, 120, 136–137
Differing site conditions, 294, 319
  category I, 319, 320
  category II, 319
Directed verdict, 235
Direct examination, 234
Discharge, contract, 272–276
  breach of contract, 273–274
  contractual provision, 273
  impossibility, 274–275
  impracticability, 275–276
  mutual agreement, 274
  operation of law, 276
  satisfactory performance, 272–273
Discipline, 43, 141
Discovery, 234, 441, 443
Disney World, 58
Dispute resolution, alternate forms, 247–251
  arbitration, 249–251
  mediation, 247–248
  mini-trial, 248–249
  negotiation, 247
Dispute review board, 248
Dissatisfiers ("hygiene factors"), 125
Disseminator, 132
Dissolution (of corporation), 343
Distribution of public services, 412
Disturbance handler, 132
Divisible (severable) contract, 257, 308
Doing, 57
Domestic corporation, 336
Donee beneficiary, 271
Don'ts for Expert Witnesses, 444–447
Double taxation, 339
Douglas and Douglas, 161
Dow Corning Corporation, 72
Downward communication, 148
Drake, R.L., 133
Drawee (of bill of exchange), 259
Drawer (of bill of exchange), 259

Drawings, 287, 291, 311–312, 315
  contract, 311
  design, 311
Drucker, P., 130, 180
du Pont de Nemours and Company, E.I., 195
Duration estimates (activities), 199
Duress, 258
Duty ethics, 386
Duty to stay informed, 363

EAC (Estimate at completion), 218
Early event time, 205, 206
Early finish, 201, 206, 207, 214
Early start, 201, 206, 207, 214
Earned value analysis, 216–219
Eastman Chemical Company, 98, 99
Eastman Kodak, 100
Economic loss, 361, 383
  in strict liability, 366
Edgerton, V., 403
Education, 426
  formalized graduate, 136
Electronic communication, 158–159
Elements, contract, 259–266
  competent parties, 260–261
  consideration, 264–265
  form, 265–266
  proper subject matter, 261
  meeting of the minds, 261–264
Employee fitness, 142
Employee loyalty, confidentiality and, 401–404, 413–414
Employee performance, 120
Employee safety and health, 142
Employer, duty to, 397–399
Employer/employee relationship, 239
Employment, 395
Employment accidents, 357
End-result contract, 301
Enforceable contract, 258, 266
Enforcement (of codes of ethics), 396–397
Engineer Ambassador program, 437
Engineering, political, 404
Engineering, versus management, 4
Engineering and architectural services contracts, 285–290
  contract elements, 286
  fee arrangements, 288–289
  relationships with other parties, 289–290
  scope of services, 286–288
  standard documents, 290
Engineering and management, similarities, 7
  differences, 10
Engineering as a career, 2–4, 425

Engineering as a profession, 425
Engineering disciplines, 2
Engineering employment, 2–3
Engineering ethic, 447
Engineering Ethics Update, 415
Engineering internship, 429
Engineering management, definition, 17
  distinguished from industrial engineering, 18
  distinguished from other management, 18
Engineering management education programs, 432
Engineering management functions, 14
  distinguished from other management functions, 18
Engineering manager, ethics and, 414–415
Engineering manager (transition to), 7
  definition of, 7
  disadvantages, 10
  scope of general problems, 9
Engineering profession, 2–4, 426
  definition, 426
Engineering Workforce Commission, 3, 9
Engineer-in-Training, 136
Engineers as social experimenters, 411
Engineer's Creed, 447, 448
Engineers' Week, 437
English law, 224
Entire contract, 257, 308
Entrepreneur, 132
Environmental education (of managers), 12, 20
Environmental ethics, 385, 386, 411–412, 413–414
Environmental Protection Agency, 226
Equal employment opportunity, 142
Equal Employment Opportunity Act, 246
Equity, 228, 234
Equity remedies, 278–279
  injunction, 279
  specific performance, 279
Equity securities, 338–339
Ernst & Young, 113
Error of law (in court appeal), 235
Errors and omissions insurance, 373, 376
Esteem needs, 124
Estimate at completion (EAC), 218
Estimation, construction costs, 287
Estoppel, agency by, 239
Ethical issues in engineering, 385, 398–414
  confidentiality, 401–404
  conflict of interest, 399–401, 413–414
  contributions, 404–405
  employee loyalty, 401–404, 413–414
  global issues, 411–413
  kickbacks, 404–405, 413

professional conduct, 410–411, 413–414
  whistleblowing, 405–410, 422
Ethical judgment, 384
Ethical rankings of professions, 384
Ethics, 385–415, 425
  codes of, 387–399, 415
  conflicts, 397–399
  definitions, 385–386
    engineering, 386
  duty, 386
  education and information exchange, 415
  and the engineering manager, 414–415
  environmental, 385, 386, 411–412, 413–414
  issues in engineering, 399–414
  rights, 386
  setting the example, 415
  utilitarianism, 386
  virtue, 386
European Common Market, 71
European Economic Community, 50, 72, 427
Evaluation and award, construction bid, 306–307
Events (in activity-on-arrow network), 203, 205
Evidence, 235–237, 267
  best evidence rule, 236–237
  hearsay, 237
  oral, 267
  parol, 237, 267
  real, 236
  testimony, 236
Excusable delays, 316
Executed contract, 257
Executory contract, 257–258
Expenses plus professional fee (fee basis), 289
Expert witness, 225, 237, 426, 439–443
  advice for, 443, 444–447
  functions, 441
Express contract, 255
Express warranty, 364

Factors, motivational, 122–125
False imprisonment, 350
Fast track construction contract, 301
Favorable variance, 216
Fayol, H., 15, 19
Featherbedding, 244
Federal Arbitration Act, 249
Federal Aviation Administration, 245, 246, 408
Federal Communications Commission, 226
Federal court system, 229–230
  courts of appeals, 230
  district courts, 230, 242
  special courts, 230
  Supreme Court, 229, 230, 396
Federal government, 430

Federal Labor-Management Relations Act of
  1947, 244, 425
Federal Quality Institute, 76
Fee arrangements, engineering and architectural
  services contract, 288–289
Feedback, 137, 167
Feigenbaum, A.V., 53, 64, 69, 70, 83
Fessler, D.W., 256, 258, 262, 264
Figurehead, 132
Final inspection, 310
Firmage, D.A., 250, 311, 397
Fishbone diagrams (Ishikawa or cause and effect
  diagrams), 65, 93, 94, 107, 108
Fitness, employee, 142
Fixed fee, 288, 289
Fixed price contract, 293–294, 303, 306, 308
Float (slack), 202, 207
Florida Power & Light, 100
Flowchart(ing), 93–94, 95, 107, 108, 112, 114
Fluor Daniel, Inc., 138
Force field analysis, 94–96
Ford Motor Company, 404
Foreign corporation, 336
Foreign countries, conducting business in, 413
Form (of contract), 265, 286, 293
Formal organization, 42–43
Formal structure (hierarchy), 26
Formation principles (contract), 259–266
  competent parties, 260–261
  consideration, 264–265
  form, 265–266
  proper subject matter, 261
  meeting of the minds, 261–264
Forward pass, 199, 201, 205
Foster Wheeler, 276
Founder societies, 434
Franchise, 345
Frank, W.O., 103
Fraud, 228, 258
Frivolous suits, 378
Function contract, 301
Functional manager, 170, 175, 188
Functional organization, 28–29, 30, 34, 175,
  176–177, 178, 182
  advantages and disadvantages, 180
Functions (of engineering management), 14
  functional categories, 14
Fundamental canons, 390
Fundamentals of engineering (F.E.) examination,
  428

Gantt, H.L., 193
Gee & Jenson, 100–103, 104, 105
  14 obligations of management, 101, 102

Gellerman, S.W., 127
General Accounting Office, 400
General conditions, 287, 291, 312
General Dynamics, Convair Division, 408–409
General Electric, 64
General layouts, 287
Geographical organizational structure, 39–40
Global competition, 50–53
Global economy, 50, 53
Global issues (in engineering ethics), 385, 411–413
   business in foreign countries, 413
   distribution of public services, 412
   environmental ethics, 412, 413–414
   national defense and weapons development, 412
Gobeli and Larson, 35
Going bare, 376
Golden limits, 172
Golden Rule, 385
Goodman, R.A., 186
Goodrich, B.F., Company, 401–402, 406
Gossip, 150
Government (employment in), 3
Governmental regulation of business, 142
Governor's Quality Award (Arkansas), 76
Grapevine communication, 152, 186
Graphical output (project schedule), 209
Gray, I., 7, 10, 12
Grievances, 141
Gross negligence, 354
Guidelines to Professional Employment for Engineers and Scientists, 389, 395

Half, R., 158
Hammurabi, 224, 357
Hardship or severe inconvenience, 275
Hawthorne studies, 126
Hayden, W.M., 78
Hearsay evidence, 237, 250
Heartland Award of Excellence, 106
Heartland Professional Construction Services, 103, 105–107, 108, 109, 110
   Award of Excellence, 106
   Vision and mission statement, 106
Hensey, M. 7, 8, 22, 23, 27, 120, 127, 137, 161
Herzberg, F., 124–125
Hierarchy, human need, 124
Hierarchy of documents, 268, 269, 315
Histograms, 86–87, 89, 90
Hoffman, G.C., 11, 12, 20
Holder in due course, 259
Hold harmless clauses, 316

Holmes, O.W., 224, 277
Holt, I.L., 385
Hoover, H., 443–444
Hopper, G.M., 130
Hughes, R.K., 248, 249
Human element, 120–142
Human need hierarchy, 124
Human resource management, 44–45
Hunt, V.D., 69
Hyatt Regency, 369, 370

IBM Corporation, 56
Illegal contract, 291
Implied-in-fact contract, 255
Implied-in-law contract (quasi-contract), 256
Implied warranty, 359, 363, 364, 367
   of fitness, 364, 367
   of merchantability, 364
Impossibility of (contract) performance, 274–275
   death or disabling illness, 275
   destruction of subject matter, 275
   hardship or severe inconvenience, 275
   labor strikes, 275
   objective, 274–275
   subjective, 274
Impoverished management, 24
Impracticability of (contract) performance, 275–276
Incentives, non-monetary, 138
Incentive systems, 137–139
   dual, 139
Incidental beneficiary, 271–272
Income taxes,
   corporate, 339–340, 341, 342
   personal, 339
Incorporation, 376
Incorporation by reference, 315
Indemnity, 316, 317
Individual work motivation, 129
Industrial exemptions, 430
Infant (contract), 260
Infliction of emotional distress, 351
Informal communication, 26, 28, 152
Informal conversation, 159, 160
Informal organization, 40–41, 42–43, 186
In-house training, 431
Initiative (as an engineering management behavioral trait), 11
Injunction, 228, 229, 279
Injuries, 383
Innovation, management of, 18
Insane person, 260
Inspection, 101, 288, 290, 307, 308

Institute for Engineering Ethics, 415
Institute of Electrical and Electronics Engineers, 388, 403, 434
Instructions to bidders, 313
Intellectual property, 225, 241–243
    copyrights, 243
    patents, 241–242
    trademarks, 242–243
    trade secrets, 243
Intended beneficiary, 271
Intentional interference (intentional tort), 350
Intentional torts, 350
    Wrongs against the person, 350
    Wrongs against property, 350
Internal and external customers, 110
Internal Revenue Code, 339
Internal Revenue Service, 245, 328
International Conference of Building Officials, 435–436
International Harvester, 364
International Latex Corporation, 402
International Organization for Standardization (ISO), 71
Internship, engineering, 429
Interpretation of written contracts, 266–269
Interrogatory (interrogatories), 234, 443
Interstate Commerce Commission, 226
Intoxicated person, 260
Invasion of privacy, 351
Inventory, management function, 27
Invitation for bids, 305
Invitee, 355–356
Ishikawa, K., 51, 52, 65, 83, 93
    diagram, 65
Ishikawa diagrams (cause and effect or fishbone diagrams), 93, 94
ISO 9000, 71–72
ITT, 62

Jervis and Levin, 304, 314
Jessup and Jessup, 305
Job design, 26
Job enrichment, 128
Job site safety, 288, 360, 369–371
Joint adventurer, 343
Joint and several contract, 256–257
Joint and several liability, 329, 377
Joint contract, 256
Joint intent, 266
Joint liability, 328
Joint venture, 343–344
    coventurers (joint adventurers), 343
    sponsoring coventurer, 344

Joke contracts, 262
Judgment, court, 229
Judgment, ethical, 384
Judgment n.o.v. (non obstante veredicto), 235
Judicial dissolution, 343
Judicial precedent, law of, 224, 227
Juran, J., 51, 61, 65–66, 87
Juran and Gryna, 83
Jurisdiction (court's authority), 231–233
    venue, 232–233
Jury, 234, 235, 439

Kammerer, A., 404
Katz and Kahn, 130
Kerzner, H., 170, 172, 190, 191
Key Contact program, 437
Kickbacks, contributions and, 404–405, 413
Kimmons, R.L., 431–432
Koza and Richter, 9, 10

Labor regulation, 244
    collective bargaining, 244
    featherbedding, 244
    National Labor Relations Act of 1935, 244
    Taft-Hartley amendment of 1947, 244, 425
    strike, 244
    union, 244
Labor relations, 43–44
Labor strikes, 275
Lag time, 206
L'Ambiance Plaza, 371, 372–373
Lammie, L.L., 415
Lanham Trademark Act of 1946, 242
Larson and Gobeli, 180, 187
Last clear chance theory, 354
Late event time, 205, 206
Late finish, 201, 202, 206, 207, 214
Late start, 201, 202, 206, 207, 214
Lateral communication, 148
Law, definitions, 224
Lawson, J.W., 35
Layoff, 141
Lead time, 206
Leader, 132
Leadership, 11, 22, 38, 42, 56, 61, 63, 64, 110, 120, 128–133, 190
    characteristics, 131
    definitions, 129–130
    qualities, 132–133
    roles, 132
    versus management, 130
Legal department, 44

Legal duty to care (in negligence), 351, 362, 370
Legally protected interest (in negligence), 352
Lex non scripta, 227
Lex scripta, 227
Liability, 383, 425
  joint, 328
  joint and several, 329
  limited (of corporate owners), 335, 341, 342
  personal, 326, 327, 328, 330
  professional, 288
  without fault, 356
Liaison, 132
Liaison office, 176, 177
Libel, 349
Licensee, 355–356
Licensing, *see* Registration, professional
Licensing boards, 226
Life cycle, project, 172–174
Limited liability company, 345
Limited liability of corporate owners, 335, 341, 342
Limited partner(ship), 332, 333
Line-staff group, 177
Line-staff relationships, 26
Ling-Temco-Vought, 406
Links, communication, 148–151
Liquidated damages, 278, 312, 316–317
Listening, 155–156
Litigants, 229
Litigation, 247
Local area networks, 159
Logistics, 43
Loulakis, M.C., 361
Loulakis and Ingberg, 316
Low bidder, 303, 305, 306
  apparent, 305
Lower control limits, 84–86
Lowest responsible bidder, 303
Lump sum contract, 299

Mackenzie, R.A., 134, 135
MacMillan, Lord, 224
Maguire Group, The, 397, 398
  Code of Ethics and Business Guidelines, 397, 398
Mail box rule, 263
Malicious prosecution, 351
Management,
  as a process, 4
  definitions, 4
  versus engineering, 4
Management and engineering, similarities, 7
  differences, 10

Management function inventory, 27
Management involvement and leadership (in quality management), 60–61, 110
Managerial Grid, 131
Manager training, 11
Manufacturing, 42
  design for, 67
  employment in, 3
Marketing and sales, 41
Marshall, J., 334
Martin and Schinzinger, 388, 406–407, 412
Martin Corporation, 62
Martin Marietta Corporation, 415
Maryland Society of Professional Engineers, 383
Maslow, A., 124, 128
Master of Business Administration, 432
Master's hearing, 440
Material breach (of contract), 273
Material samples, 307
Material suppliers, 307
MATHCOUNTS, 437
Matrix organization, 34–36, 40, 178, 180–182, 184, 185
  advantages and disadvantages, 180, 182, 183
  balanced, 183, 187
  effectiveness, 180
  functional, 183, 187
  keys for success, 35
  project, 183, 187
  strong, 178, 187
  weak, 178, 187
McCanse, A., 24
McClelland, D.C., 127
McDonald's, 58
McDonnell Douglas DC-10, 408–409
McGregor, D., 125–126
Measuring, 14, 16
Media, 46
Mediation, 247–248
Mediator, 248
Medicare, 245
Meeting of the minds, 261–264, 286, 291–292
Meetings, 22, 161–162
  guidelines, 161–162
  minutes, 162
  weekly staff, 162
Mental infirmity, 258
Meredith and Mantel, 169, 171, 176, 189, 207, 208, 219
Microsoft Project for Windows, 210
Miller, R.M., 100, 103
Mini-trial, 248–249
Minor person (infant), 258, 260
Mintzberg, H., 132

Mirror-image acceptance rule, 263
Misconduct, 235
Mission statement, 105, 106, 138
Missouri Board of Architects, Professional Engineers and Land Surveyors, 355
Mistake, 258
Mistakes in bids, 263–264, 304
Mitigation of damages, 277
Moder, Phillips, and Davis, 207
Monitor, 132
Monitoring project progress, 209, 375
Monthly progress payments, 308
Morality and ethics, 385–387
  definitions, 385–386
Morals (morality, moral values), 385–387
  definitions, 385–386
  personal sense of, 415–416
Morrison, P., 5, 6
Morton Thiokol, 407
Motion to dismiss (nonsuit), 235
Motivating, 14, 15
Motivation, 120–128, 137
  astronauts, 128
  factors, 122–125
  individual work, 129
  technical professionals, 127
Motivators, 125
Motorola Corporation, 56
Mouton, J., 24
Multinational corporations, 413
Multiple of direct hourly expense (fee basis), 289
Mutual agency, 328, 330
Mutual agreement (to terminate contract), 274
  novation, 274
  recission, 274
  waiver of performance, 274
Mutual assent to the (contract) terms, 261

NASA, 407
National Association of Legal Assistants, 388
National Council of Examiners for Engineering and Surveying, 427, 428–429, 433, 434
National defense and weapons development, 385, 412
National Electrical Manufacturers' Association, 315
National Highway Traffic Safety Administration, 245
National Institute for Engineering Management and Systems, 54, 58
National Labor Relations Act of 1935, 244
National Labor Relations Board, 246

National Quality Forum, 53
National Society of Professional Engineers, 54, 61, 63, 77, 100, 155, 290, 334, 384, 388, 389, 397, 416, 434, 436, 437, 447
  Engineer Ambassador program, 437
National Technological University, 13, 137
Necessity, agency by, 239
Need satisfaction, 124
Negligence, 228, 239, 240, 350, 351–356, 360, 361, 362, 364
  comparative, 354
  contributory, 353
  gross, 354
  in services liability, 366
  job site safety, 370
  per se, 353
  shop drawing review, 369
Negligence per se, 353
Negotiable instruments, 259, 265
Negotiation, 162–163, 247
Negotiator, 132
Nelson and Peterson, 399
Network preparation, 196
Network schedule, 309
Network schedule analysis, 193, 195–210
New trial, motion for, 235
No damages for delay clause, 318
Noise, 147–148
Nominal damages, 277
Nominal group technique, 96–97, 99
Nonconformance, cost of, 70
Noncumulative preferred stock, 339
Noncumulative voting, 337
Noneconomic damages, 377
Non-material breach (of contract), 273
Nonnumeric tools, 92–97, 98
  brainstorming, 92–93
  cause and effect diagrams (Ishikawa or fishbone diagrams), 83, 94
  flowcharting, 93–94, 95
  force fields, 94–96
  nominal group technique, 96–97
Nonresponsive bids, 306
Nonsuit, 235
Nord, M., 238, 259
North American Free Trade Agreement, 427
Notes, corporate, 339
Not-for-profit corporation, 344–345
Notice to proceed, 307
Nuclear Regulatory Commission, 245
Nuisance, 349
Numeric(al) tools, 82–92, 98, 103
  control charts, 82–86, 87, 88
  histograms, 86–87, 89, 90

Pareto charts, 87, 89–91, 97
run charts (timeline charts), 91
scatter diagrams, 91–92

O'Kon, J., 372
O'Kon and Company, 372–373
O-ring, 407–408
Objective impossibility, 274–275
Obligations, conflicting, 397–399
Obligor, 270
Observation of work, 288
Occupational Safety and Health Administration,
    (OSHA), 226, 245, 246
Offer(s), 259, 261, 291
    acceptance of, 262
    and acceptance, 261
    counteroffer, 262
    error, 263
    termination, 262
Officers, corporate, 337
Oldenquist and Slowter, 389
One-way communication, 154–155
On the job training, 136, 432
Operation of law (in contract discharge), 276
Operational aspects (of management), 5
Opinion evidence, 237
    expert witness, 237
    ordinary witness, 237
Oral communication, 152–156, 443
    speech suggestions, 153–154
    visual aids, 154–157, 441, 442
Oral contract, 298
Oral evidence, 267
Ordinances, 227
Ordinary partnership, 326–332
Ordinary witness, 237
Ordinary words, 267
Organization,
    relationships outside, 45–46
    relationships within, 41–45
Organizational charts, 28
Organizational design, 27
Organizational effectiveness grid, 24
Organizational structure, 22, 28–41, 171, 174
    clusters, 38–39
    functional, 28–29, 30, 24
    geographical, 39–40
    hierarchical, 151
    informal, 28, 40–41
    matrix, 34–36
    product, 29, 31, 32
    project, 31, 33, 34, 35
    self-directed teams, 36–38
    throwaway, 40

Organizational structure principles, 25–27
    authority and power, 25
    coordination and communication, 26
    division of labor and specialization, 26
Organizing, 14, 15

Parallel redundancy, 163
Pareto, V., 87
Pareto chart(s), 87, 89–91, 97, 103
Parol evidence rule, 237, 267
Parsons Brinkerhoff, Inc., 415
Partial performance, 273
Partnering, 45, 46, 332–334
Partnership, 240, 260, 326–334, 336, 340, 342,
    344, 346
    advantages and disadvantages, 330
    agreement, 328, 329
    articles of, 329
    corporation, compared with, 342
    distribution of property, 330–332
    formation, 328, 342
    general, 326–332
    limited, 332
    subpartnership, 332
    taxation, 328, 342
    termination, 329–330
        dissolution, 329–330
        winding up, 330–332
Pascal, B., 158
Patent and Trademark Office, 242
Patents, 241–242
    design, 241
    letters patent, 242
    search, 242
Payee (of bill of exchange), 259
Payment, for extra work, 309
Payment bond, 314
Payments, periodic, 312
Peer review, 374
Penalty, 316
Penney, J.C., 52
Peppercorn theory, 265
Percentage of construction cost (fee basis), 289
Performance, employee, 120
Performance (of contract),
    complete, 272
    partial, 273
    prevention of, 274
    satisfactory, 272–273
    substantial, 272–273, 309–310
    tender of, 273
Performance appraisal, 140
Performance bond, 314
Performance measurement, 18

Periodic payments, 312
Personal property,
  conversion of, 351
  trespass to, 351
Personnel administration function(s), 120, 139–
  142
Personnel department(s), 139, 140
Personnel development, 136–137
PERT (Project Evaluation and Review Tech-
  nique), 195, 196
PERT/COST, 196
Peters and Waterman, 34–35, 180
Physiological needs, 124
Piercing the corporate veil, 335
Pinto, 403–404, 405
Placing personnel, 140
Plaintiff, 233
Planning, 14, 57
Planning phase, 287, 289, 291, 299, 302, 315
Plans, 311–312
Polaris Program, 195, 203
Political action committees, 437
Political engineering, 405
Pound, R., 425
Power,
  principle of organizational structure, 25, 26
Pozner, B.Z., 190
Precedence networking, 196, 206–207
Precedent, 251
Preferred stock, 338–339
Prequalification (of bidders), 291, 303–304
President's Award for Quality, 76
Pretrial conference, 234
Pretrial procedure, 233–234, 236
  admissions, 234
  complaint (filing), 234
  demurrer, 234
  deposition(s), 234, 443
  discovery, 234, 441, 443
  interrogatories, 234, 443
  pretrial conference, 234
  summary judgment, 234
  summons, 234
Prevention (of contract performance), 274
Principal, 238–239
Principal/agent, 260
Principles and practice (P.E.) examination, 429
Private corporations, 334, 335–343
Privity defense, 290
Privity of contract, 358, 359, 360
  demise of privity defense, 361–362
Problem causes, in engineering management, 16
Process, construction contract, 302–311
  advertising, 302–303

bid opening, 305–306
completion, 310–311
evaluation and award, 306–307
proposals, 303–305
the work proceeds, 307–310
Process analysis, 112
Product liability, 357, 364–366, 375
  statutes of limitations and, 368
Product organization, 29, 31, 32
  pure, 178
Productivity, 121
Profession,
  characteristics, 426
  definitions, 425–426
  duty to, 397
Professional competency, 374
Professional conduct, 410–411, 413–414
  computer-aided design and drafting, 411
  criticizing another's work, 411
  sealing drawings, 411
Professional development, 7, 8, 395
  four stages, 7, 8
Professional employee, definition, 425
Professional engineer(s), 429, 430, 448
Professional engineering societies, 395, 426,
  434–439
  conferences, 437
  founder societies, 433
  membership, 436
  political action committees, 437
  professional conduct matters, 437, 439
  publications, 437
Professional ethics, 383–416
  codes of ethics, 387–399, 415
  conflicting obligations, 397
  engineering manager, 414–415
  issues in engineering, 398–415
  morality and ethics, 385–387
Professionalism, 426
Professional liability, 225, 288, 376
  insurance, 376
Professional licensing (registration), see Registration, professional
Professional obligations, 392–394
Professional practice standard (of care), 362, 363
Professional registration, see Registration, pro-
  fessional
Professional status, 121
Professions, ethical rankings, 384
Program (as distinct from project), 170
Progress payments, periodic, 308
Project,
  characteristics, 170
  definition, 169–170

Project budget(ing), 210, 213–215, 375
Project completion (date), 162, 312
Project control, 190, 209, 219
Project cost control, 215–219
Project cost estimating, 210, 213
Project Evaluation and Review Technique (PERT), 195
Project feasibility, 287
Project leader, 176, 177
Project life cycle, 172–174
Project management, 169–219, 374
  definition, 170
Project management approach, 175–176
  advantages, 175
  disadvantages, 175
Project Management Institute, 436
Project manager, 170, 175, 186–190, 219
  attributes, 189–190
  authority areas, 186
  reporting levels, 187
  roles and responsibilities, 188
    parent firm, 188
    project, 189
    project team members, 189
    skills, 190, 191
Project manual, 183, 311
Project organization(s), 31, 33, 34, 35, 174–187
  role conflict in, 182–183, 185
Project planning, 190–210
Project schedule (scheduling), 190–210, 308–309, 375
Project scope, 299
Project start date, 312
Projectized organization, 31, 40, 177–178, 179, 182
  advantages and disadvantages, 180
Promise(s), 254, 255, 264, 286
Promisee, 254, 270, 271, 272
Promisor, 254, 270, 271, 272
Promissory note, 259
Promotion, 140
Proper subject matter, 261, 286, 291
Property damage, 383
Proposals, construction, 303–305
  competitive negotiation, 305
  cost-plus-time, 304
  lane rental, 304
  technical merit, 304
Proprietorship, sole, 325–326, 327
  advantages and disadvantages, 327
  formation, 325
Proximate cause, 352, 353, 365
Proxmire, W., 406
Public, duty to, 397–399

Public corporations, 344
Public health, safety and welfare, 395
Public organizations, 45–46
Public policy, 261
Public relations, 45
Public service, 425, 426, 447
Public speaking, 152–154
  speech suggestions, 153–154
  visual aids, 155–157, 441, 442
Publicly held corporations, 344
Punch list, 310
Punitive damages, 277, 377
Purchasing and expediting, 41–42

Qualities of effective leaders, 132–133
Quality, 49
  control, 66
  cost of, 52, 62, 69–71
  culture, 76
  definitions, 54–55
  improvement, 66
  management of design organizations, 68
  partnerships, 60
  planning, 65
  trilogy, 65
Quality department, 43
Quality function deployment (QFD), 65
Quality Improvement Potential (QIP) index, 53, 70
Quality improvement program, 101
Quality Management Maturity Grid, 53
Quality partnerships, 60
Quasi-contract (implied-in-law contract), 256

R chart, 83–86
Ratification, agency by, 239
Reading, 137
Real evidence, 236
Receiver, 147–148
Recission (of a contract), 228
Recognition, 121
Recognition systems, 120, 137–139
  innovative, 139
  tangible, 138
Re-cross-examination, 235
Recruitment, 140, 395
Redirect examination, 235
Reformation (of instruments), 229
Registered professional engineer, 325
Registration, professional, 137, 261, 286, 372, 373, 395–396, 414, 426, 427–431, 432, 440
  boards, 395–396, 414, 426, 427, 429
    conduct enforcement, 429

comity, 429
conduct, 429
continuing professional competence, 431
examinations, 427
history, 427
industrial exemptions, 430
proportion registered, 427, 428
requirements, 428–429
tripartite model, 428
special, 429
Regression analysis, 91, 92
Regulation of business, governmental, 142
Regulations, 226
enforcement, 226
establishment, 226
Regulatory compliance, 225, 244–247
hypothetical construction company, 245–247
labor, 244
workers' compensation, 244
workplace safety, 245
Regulatory standard (of care), 363
Reilly, L.P., 162–163
Reimbursable costs, 307
Rejection, bids, 305
Relationships with other parties (design services), 289–290
Remand (of court case), 235
Remedies for breach of contract, 276–279
damages, 277–278
actual, 277
compensatory, 277
liquidated, 278
mitigation of, 277
nominal, 277
punitive, 277
equity, 278–279
injunction, 279
specific performance, 279
restitution, 278
Repetition, 151
Reporting levels, project manager, 187
Repudiation, 273–274
Res ipsa loquitur, 352, 359
three required conditions, 352
Research and development, 42
Resource allocator, 132
Resource analysis (in project schedule), 207–208, 209, 212
Responsibility (responsibilities), 25, 27, 134, 178
of project manager, 188–189
Restitution, 278
Restraining order, 229
Retainage, 308
Revocation (of offer), 262

Reward system(s), 42, 137–139
team rewards, 139
Rights ethics, 386
Risk taking, 11
Roadstrum, W.H., 134
Rocket seal, 407–408
Role conflict (in project organizations), 182–183, 185
Roles,
of engineers, 5, 6
leadership, 132
decision, 132
informational, 132
interpersonal, 132
of managers, 4, 6
of project managers, 188–189
Rosenau, M.D., 171, 172, 196, 208
Rossini, F.A., 150
Rules of practice, 390–392
Run chart(s) (timeline charts), 91, 107, 109, 110, 113

S corporations, 339–340, 342
Safety, health and welfare of the public, 389, 390, 427
Safety and health, employee, 142
Safety needs, 124
Sarchet, B.R., 38
Satisfaction of needs, 124
Satisfactory performance, 272–273
Satisfiers ("motivators"), 125
Scalar principle (of organizational structure), 25
Scatter diagrams (XY plots), 91–92
Schedule updates, 309
Schedule variance, 216, 217
Schoumacher, B., 239, 241, 250, 344
Schultz, W.N., 133
Scope of services, engineering and architectural services contracts, 286–288
construction phase, 288
design phase, 287
planning phase, 287
Scott, H.A., 49
Screening personnel, 140
Securities and Exchange Commission, 245
Selecting personnel, 140
Self-actualization needs, 124
Self-directed teams, 22, 36–38, 59
job assignments within, 37
Self insurance, 376
Sensitivity, 190
Serial redundancy (repetition), 151, 163
Service industries, 66–69
Services (employment in), 3

Services contracts, 286
Services liability, 366–368
Severable (divisible) contract, 257, 308
Several contract, 256
Shannon, R.E., 14, 20, 34, 40
Shareholders, 337
Share-office arrangement, 345–346
Shewart, W., 63
Shewart cycle, 57
Shop drawing(s), 288, 307, 369, 372, 373, 375
 review, 288, 369, 373, 375
Short courses, 137
Signal-to-noise ratio, 163
Silverman, M., 10, 19, 134, 161, 170, 172, 183,
 187, 188
Simon, M.S., 247, 250, 251, 304, 309, 310, 316
SKANSKA, 192–193
Skills, project management, 190, 191
Skills required (of engineer and engineering
 manager), 10
Skinner, B.F., 127
Slack (float), 202, 206, 207, 214
Social experimenters, engineers as, 411
Social justice, 384
Social needs, 124
Social security tax, 245
Society of American Military Engineers, 434
Society of Automotive Engineers, 434
Society of Manufacturing Engineers, 434
Society of Plastics Engineers, 434
Sole proprietorship(s), 325–326, 327, 334, 340
 advantages and disadvantages, 327
 formation, 325
Sommers, C.H., 387
Space shuttle *Challenger*, 407–408
Span of control, 26
Special causes, 82
Special conditions, 287, 291, 312, 315
Specialty conferences, 437
Specifications, construction, 159
Specifications, technical, 287, 291, 315
Specific performance, 228, 279
Spokesperson, 132
Sponsoring coventurer, 344
Staffing, 14, 15
Standard documents, 290
Standards of care (in liability), 349, 362, 367
 duty to stay informed, 363
 implied warranty, 363
 professional practice, 362, 363
 strict liability, 363
 regulatory standard, 363
Stare decisis, 227
STAR structure, 36–37

Start date, project, 312
State court systems, 231
 Alaska, 231, 232
 appeals court, 231
 minor judiciary, 231
 supreme court, 231
 trial courts of general and original jurisdic-
  tion, 231
Statistical control, 82, 83
Statistical process control, 51, 101
Statute law, 226–227, 251
Statute of frauds, 258–259, 265–266, 286, 293
Statute of limitations, 276, 368, 378
Statute of repose, 368
Stewart, J.M., 171
Stock, corporate (equity securities), 338–339
 common, 338
 preferred, 338–339
  cumulative, 338–339
  noncumulative, 339
Storyboarding, 93
Straight voting, 337
Strategies for protection (against liability suits),
 373–376
 administrative strategies, 374–376
 legal strategies, 376
 technical strategies, 374
Strict liability (liability without fault), 350, 356–
 357, 363
 in product liability, 364, 365
 in services liability, 366, 368
Strike(s), 244, 275
Structured settlements, 377
Stuckenbruck, L., 149, 170
Subcontractors, 307
Subjective impossibility, 274
Subpartner(ship), 332, 333
Subrogation, 229
Substantial completion, 310, 312, 316
Substantial performance, 272–273, 309–310
Summary judgment, 234
Summons, 234
Supervening illegality, 240, 276
Supervision of (construction) work, 360
Supplementary conditions, 312
Supplementary general conditions, 312
Supplier(s), 45, 59
Supplier-customer relations, 59
Supreme court,
 federal, 229, 230
 state, 231
Surety, 314
Sweet, J., 231, 233, 240, 241, 243, 267, 285,
 304, 314

Taft-Hartley Act (amendment of 1947), 244, 425
Taguchi, G., 66
Tangible recognition, 138
Target price contract, 294
Task force, 26, 176, 177
Task specialization, 26
Tatum, C.B., 431
Taxation, corporate, 339–340, 341, 342
  double taxation, 339
Taxes, corporate income, 339–340
Taxes, personal income, 339
Team building, 11
Team management, 24, 25
Team(s), 22, 56, 59, 106, 139
  cross functional, 110, 111
  self-directed, 22
Teamwork, 22
Technical competence (maintenance of), 13
Technical society (societies), 13, 432
Technical specifications, 312–313, 315
Ten Commandments, 224
Tender of performance, 273
Tennessee Eastman Company, 77
Termination, of agency, 240
Termination, construction contract, 318–319
  for convenience, 319
  for default, 319
Termination, corporation, 343
Termination, partnership, 329–330
  dissolution, 329–330
  winding up, 330–332
Termination for convenience, 273
Termination (of contract), 268
Termination (of employment), 141, 396
Termination (of offer), 262, 263
Testimony, 236
Texaco USA, 107, 110–113, 114, 115
Texas Instruments, 67, 76
Thamhain and Wilemon, 219
Theory X, 125–126
Theory Y, 126
Third-party beneficiaries, 270–272
  creditor, 271
  donee, 271
  incidental, 271–272
  intended, 271
Third-party liability, 358, 361
  in service contracts, 359
Third-party rights, 269–272
  assignment, 269–270
  third-party beneficiaries, 270–272
Thomsett, R., 150
Three-legged stool, 171, 172
Throwaway organization, 40

Time extension, 318
Timeline charts (run charts), 91
Time of completion, 312, 316–317
Time of payment, 268
Time of the essence, 273, 316
Tirella and Bates, 163
Toffler, A., 40
Tort liability, 326
Tort reform, 377–378
Tort(s), 225, 261, 328, 349–357
  bases for tort liability, 350
Tortum, 349
Total quality culture, 76
Total quality management, 49–78, 82–115, 142
  applications in engineering, 100–113
  history of, 50–54
  definition, 55–56
  design organizations, 68
  gurus, 62–66
  implementation, 76–77
  Maturity Grid, 53
  nonnumeric tools, 92–97
  numeric tools, 82–92
  standards of excellence, 71–76
Total variance, 216, 217
Townsend, R., 162
Tozer, F.L., 439
Trade-off management, 172, 188
Trademarks, 242–243
Trade secrets, 243
Trade terms, 267
Training, 22, 63, 102, 106, 111, 431–432
  in-house, 431
  of competent managers, 11
  on-the-job, 136, 432
Traits of effective leaders, 132–133
Transfer, 140, 396
Transmission medium, 147–148
Transmitter, 147–148
Trespass, 349
  personal property, 351
  real property, 351
Trespasser, 355
Trial, 225, 440
  appellant, 235
  appellate court, 235
  appellee, 235
  closing arguments, 235
  cross-examination, 234
  damages, 235
  direct examination, 234
  directed verdict, 235
  error of law (in court appeal), 235
  judgment n.o.v. (*non obstante veredicto*), 235

jury, 234, 235
  misconduct, 235
  motion to dismiss (nonsuit), 235
  new trial, motion for, 235
  nonsuit, 235
  process, 441
  re-cross-examination, 235
  redirect examination, 235
  remand (of court case), 235
  verdict, 235
Trial procedure, 233–237
  evidence, 235–237
  pretrial, 233–234, 236
  trial, 234–235, 236
Triangle, agency, 240
Tribus, M., 61, 87, 93, 96
Trilogy, quality, 65
Triple constraint, 171, 172, 188, 219
Turnkey contract, 299
Two-party contracts, 254
Two-way communication, 154–155
Types of construction contracts, 293–301
  construction manager, 300, 318
  cost plus, 297–298, 306, 308
  design build, 298–299
  end result, 301
  fast track, 301
  fixed price, 293–294, 299, 306, 308
  turnkey, 299
  unit price, 294–297, 299, 306, 308, 309

U.S. Army Corps of Engineers, 246, 248, 332, 334, 412
U.S. Bureau of Reclamation, 332–333
U.S. Constitution, 427
U.S. Department of Defense, 216, 246
U.S. Department of Justice, 396
U.S. Department of Labor, 245, 246
U.S. Environmental Protection Agency, 363
U.S. Fish and Wildlife Service, 246
U.S. General Services Administration, 409
U.S. Navy Fleet Ballistics Missile Program (Polaris Program), 195, 203
U.S. Supreme Court, 397
Ultra vires contract(s), 260, 336
Unemployment insurance, 245
Unenforceable contract, 258, 292
Unfavorable variance, 216, 217
Unified risk insurance, 376
Uniform Arbitration Act, 249
Uniform Building Code, 246
Uniform Construction Index, 313
Uniform Partnership Act, 328

Unilateral contract, 255
Unincorporated association, 345
Unintentional tort (negligence), 351
Union, 244
Union of Japanese Scientists and Engineers, 73
United States Constitution, 229
United States National Quality Award (Baldrige Award), 72
Unit price contract, 294–297, 306, 308, 309
Unity of command, 25
Unjust enrichment, 256
Unlimited personal liability, 326, 327, 328, 330
Upper control limits, 84–86
Upward communication, 148
Utilitarianism, 386

Vandiver, K., 406
Variances (in project cost control), 216
Vaughn, R.C., 241, 244, 251, 264
Venue (for court proceedings), 232–233
Verbal communication, 152–153
Verdict, 235
Vesilund, P.A., 387
Violation of statutory duty, 352
Virtue ethics, 386
Vision statement, 105, 106
Visual aids, 155–157, 441, 442
  guidelines, 156, 157
Visual communication, 152–153, 443
Vocal communication, 152–153, 443
Voidable contract, 258
Void contract, 258
Voluntary disability, 274
Voluntary dissolution, 343
von Braun, W., 41
Voting, corporate, 337, 338
Vroom, V.H., 127
VTT, the Technical Research Center of Finland, 53, 90

Walton, S., 60
Warranties, 376
Weinert, D.W., 434
Wellins, Byham, and Wilson, 36, 38
West Publishing Company, 237
Wetlands, 246
Whistleblower Protector Act of 1989, 410
Whistleblowing, 385, 399, 405–410, 422
White, E.B., 156–157
Williams, L.P., 133
Williamson, M., 17
Willis, H.E., 224

Wilson, L.P., 444–447
Wisely, W.H., 426, 447
Witness,
  expert, 237
  ordinary, 237
Wohlgemuth, D., 401–402
Words versus figures, 267
Work breakdown structure, 191–193, 196, 197,
  213, 214
Workers' compensation, 244, 357, 378
  insurance, 246
Workplace issues, 384
Workplace safety, 244
Workshops, 137
Work team,
  effectiveness, 22
  development, 22, 23
    stages of, 23

Wright, S., 276
Wright brothers, 294
Written communication, 156–159, 161
  ambiguous wording, 159
  careless modifiers, 158
  careless pronouns, 158
  construction specifications, 159
  electronic, 158
  emphasis, 158
  principles, 156, 159

x-bar chart, 83–85
Xerox Corporation, 98
XY plots (scatter diagrams), 91–92

Zero defects, 62
Zoning and land use, 246

CPSIA information can be obtained
at www.ICGtesting.com
Printed in the USA
BVOW06s2133151017
497435BV00026B/60/P

9 780471 593294